Kohlert / Chistyakova

Innovations in Film Manufacture

Christian Kohlert
Tamara Chistyakova

Innovations in Film Manufacture

Exemplified by Calender Technology

HANSER

Print-ISBN: 978-1-56990-440-4
E-Book-ISBN: 978-1-56990-442-8
E-Pub-ISBN: 978-1-56990-958-4

Bibliographic information of the German National Library:
The German National Library lists this publication in the German National Bibliography; detailed bibliographic data are available on the Internet at http://dnb.d-nb.de.

Vilshofener Straße 10 | 81679 Munich | info@hanser.de
www.hanserpublications.com
www.hanser-fachbuch.de
Editor: Dr. Mark Smith
Production Management: Cornelia Speckmaier
Cover concept: Marc Müller-Bremer, www.rebranding.de, Munich
Cover design: Max Kostopoulos
Cover picture: © stock.adobe.com/edojob
Typesetting: le-tex publishing services GmbH, Leipzig

Opening Remarks

Christian Kohlert

This book was developed by specialists in plastics technology, especially calendering technology, and is based on around 20 years of professional experience and scientific collaborations with international universities and research institutions. Results are presented based on around 20 doctoral theses and 200 diploma and master's theses, which were supervised by the authors.

The aim of this book is not a general presentation of the calendering process, as this is already sufficiently contained in the books by Kopsch and Kohlert. It is also not intended to convey knowledge that can be easily researched on the internet. Instead, the aim is to present innovations and practical experiences in plastics processing—using the example of the calendering process—that can also be used in other plastics technologies, such as extrusion, injection moulding, or coating. All calculation methods presented are applicable to other plastic technologies, or can be transferred to them.

The images used in the book are taken from the authors' working lives, unless otherwise stated in the image caption or at the end of the section. We recommend that anyone interested in re-using the images contact the authors' collective at info@kohlert-consulting.de.

The book was written by Prof. Dr. Christian Kohlert and Prof. Dr. Tamara Chistyakova, with the participation of Helmut Welbers, Dr. Andrey Polosin, Dr. Michael Kohlert and Prof. Dr. Frank Kleinert.

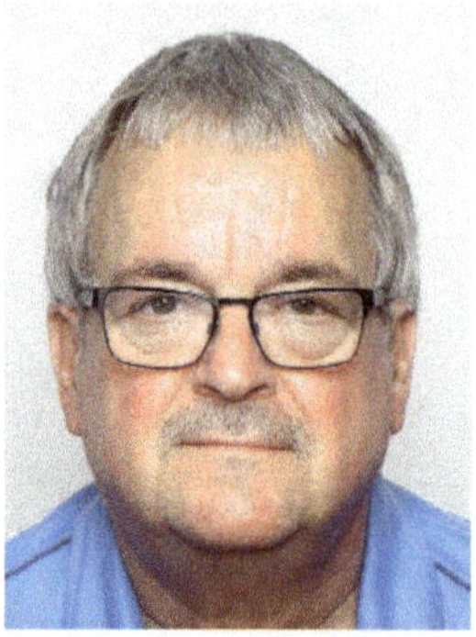

Prof. Dr. Christian Kohlert

Prof. Dr. Tamara Chistyakova

Helmut Welbers

Dr. Andrey Polosin

Dr. Michael Kohlert

Prof. Dr. Frank Kleinert

Contents

1 Introduction

Christian Kohlert

Via AI you will find the following definition of “calendering”, and this underlines the importance of our focus on innovations and similar evaluations in this book:

> Calendering is a process in the plastics and rubber industry in which materials, especially plastics or rubber, are processed into thin layers or films by rolling. This process is primarily used for the production of flat materials such as films, panels, and sheets.

Calendering involves four main steps:

1. **Heating:** The raw material (plastic or rubber) is heated to a high temperature to soften it and make it malleable.
2. **Rolling:** The heated material is passed through a series of rollers that are coordinated to maintain the correct pressure and temperature. As it passes through the rollers, the material is formed into a thinner layer.
3. **Cooling:** The thin layer or film is cooled to achieve the desired strength and thickness.
4. **Slitting:** The cooled material is slit to the desired dimensions.

Calendering can also be used to bond different materials together, for example to produce a composite of two or more plastic or rubber layers. The process parameters, such as temperature, pressure, and roll speed must be carefully controlled to achieve the desired material properties and dimensions.

Calendered products are used in a variety of applications, such as the packaging, automotive, construction, and electrical industries. Calendered plastic films play a major role as homopolymer films, as multilayer films, or as mixed films in the packaging industry, as well as in technical and agricultural applications, and they are an integral part of polymer value chains.

Of course, the type of film, material, and property profile, as well as new innovations, play a major role in the optimization processes, which in turn affect the price–property ratio. Many innovations have been incorporated into polymer production processes in recent years, especially sensors (driven by advances in microelectronics) and in data processing, and these can generally be used in any polymer film manufacturing process.

As shown in Figure 1.1, every film manufacturer faces market tensions, and needs to find a system that is optimal for them. Raw material suppliers are concentrating on the production of cheaper raw materials and have a better market position; customers are becoming more demanding regarding quality and property profiles; and new competitors from outside Europe are striving to enter the market, and in so doing, substitute products with their own polymer bases.

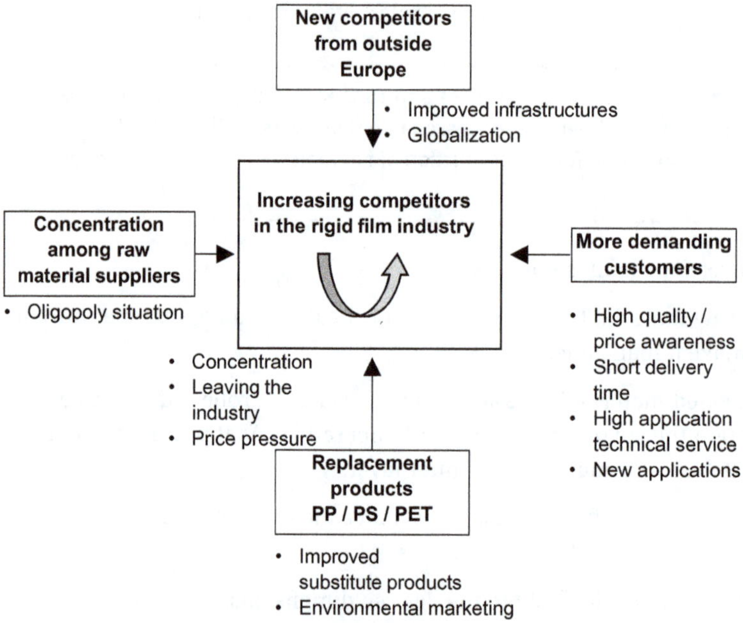

Figure 1.1 Market tensions in the production of polymer films

Polymer films are usually produced using three manufacturing processes (Figure 1.2). Of these, the calendering process is the most productive process in terms of tonnage, thickness tolerance, and quick changeover times (see Figure 1.3). The basics of calendering are covered in the books by Kopsch and Kohlert [1, 2]. Extrusion has its advantages in terms of the investment amount, the number of workers, and the colour change. Spreading and laminating several films can be used for finishing after any film production process.

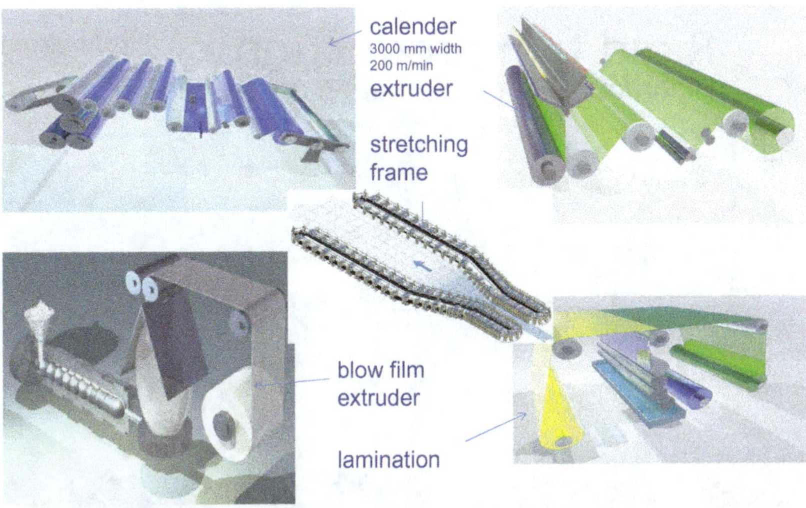

Figure 1.2 Manufacturing process for polymeric films and their finishing (Source: [3])

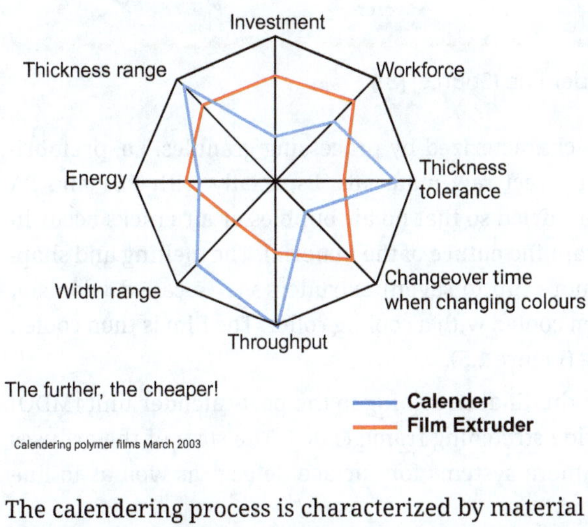

Figure 1.3 Comparison of calendering and extrusion

The calendering process is characterized by material storage and preparation, dosing and mixing, the fluxing process in an extruder or co-kneader, the actual calendering process by molding between several rollers, the after-treatment for the final adjustment of the film thickness and film shrinkage, as well as the temperature control for cooling and the winding up.

For customer requests, a winder or format cutter is usually used for final packaging (Figure 1.4).

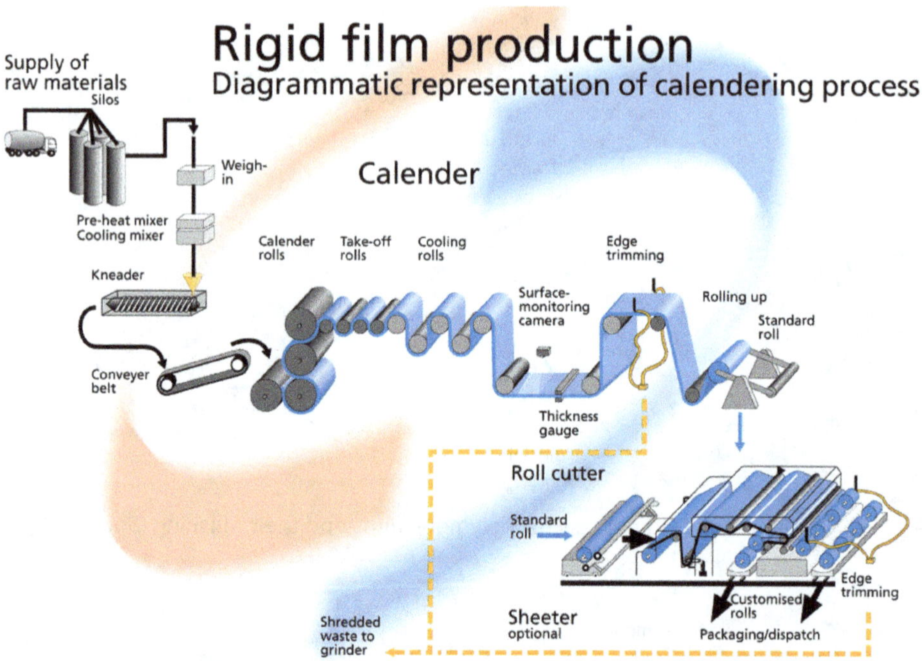

Figure 1.4 Schematics of a calender line (Source: [3])

The extrusion process is usually characterized by processing granules, i.e. prefabricated material particles and not direct raw materials. Especially with PET and PA granules, the material must be pre-dried so that no air bubbles or air cracks occur in the finished film, due to the hydrophilic nature of the material. The melting and shaping then takes place in one or more (up to seven) extruders via so-called wide slot dies, and the emerging film is then cooled with a cooling roller. The film is then cooled and assembled, as in calendering (Figure 1.5).

The finishing options include longitudinal stretching in the post-calender unit (MDO) and transverse stretching in a wide stretching frame (TDO). The state of the art uses thickness measuring systems, camera systems for surface defects, as well as in-line measurements of colour, gloss, and haze.

Film thicknesses between 20 and 1600 µm and widths of up to 2.2 m are achieved, and up to 6.2 m when stretched. By stretching and using a suitable T-guide, shrinkage values of up to 80% can be achieved.

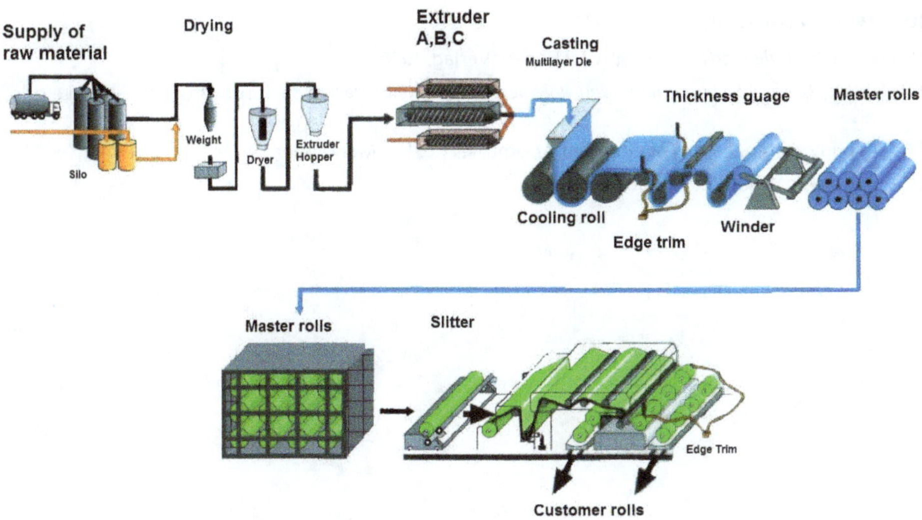

Figure 1.5 Schematic of a film extrusion line (Source: [3])

When extruding blown film, the prefabricated granules are formed into a multilayer film in the range of several µm per layer, possibly with pre-drying, using several extruders of different sizes in a mixing nozzle and expanded by several air bubbles. This allows the desired film thickness and width, as well as the corresponding shrinkage, to be achieved (Figure 1.6).

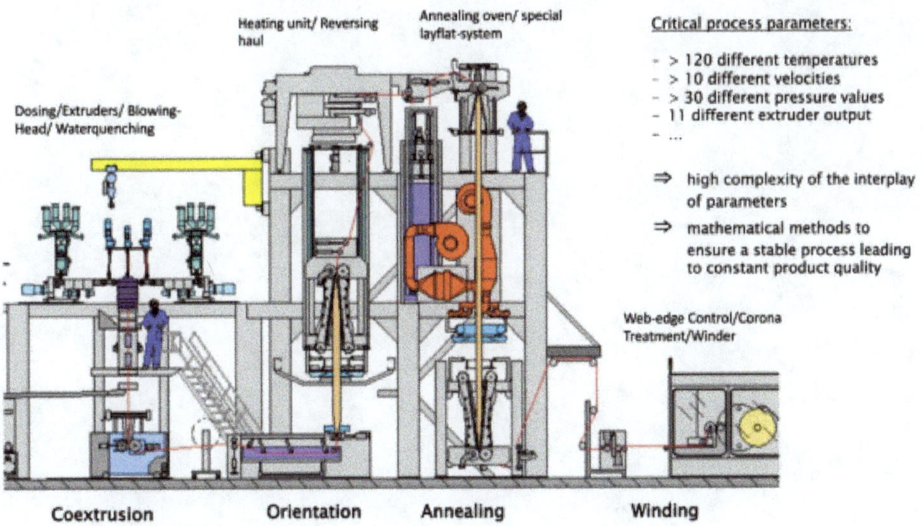

Figure 1.6 Schematics of a blown film extrusion line (Source: [4])

References for Chapter 1

[1] Kopsch, H.: *Calender Technology*, Carl Hanser Verlag, 1978

[2] Kohlert, C. et al.: *Kalandrieren von Polymeren*, Deutscher Verlag für Grundstoffindustrie, Leipzig, 1992

[3] Company material, Klöckner Pentaplast GmbH & Co. KG, Montabaur

[4] Company material, Maria Soell GmbH

2 Production of Calendered Films

2.1 Calendering

Christian Kohlert

Calendering is the process of forming polymeric masses in the nips of a calender. During calendering, the processing material passes only once through each nip, whereas during rolling, the material passes through the nip several times until the desired homogenization is achieved.

The aim of the calendering process is to achieve a uniform thickness distribution for a homogeneous polymer mixture at high film speeds in the longitudinal and transverse directions. In addition to the film thickness, also important are color consistency, uniform shrinkage, and strength values, as well as the avoidance of film defects such as black dots, specks, wrinkles, etc.

In practice, 4- and 5-roll L-shaped calenders with film guides from bottom to top have become established for hard PVC, as this makes it easier to manage the cleanliness of the rolls. For soft PVC, 4-roll calenders in F and Z shapes are used because of better suction of the vapors created by the plasticizers. Common calender designs in current use are shown in Figure 2.1 and Figure 2.2.

design		***general areas of application***
	2-roll-calender; I-shape	floor coverings, vinyl asbestos tiles, satin finishing of IT panels, rough rubber plates, sole plates
	3-roll-calender; I-shape	floor coverings, rubber plates, friction
	4-roll-calender; I-shape	rigid-PVC-film, rubber plates, covering fabrics and cord threads stripes on one and both sides, friction
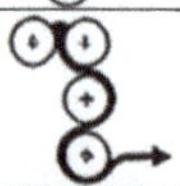	4-roll-calender; F-shape	soft-PVC-film, rubber plates, covering fabrics and cord threads stripes on one and both sides, friction, duplicate
design		***general areas of application***
	4-roll-calender; F-shape with rear breast roller	soft-PVC-film, rubber plates, covering fabrics and cord threads stripes on one and both sides, friction, duplicate
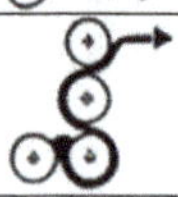	4-roll-calender; L-shape	rigid-PVC-film
	5-roll-calender; L-shape	rigid-PVC-film
	4-roll-calender; Z-shape	soft-PVC-film, rubber plates, covering fabrics and cord threads stripes on one and both sides, friction, duplicate
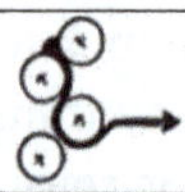	4-roll-calender; S-shape	rigid and soft-PVC-film, rubber plates, covering fabrics and cord threads stripes on one and both sides, friction, duplicate
Special design		***general areas of application***
	2-roll-calender; horizontal-shape	rough rubber plates, floor covering
	2-roll-calender; inclined-shape	rough rubber plates, linoleum
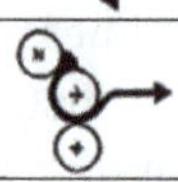	3-roll-calender; inclined-shape	rubber plates, covering fabrics on one sides, friction
	3-roll-calender; A-shape	pre-film for rigid PVC pressed sheet
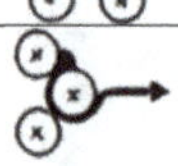	3-roll-calender; triangular-shape	rubber plates, covering fabrics on one sides, friction
	4-roll-calender; Z-shape with raised front roller	rubber plates, covering fabrics and cord threads stripes on one and both sides, friction, duplicate
	4-roll-calender; I-shape with inclined top roller	rubber plates, covering fabrics and cord threads stripes on one and both sides, friction, duplicate

Figure 2.1 Calender designs (Source: [1])

Figure 2.2 shows the most important parts of a calender, namely the rollers, roller bearings, stands, and post-calender unit. In the early days of calender technology, bottle neck-drilled rolls were used, with heating taking place by electrical electrolytes in chambers filled with water. Although this was an inexpensive technology, it required long changeover times for the roll temperatures and resulted in a temperature gradient from the centre of the roll to the edge of the roll. With the desire to reduce changeover times, peripherally drilled rollers were used (Figure 2.3). Here, externally heated oil or water in the printing circuit made it possible to achieve the necessary surface temperatures more quickly for different recipes and film thicknesses. The disadvantage of filling with oil is the contamination when changing the rollers and the oil costs.

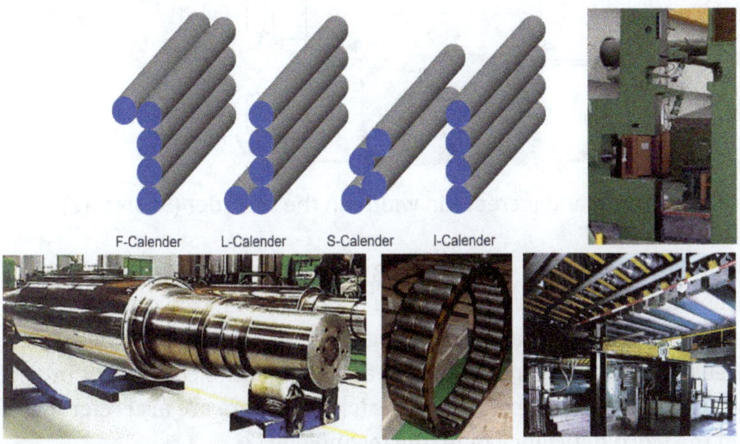

Figure 2.2 Calender designs (Source: [2])

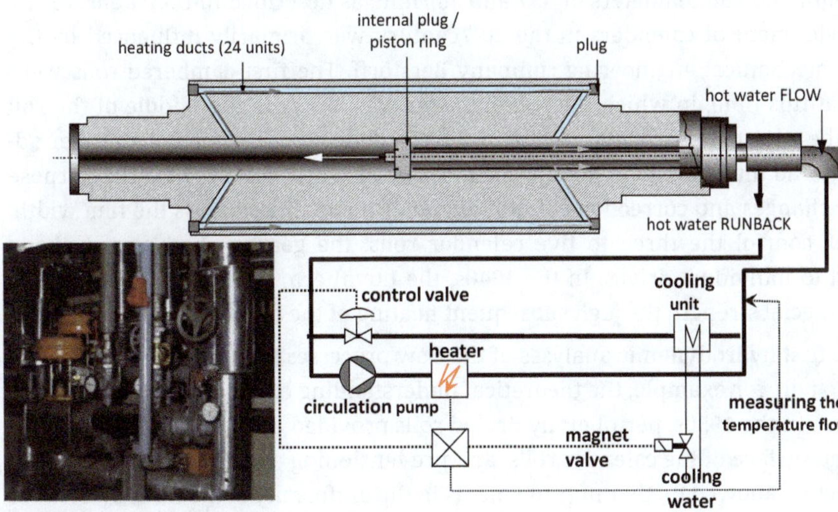

Figure 2.3 Roller structure (Source: [2])

Using the example of a 4-roll L-calender, the thickness reduction and increase in width between the individual rolls are shown in Figure 2.4. The higher the viscosity and the narrower the roll gap, the greater the increase in width. This must be taken account of by adjusting the input width of the material in the first gap.

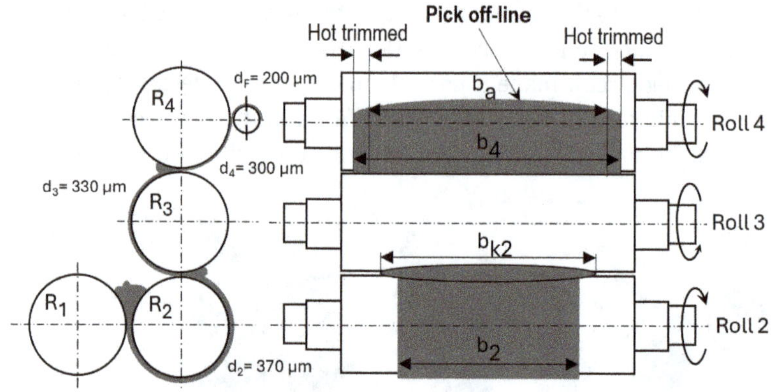

Figure 2.4 Thickness reduction and increase in width on the calender (Source: [2])

Calendering paper and plastics are an old technology that appears not to have evolved significantly in recent decades. However, when you take a closer look, you notice endless technical progress in machines and technology. The word "calendering" could come from the Greek word for "cylinder", and was first mentioned in 1798 in the context of a smoothing calender for textile surfaces (cf. [1]). There are also references to the naming from the word deficiency of textile technology [3].

Figure 2.5 shows the first patented calender for rubberizing fabrics, with a roll width of 1830 mm and roll diameters of 457 and 305 mm, as described in US Patent No. 16. The development of calenders in the 20th century was primarily influenced by the German mechanical engineering company Berstorff. The first cambered rolls were created at this time, in which the roll diameter was larger in the middle of the roll than at the edge to absorb a large part of the roll deflection. Electromotive roller adjustments and tilt adjustment options for the central rollers followed for the purpose of faster changes and corrections of film diameter inequalities across the film width. To better control the three to five calender rolls, the gearbox had already been switched to individual drives. In the 1940s, the Luvitherm process was patented to achieve special strength through subsequent heating of the film leaving the calender.

With the first hydrodynamic analyses of the flow processes in the roll gap by Alexandre Ardichvili, for example, the theoretical understanding of the effects of a calender gap began. In the 1950s, peripherally drilled rolls provided a better temperature profile on the surface of the calender rolls, and pre-tensioning and roll bending as counter-deflection allowed further improvements in the uniformity of the films in terms of thickness and shrinkage tolerance. Around 1960, people began to switch to continuous pre-plasticization for higher throughput and to use larger roller widths (up to

2000 mm) and higher speeds in new buildings. At that time, it was already possible to achieve calendering speeds of 60 m/min for hard PVC and up to 150 m/min for soft PVC. Figure 2.6 shows a modern calender line starting from fluxing to reel slitting.

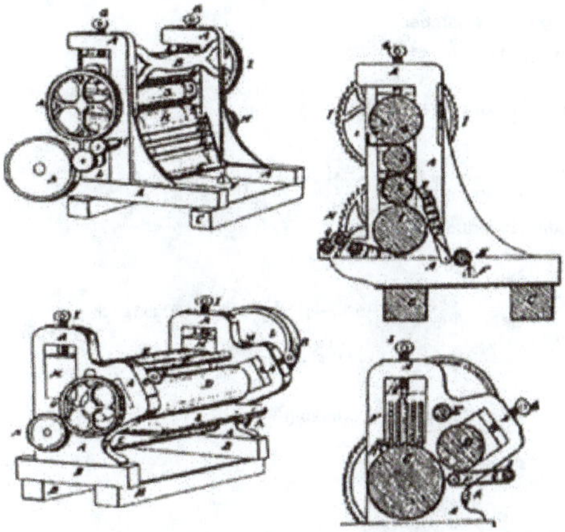

Figure 2.5 First patented calender variants (Source: [1])

In the 1970s, the first calculation models for throughput and pressure (for example, by J. S. Chong) succeeded in bringing the roll shapes (camber or crowning) so close to reality that 80–90% of the roll deflection due to material pressure could be compensated for (Figure 2.7).

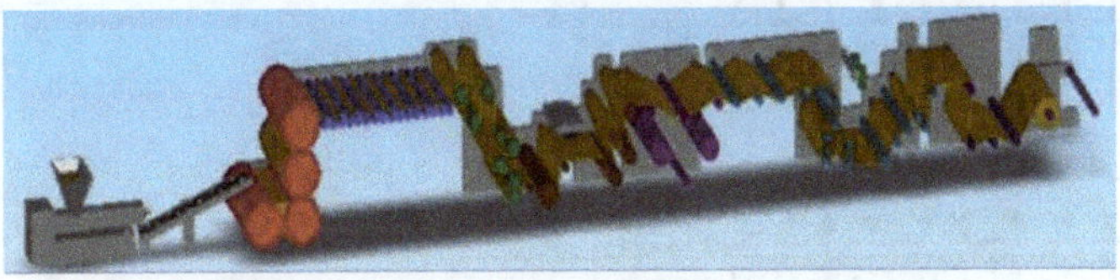

Figure 2.6 Modern 4-roll L-calender line (Source: [2])

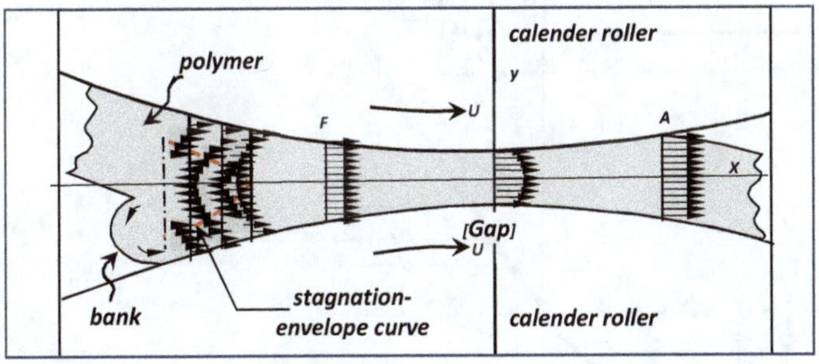

Figure 2.7 Speed distribution in the roller gap (Source: [1])

Figure 2.8 shows the basic equations for calculating the bending under camber, roll bending, and axis crossing. The first calculation models for roll deflection and compensation were based on this (Figure 2.9). Initial process optimizations (for example by Ernst-Otto Reher and Vladimir Krasovski) made it possible to achieve accuracies in the µm range for PVC rigid films.

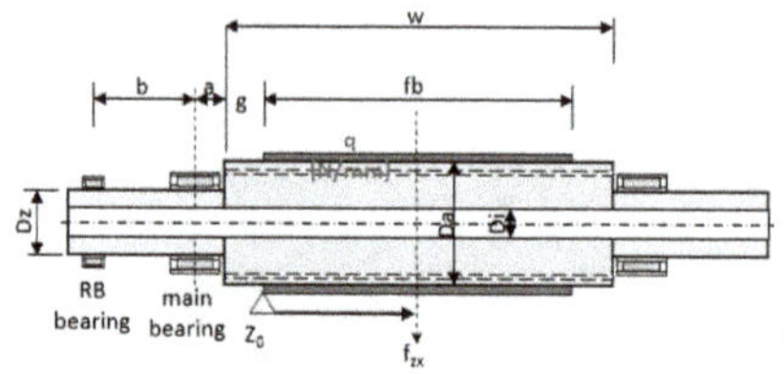

rolling hide width on the roller	fb	mm
specific gap load (if known)	Q	N/mm
axis crossing	Sb	Mm
roll bending force (N) or, ggf, pressure in the cylinder	Rb	N
camber of the roller (if known)	H	mm
roller material (steel or cast)	St/GGG	
roller diameter	Da	mm
face width, (bowl width)	w	mm
drilling diameter in the roller center	Di	mm
neck diameter main bearing	Dz	mm
distance main bearing > roller edge	A	mm
distance main bearing > roll bending bearing	b	mm

calculation of the camber

$$\alpha = \frac{\pi}{180} * 20 + \frac{7}{18} * \pi * \frac{w - fb}{w} \quad (0.9)$$

$$\beta = \frac{\pi}{180} * 160 - \frac{7}{18} * \pi * \frac{w - fb}{w} \quad (0.10)$$

$$bb_z = \frac{h}{1 - \sin\alpha} * \left[\sin * \left[\alpha + \frac{\beta - \alpha}{w} * \left(z + \frac{w - fb}{2} \right) \right] - \sin\alpha \right] \quad (0.11)$$

roll bending line due to gap load

$$g = \frac{w - fb}{2} \quad (0.12)$$

$$f_z = \frac{q * fb}{24 * e * Jb} * \left[\begin{matrix} \frac{z^4}{fb} - 2 * z^3 - 6 * (a + g) * z^2 + (6 * fb * a + 6 * fb * g + fb^2) \ldots \\ * (z + g + a) + 4 * \left(g^3 + 3 * a * g^2 + 3 * a^2 * g + \frac{Jb}{Jz} * a^3 \right) \end{matrix} \right] * 1000 \quad (0.13)$$

gap deformation through axis crossing

$$s = \sqrt{\left(sb * \frac{fb}{w} \right)^2 + Da^2} - Da \quad (0.14)$$

$$s_z = s * \left(1 - 2 * \frac{z}{fb}\right)^2 \quad (0.15)$$

gap deformation through roll bending

$$rb_z = Rb * \frac{b}{2 * e * Jb} * \left[z^2 - z * fb - a * \left(fb + \frac{Jb}{Jz} * a \right) \right] \quad (0.16)$$

Figure 2.8 Roll calculation (Source: [2])

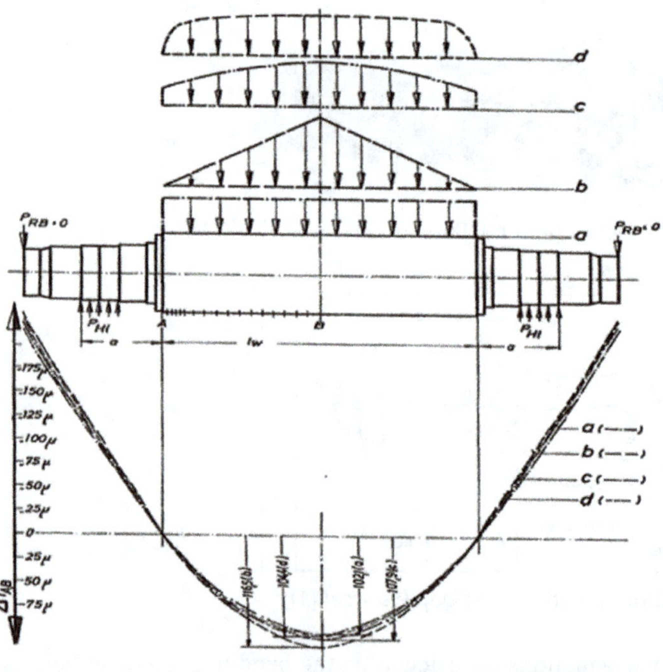

Figure 2.9 First calculation models for throughput and pressure (Source: [1])

In Figure 2.10 are shown different flow patterns depending on the friction. This also required a development of measurement technology in which these accuracies could also be practically assessed using radiometric thickness measuring systems. In the 1980s, the calender was supplemented with wedge-shaped internals (after Heinrich Collin and Christian Kohlert [4]), which were intended to improve the mixing process in the calender gaps. The first major automations and controls of temperatures and speeds led to increased system analyses of the influencing parameters and their connections in relation to film qualities (after Klaus Hartmann).

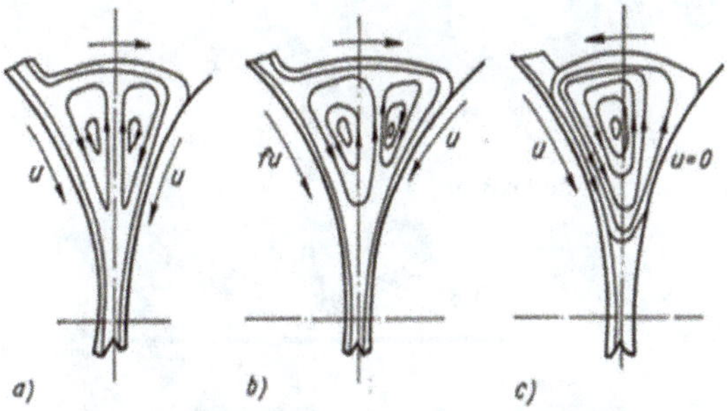

Figure 2.10
Quantitative image of the flow of a Newtonian fluid in the gap, with (a) $f = 1$, (b) $f = 1.3$, and (c) $f = \infty$ (Source: [1])

With the advent of microcomputer technology, in-line measurement technology was placed at the forefront of developments, especially in the 1990s. As a result, the quality of the film features could be measured at the point of creation, and influenced by the calender operator. This is how the defect inspection systems for evaluating surface quality, shrinkage measuring systems using polarized light, level measuring systems using ultrasound or light speed measurement, as well as in-line color and gloss devices were created.

The transition to modern techniques around the turn of the century was characterized by further strong development of in-line measurement technology, as well as automation solutions from dosing to winding technology. A modern calender line from the company Klöckner Pentaplast is shown schematically in Figure 2.11.

Film properties that need to be considered during calendering are the thickness distribution in the longitudinal (MDO) and transverse directions (TDO), the shrinkage in MDO and TDO, surface defects (black specks, gels, fibres, holes, etc.), surface roughness (glossy or matt), barriers against water vapor and gas, thermoformability, printability, color fastness, and mechanical and thermal film properties.

Calendering technology is divided into the section material storage, dosing, hot and cold mixers, fluxing, calendering, pick-off rolls, take-off-rolls, cooling rolls, measurement technology, winding and packaging using slitters or sheet cutters.

Today there are complex calender lines for film thicknesses from 25–1200 µm, with roll speeds of up to 200 m/min and roll widths of up to 3500 mm. Shrinkage values of 0–60% are possible in MDO.

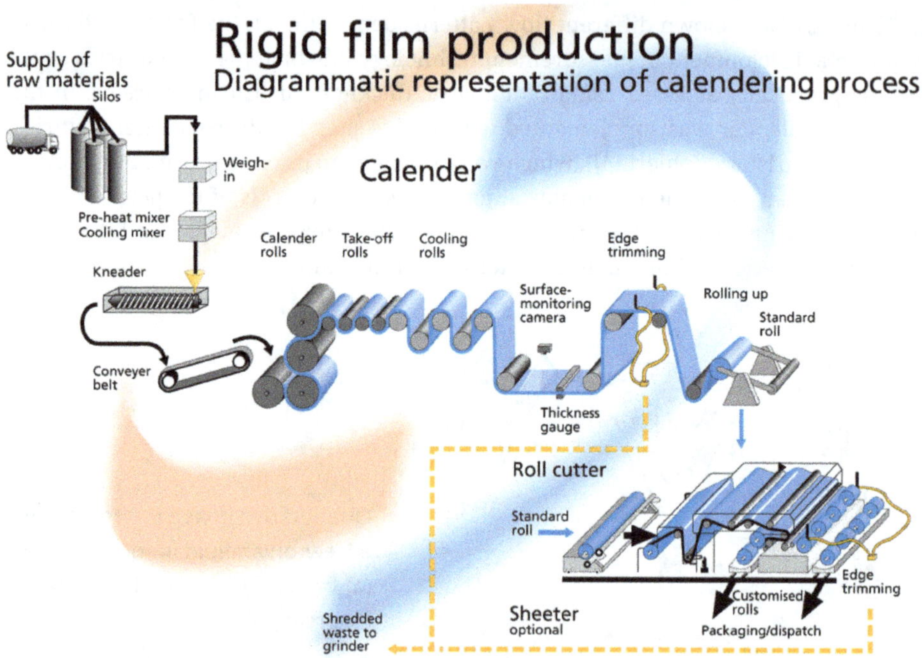

Figure 2.11 Modern rigid-PVC calender (Source: [2])

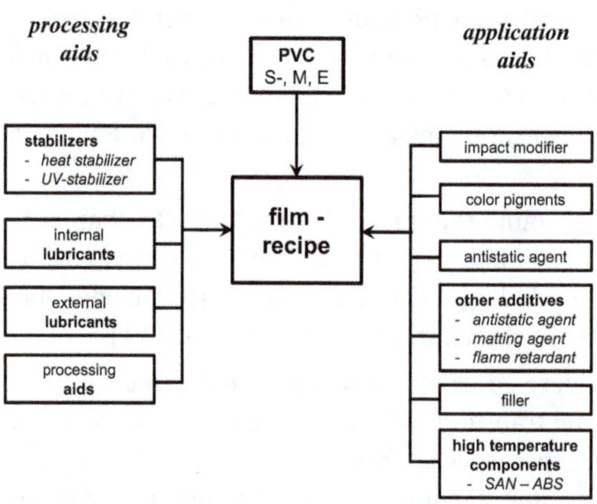

Figure 2.12 Film recipe for PVC rigid films (Source: [2])

For the main product of a plastic calenders, a basic recipe (for PVC rigid films) is shown in Figure 2.12. It is assumed that around 80% of the recipe consists of PVC and around 10% each is made up of processing aids and application aids.

The processing aids include stabilizers, internal and external lubricants to reduce temperature peaks and reduce the adhesive effect on the calender rolls, as well as calendering aids to influence the viscosity of the polymeric mass to be processed. The application aids include impact modifiers to influence film strength, color pigments, and antistatic agents to prevent electrical charging, as well as other additives and fillers for special applications or reducing recipe costs.

This allows throughputs of up to 4 t/h to be achieved. This is possible through automated dosing of the individual components, computer-aided control, process monitoring with control programs such as WinCC, trend analyses, and statistical process control. In-line and off-line quality control as well as energy efficiency measures complement this modern technology. Certifications according to ISO 9001 (quality), GMP, ISO 14001 (environment), ISO 50001 (energy), and hygiene are now standard in calendering companies.

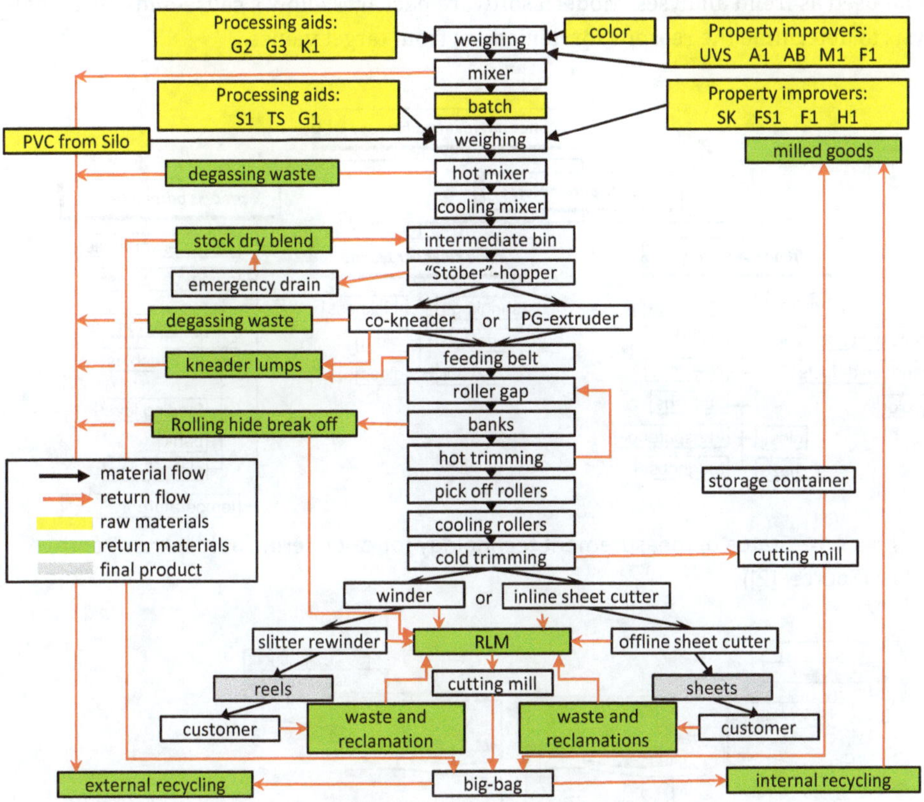

Figure 2.13 Material flow diagram in a calender line (Source: [2])

The material flow diagram of a calender line is shown in Figure 2.13. Many returns to the output flows can be seen as internal recycling. In modern calendering systems, an average yield of 85% is obtained directly from the raw material to the film for the customers. Around 15% is returned to the cycle or stored in big bags as granulated material for later further processing. Material that can no longer be processed is created through combustion or pollution, and is usually sold as down-recycling material in the range of 0.1–0.3%.

Figure 2.14 and Figure 2.15 show the status of measurement technology and data linkage at Klöckner Pentaplast. The aim of the development of measurement technology in recent years has been to capture the desired values in-line, since the operator can intervene immediately in the event of deviations from the target value. A later laboratory assessment result in ever larger unusable shortages if deviations are identified. Modern calender lines now have traversing in-line thickness measuring systems with shrinkage measuring systems, line cameras for surface evaluation, as well as color and gloss measuring systems if the film quality requires this (Figure 2.14). The results of these measuring systems are displayed directly on the calender to provide the operator with information about the film qualities, and are collected in data systems and used as trend analyses. Modern software packages allow a cause-and-effect analysis to determine the reasons for deviations from target values.

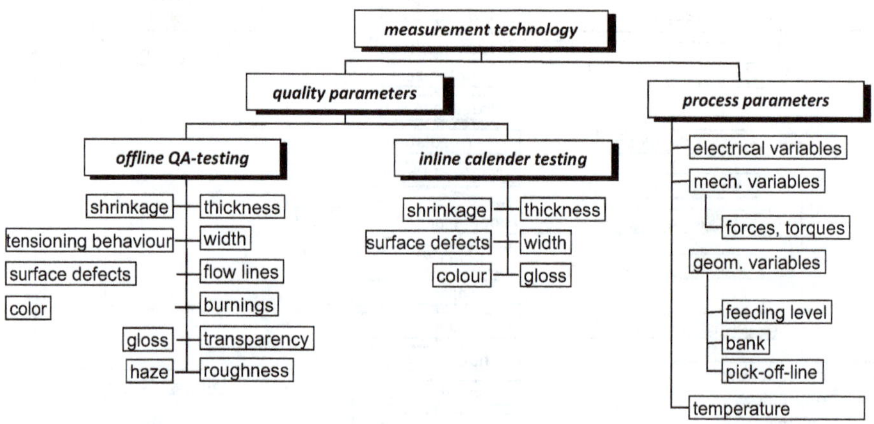

Figure 2.14 Status of measurement technology for calendering at Klöckner Pentaplast (Source: [2])

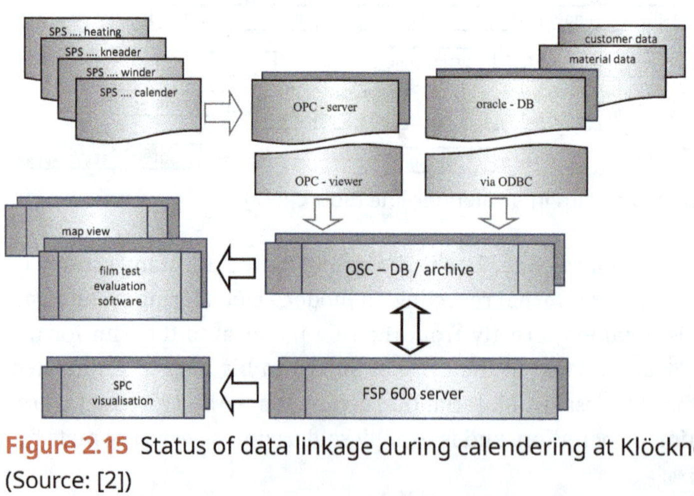

Figure 2.15 Status of data linkage during calendering at Klöckner Pentaplast (Source: [2])

Options for surface treatment are shown in Table 2.1, with costs. Figure 2.16 shows an example of a surface treatment for better printability. If a normal corona treatment loses its increased surface tension values after 6–10 weeks, a corona/aerosol treatment is much more effective in the long term and allows the buyer to process the film for printing purposes later.

Table 2.1 Fixed Costs for Surface Treatment of Films (as of 2010), in €

Corona	Corona aerosol	Rotary spray systems	Plasma	Corona gas
~80,000	~160,000	~130,000	>500,000	>280,000

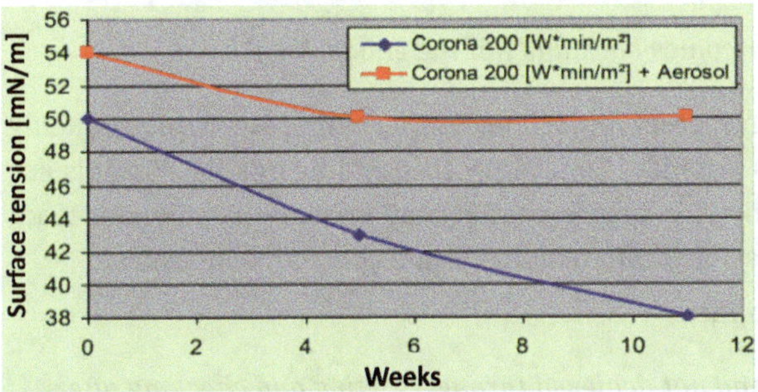

Figure 2.16 Improvement of surface tension through corona treatment films (as of 2010) (Source: [2])

Historically, the development of calender technology can be broken down into the following steps: textile calender, rubber and plastic calenders, modernization of construction and electrical engineering, coating technology and new calenderable polymers (Figure 2.17).

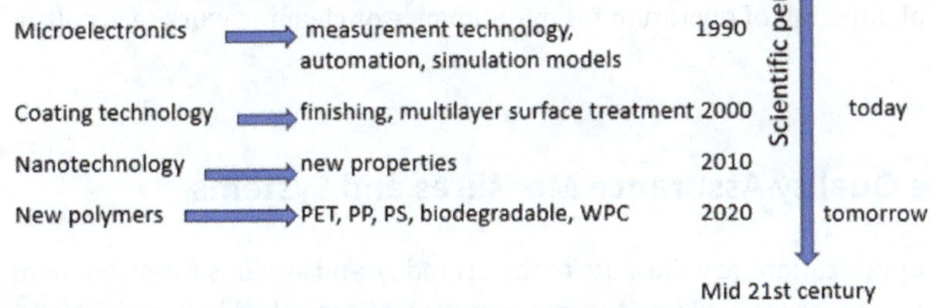

Figure 2.17 Development of calender technology

2.2 Film Errors and Possible Compensation Measures

Helmut Welbers

2.2.1 Introduction

Every manufacturer of calendered films is aware that the customer needs to be supplied with goods that they can process without any problems. The customer can expect to receive the delivery at the agreed time in the ordered quantity and in a quality that meets their needs, in order to provide the customer or the end user with satisfactory goods.

The following principle always applies:

> *"It's better that the customer comes back and not the goods!"*

When producing calendered films, great importance is therefore placed on producing goods that are as defect-free as possible. To achieve this, in-line quality controls are carried out, which reliably report any problems that may arise to the calender team at an early stage and, where possible, initiate appropriate corrective measures.

A further principle therefore applies:

> *"Quality must be produced and not achieved through sorting and checking afterwards."*

However, this does not mean that there is no final inspection, but this can be largely limited to checking the records of the in-line quality tests. A well-coordinated calendering team knows best whether the film produced meets the specified quality requirements. To make the status of the current production on the calender visible, a "traffic light" system can be installed, for example, which signals the current situation for everyone to see.

- "Green": Everything OK; quality requirements met.
- "Yellow": Changeover (e.g. new order)
- "Red": Quality requirements not met; we are working on the solution.
- No lights: Calender out of operation for maintenance or cleaning work (e.g. polishing).

2.2.2 In-line Quality Assurance Measures and Systems

In-line quality optimizations are the state of the art today, and systems for permanent quality assurance can be divided into two main groups, as described in Section 2.2.2.1 and Section 2.2.2.2.

2.2.2.1 Systems With Direct Influence

These systems have a direct influence on film quality and optimization through automatic adjustment of the settings, and includes, for example:

Mixing:

- Control of the dosage for the individual components of the mixture
- Temperature measurement of the mixture in the hot mixer and release of the stabilizer injection
- Temperature measurement of the mixture in the hot mixer and opening the ejection flap into the cooling mixer
- Temperature measurement of the mixture in the cooling mixer and opening the ejection flap into the storage container for the gelling unit.

Fluxing unit (co-kneader; planetary gear extruder (PGE)):

- Ampere recording and speed (this system must interface with the level control in gap 1).

Feed belt (either to the rolling mill or to the swing belt):

- Search coil for metal particles with ejection device (interface with gelling unit and filling level gap 1- or 2-roll mixing-mill).

Two-roll mixing-mill (if available):

- Filling height in the gap (interface with fluxing unit and calender must be present)
- Width of the feeding stripe in the calender (with interface to the speed of the belt).

Conveyor belt (swivel belt):

- Speed of the belt (with interface to the filling level in gap 1 and width of the feeding stripe).
- Swing cycle per minute (with interface to the filling level in gap 1)
- Dwell time in the turning points (with interface for distribution (left/middle/right) in gap 1).

Calender:

- Filling height control gap 1 (interface to swivel conveyor speed)
- Distribution (left/centre/right) in gap 1 (interface to wig-wag belt dwell time)
- Roller peripheral speeds: friction (guide roll can be the last or penultimate calender roll)
- Roller gap (interface with thickness measuring and control system)
- Roller bending device (interface with thickness measuring and control system)
- Roller axis crossing (interface with thickness measuring and control system).

Post-calender unit(s):

- Thickness measuring and control system (with interface to roll gap, bending device, and axis crossing adjustment).

2.2.2.2 Systems with an Indirect Influence

These systems have an indirect influence on film quality by informing the calender team, and include, for example:

Mixing:

- Database for the recipes
- Barcode on the mixing instructions to avoid input errors
- Measurement and control of the power consumption of the hot mixer and the mixing time until the desired temperature is reached
- Measurement and control of the power consumption of the cooling mixer and the mixing time until the desired temperature is reached for opening the ejection flap.

Gelling unit (co-kneader; planetary gear extruder (PGE)):

- Database for films (type/color/thickness)
- Speed of the feed hopper's screw to control the degree of filling in the fluxing unit (particularly in the co-kneader)
- Target and actual temperature of the heating zones in the housing and the screw conveyor (co-kneader) or central spindle (PGE)
- Temperature of the plasticized material.

Feed belt (either to the rolling mill or to the swing belt):

- Belt speed: 2-roll mixing-mill (if available)
- Type/color/thickness database
- Temperature of the rollers
- Speed of the rollers
- Hide thickness on the work roll (the material on a roller is called "hide")
- Measurement of hide temperature.

Conveyor belt (swivel belt):

- Database for basic settings (type/color/thickness/surface/width)

Calender:

- Roller peripheral speeds: friction (guide roll is the last calender roll)
- Roller temperatures
- Size of the banks

- Bank pressure
- Hide temperatures
- Hide monitoring
- Current consumption (amperes) of the rollers
- Lubricating oil flow in the main and auxiliary bearings.

Post-calender train:

- Type/color/thickness/surface/width database
- Roller circumferential speed (guide roll is the last or penultimate calender roller)
- Roller temperatures
- Current consumption (amperes) of the rollers.

2.2.2.3 Additional Systems

There are also additional systems for quality assurance within the post-calender train that (usually) do not have a direct influence on the calender settings, but instead come with a warning signal:

- Surface inspection: to detect film defects
- Shrinkage measurement
- Gloss measurement
- Color measurement/thermal degradation of the film
- Winding tension
- Film temperature before winding.

2.2.2.4 Auxiliary Tools

The systems described so far for avoiding or providing quick information about film errors that have occurred cannot (yet) provide reliable, automated information about the sources and, in some cases, not about the causes of the detected errors. Sometimes human intuition and a detective approach are required to get to the bottom of the problem, especially since the aforementioned systems are not all available, especially for older systems. Therefore, the special attention of the calendering team is required.

The following still applies:

> *"Calendering is as easy as playing the piano. All you must do is press the right button at the right time for the right amount of time.*
> *The environment must also be appropriate to the event (when playing the piano, you must attend to the acoustics; when calendering, you must attend to the cleanliness)."*

Should there be further automation of the calender systems in the future, including autonomous calender production, at least the above systems will be required. But this does not mean that people can then be completely dispensed with. For example, quality defects caused by foreign inclusions must be processed and eliminated through measures outside the calender line.

This compilation is intended to aid the identification and elimination of the cause of quality problems in an organized and controlled manner. The following devices should be available as aids for this:

- Infrared spectrometer to check for incorrect dry blends
- Vibrating channel for inspecting the delivered PVC powder for black particles or gel particles
- Laboratory rolling mill for testing stabilization in the event of color degradation
- Color measuring device for analysing wrong colors
- Gloss meter.

And finally, because it is essential for troubleshooting:

- Microscope
- Permanent markers
- A trouble archive!

If identification of the foreign body with the naked eye is not possible, the use of a microscope is recommended. The method of using the microscope is described below, and should be viewed as work instructions for troubleshooting in the event of external influences.

1. Mark the foreign body by circling as many defects as possible with a permanent marker.
2. Use the microscope: Switch on both light sources (top light and bottom light) and check the polarization filters. For a good result, the angle of the skylight must be set at an angle. As a rule, it is better if the light falls on the defect at an angle.
3. Set the foreign objects in sharply in focus: Use the reflected light to check the color, and use the polarizing filter to check for any lighting-up of the foreign body.
4. Compare the appearance of the foreign particle with the description in Section 2.2.8.
5. Check the trouble archive to see if a similar foreign object is already saved and look at the comment made about it.

If identification is successful, it is strongly recommended to take a sample containing the particle, clearly label it, and archive it. A supplementary description of the successful compensation measure(s) is always very helpful.

2.2.3 Influence of Mixtures on Film Defects

Inaccuracies in the component weight or the processes in the hot and cool mixer can lead to imperfections in the film, such as flow lines, friction stripes, burns, etc. The main considerations of dosage, cycle time, flowability, density and mixing tool speed are discussed below.

Dosage

The mixing process begins with weighing the mixture components into the weighing container. To achieve the lowest possible tolerance for the individual weights, a distinction is made between coarse (or gross) dosing and fine dosing (Figure 2.18). The run-out time of the dosing screw and the product still falling into the container at the time of standstill must also be considered. The settings are specified in the dosing program and optimized using appropriate software.

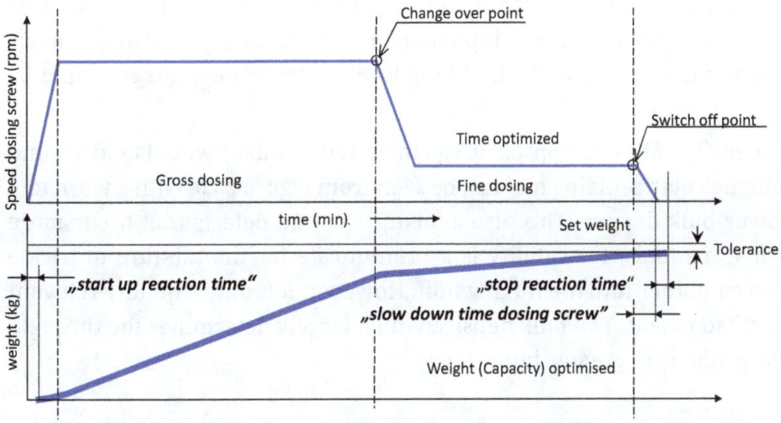

Figure 2.18 Dependence of dosing time on weight and screw speed

Cycle time

The cycle time of the mixing process in conjunction with the capacity of the mixing line determines the maximum throughput of the calendering line. When mixing, the PVC powder must be mixed with the additives, e.g. external lubricants. Due to the friction of the PVC grains, the additives heat up, melt, and can thus diffuse into the PVC grain. At a temperature of around 50–60°C (122–140°F), the liquid mixture components, such as stabilizers and liquid lubricants, must then also be injected to prevent thermal degradation during further heating. After reaching the desired temperature in the hot mixer, the mixture is emptied into the cooling mixer (Figure 2.19). There it is cooled to such an extent that no thermal damage occurs until it is plasticized, and the material remains free-flowing. When mixing "suspension PVC" (S-PVC), it is crucial to break the "skin" surrounding the grain.

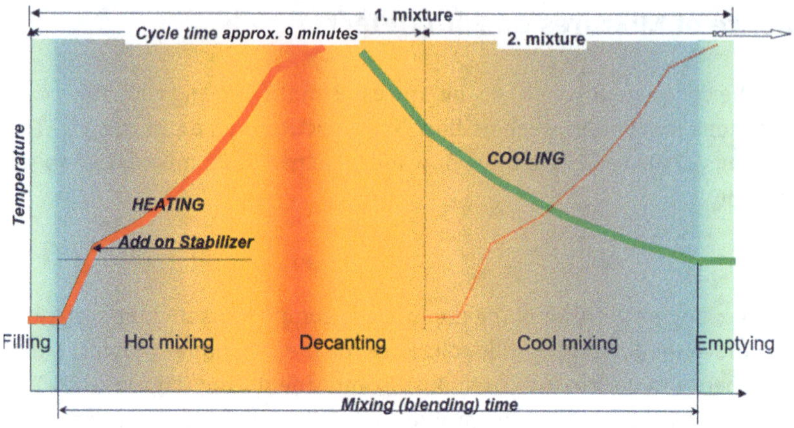

Figure 2.19 The stages of a typical mixing cycle

Flowability and bulk density [5]

A higher bulk density can be achieved by adding "micronizer" regrind, which also increases the energy input. The energy input depends on the filling level of the mixer and the condition of the mixing tool. The filling level of the heating mixer should be 85%.

Given the bulk density of PVC of approx. 0.5 g/cm^3, a 1200 L mixer would hold around 510 kg. The influence of the mixing tool can be seen from Figure 2.20, with a worn tool leading to a lower bulk density. This also explains the film defects that occur more frequently with worn tools. Flowability is a prerequisite for the mixture to trickle from the storage container into the fluxing unit. However, it is only a quality-relevant property to a limited extent. The bulk density, which largely determines the throughput in the fluxing unit, is of greater importance.

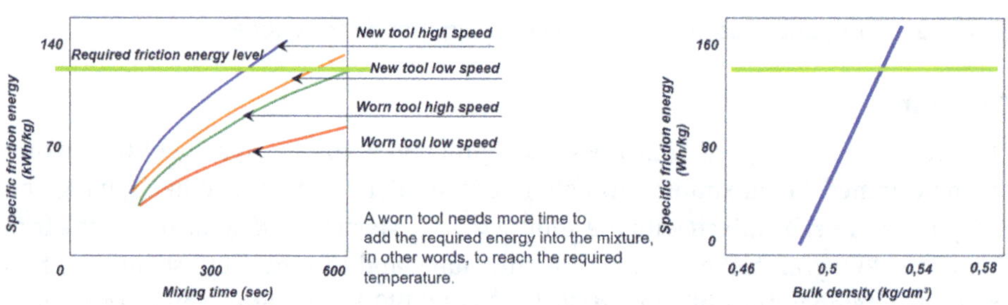

Figure 2.20 Dependence of specific friction energy on (left) mixing time and (right) bulk density. A worn tool needs more time to add the required energy into the mixture; in other words, to reach the required temperature (Source: [5])

Mixing tool speed and blending time

The peripheral speed of the mixing tool in the hot mixer must be adjusted so that a high mixing quality is achieved, and the dry blend components are mixed homogeneously without agglomeration taking place (Figure 2.21). This is achieved by selecting the speed of the mixing tool so that the time required for homogenization is reached after around 6–8 min and the temperature is not more than 130°C. As an indication, a value of 20–32 m/s can be set for the mixing tool. With a diameter of the mixing container of approx. 1.2 m, this corresponds to a speed of approx. 470 rpm.

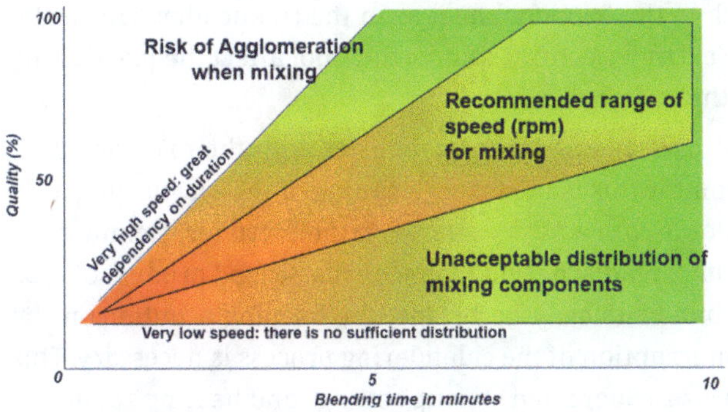

Figure 2.21
Effect of blending time upon dry blend quality

2.2.4 Basic Settings of the Calender Line as a Basis for Troubleshooting

The following information can only be a guide to checking the basic settings of the calender. These basic setting parameters are the key to troubleshooting when quality problems are observed that are not due to the influence of contamination in the area around the production line or the raw material.

The final settings depend on many different influences. Therefore, the setting parameters must always be adapted to the actual situation to achieve optimal quality.

2.2.4.1 Fluxing Unit

Co-kneader [6]

Since the co-kneaders plastify the powdery mixture mainly through frictional heat, it is very difficult to start the kneader manually. Before the plasticizing can begin, the process chamber must be completely filled. This can be seen when the electrical load (amperes) increases. At this moment, the speed of the filling screw must be reduced to avoid blocking the screw conveyor. Another option is to close the output and only open it quickly when the electrical load increases. In addition, the temperature of the heating-cooling units (HCUs) must be high to start gelling, and must be reduced immediately when the plasticizing process starts. Today, this task is taken over by the software in the startup program.

The co-kneader should only be started in automatic mode, and the screw rotation of the co-kneader should be approx. 30 rpm (the hopper feeding screw should be 1–3 rpm). The HCU temperature should be 170°C (it better that they are too hot than too cold), but not more than 190°C.

Planetary gear extruder (PGE) [7]

The plasticizing formation in the PGE is linked to the temperature in the different heating zones and not as much to the frictional heat as in the co-kneader. This is the reason why the rotation of the extruder central spindle does not affect the plasticizing of the material as much as in the co-kneader.

A type or color change can be carried out faster and easier with the extruder than with the co-kneader. The reason for this is the faster cleaning of the plasticizing area (completing the plasticization of mixture "A" by emptying the extruder, cleaning the plasticized part by blowing it out, and discharging the new plasticized product of type "B" takes about 5 min). This time can easily be bridged by forming a buffer on the belts and in gap 1, so that no interruption of the calendering process is necessary. This is important for smaller batch sizes. Suggested startup settings and heating zone temperature limits depending on the amount of polyvinyl chloride copolymer (COPO) are shown in Table 2.2:

Table 2.2 Suggested Startup Settings and Heating Zone Temperature Limits

Film type	>15% COPO	<15% COPO	Max. value
Central spindle	170°C	180°C	<190°C
Heating zone 1	180°C	185°C	<190°C
Heating zone 2	190°C	200°C	Max. 210°C

The central spindle speed must be adjusted according to the calender output. To ensure good quality, we must check the plasticization and adjust the values. If the material has non-plasticized areas, first heat up zone 2, then zone 1 and finally the central spindle; if it is too plasticized, cool down the central spindle first, then zone 1, followed by zone 2.

Two-roll mixing-mill (if any)

If the calender line does not have a rolling mill, it is important that there are no gas bubbles in the material. In this case, effective vacuum suction is required for both gelling options. This is particularly important with the co-kneader.

The rolling mill has the task of homogenizing the material emerging from the PGE or co-kneader even more evenly, especially regarding the smallest mixture components, such as colors. It also serves to degas the material. It also allows the edge stripes and a certain amount of foil to be fed directly without granulating them. Of course, you must pay attention to the same types and colors as well as absolute cleanliness.

Three criteria are important for smooth production when operating a rolling mill.

- The **hide thickness** should be set to 4–6 mm by adjusting the roller gap.
- The **roll speed** of the 2-roll mixing-mill should be in the range 35–45 m/min. The friction can have a fixed value, and is 1:1.16 up to 1:1.25.
- The **hide temperature** depends on the amount of COPO. If the COPO content is less than 15%, the temperature should be around 182°C and if the COPO content is more than 15%, it should be 178°C.

A useful tip is that the thicker the film, the more homogeneous the plasticization must be—the material in gap 1 must rotate and not too much should be present.

Feed conveyor belt and wig-wag conveyor belt

The transport of the gelled material to the first calender gap must take place as quickly as possible to avoid excessive cooling.

The belt speed must be correlated to the calender performance. The speed is normally in the range 4–40 m/min. The oscillation of the swivel belt is more important for the even filling of the first gap, and should be approximately 12 cycles/min, with a short stop at the edges.

Calender and post-calender train

In the calender, the doughy mass produced in the co-kneader or PGE (and mixed and degassed again on the 2-roll mixing-mill) is rolled out into a film. The individual components of the mixture are only mixed to a very small extent in the calender.

The setting values of the calender are crucial for the film quality, and must therefore be carefully selected and adjusted accordingly. In conjunction with the parameters of the succession, a high-quality plastic film is created.

Below is a suggestion for the basic setting of a calender. In practice, a corresponding database is very helpful for this. Despite all efforts, quality deviations can occasionally occur. Therefore, the most important parameters that need to be checked in case of difficulties are listed here:

- **Roller and hide temperature:** Measure the roller surface temperature manually and compare with the setting values. If there is a deviation of more than 3 K or a fluctuation in the values, inform the maintenance service to find the cause and eliminate it. Measuring the hide temperature also gives good indications of irregularities

- **Roller circumferential speed:** Measure the circumferential speed manually and compare with the setting values. If a deviation of more than 1 m/min or a fluctuation is detected, contact the maintenance service to restore compliance
- **Bank size:** The size of the bank in the gap is one of the most important setting parameters for a high-quality film with good flatness. It affects the flow lines, the air bubbles/craters, the friction streaks, and the burnings. These quality aspects are directly linked to the gap pressure. A bank that is too small leads to air bubbles and flow lines due to the lower pressure, while a bank that is too large leads to surface defects due to the high gap pressure.

Experience in calendering operations leads to the limits stated in Figure 2.22 (left) for the first gap between roller 1 and roller 2. Figure 2.22 (right) applies to the last gap between roller 4 and roller 5. The other gaps can be slightly larger than the last gap.

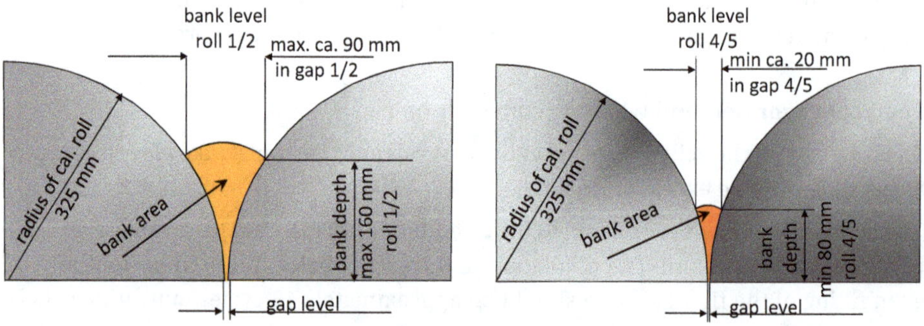

Figure 2.22 Gaps between rollers 1 and 2, as well as between rollers 4 and 5, of a 5-roll calender

It is important that all banks rotate smoothly and do not form "tongues". These cool down faster than the rest of the bank and lead to temperature differences on the edges, which in turn has a negative impact on quality. The second important relationship for the final film quality is the friction between the rolls and the relationship between the take-off speed and the speed of the penultimate calender roll.

Avoiding roller damage

Today's calenders, designed according to the state of the art, have a hydraulic fast opening of at least the gap of the penultimate and last calender rolls. This is not the case with older systems and cannot be retrofitted at a reasonable cost.

The system of the calender stands, bending and pre-stressing devices, and bearing clearance can be seen as a kind of "spring" that is pushed apart by the gap pressure (i.e. "tightened"), and that will collapse if the gap pressure is suddenly lost (i.e. "relaxed").

If the amount of “stretching” can then be calculated or estimated, conclusions can be drawn on the film thickness up to which there is risk of damage to a roll. Experience shows that these values can be up to a gap level of 450 µm. Damage can also occur between rollers 1 and 2 with thin films, but the danger ends with a gap width of approx. 250 µm. The gap between two rollers can be calculated from the friction conditions, as shown in Figure 2.23.

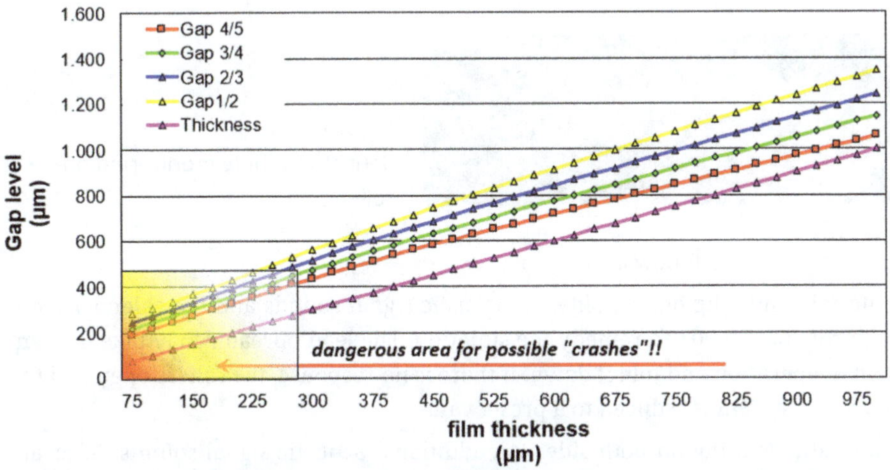

Figure 2.23 Roll gap as a function of film thickness

To recognize the potential danger even better, a further assessment can be made by calculating the dwell time in the banks as shown in Figure 2.24.

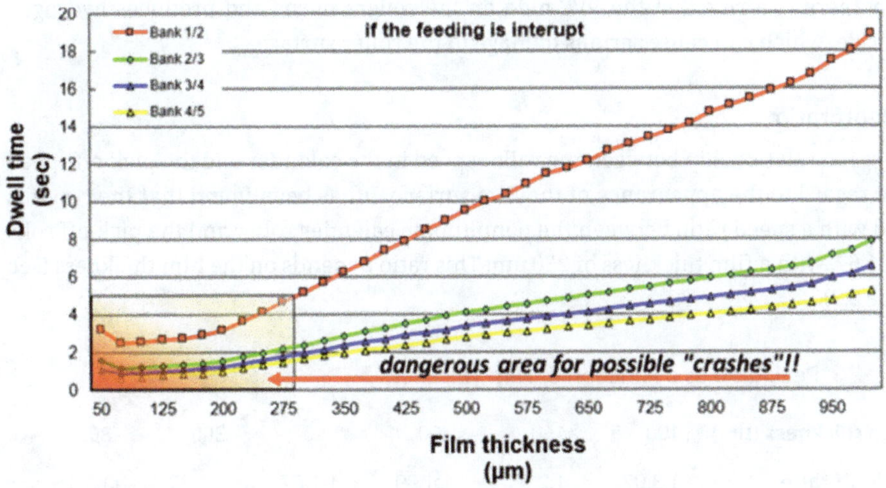

Figure 2.24 Dwell time in the roll gap as a function of film thickness

This diagram makes clear that, depending on the situation and the film thickness, the calendering team have a maximum of 5–10 s to take countermeasures. The most dangerous thing is, if the last gap begins to run empty, then only about 1–2 s will remain, making a “touchdown” almost unavoidable.

The two sensors shown in Figure 2.25 monitor the occupancy of roller 2 with material. They are positioned approximately 150–200 mm from the edge of the hide on the roller.

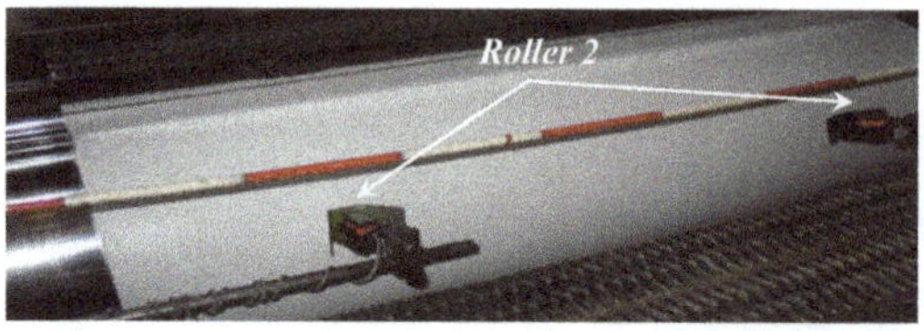

Figure 2.25 Sensors for hide monitoring on roller 2

The sensors work as follows:

If material is missing on one side, an acoustic signal sounds and the calender team has approximately 90 s to remedy the situation. The lead opreator may take the sensor out of operation and fold it down. If there is no response, the rollers open and the speed of the system is reduced to a preset value.

If material is missing on both sides, an additional acoustic signal sounds. After approx. 5 s, the rollers open at high speed and the calender line switches off. This is not an “emergency stop”; instead, it is termed “rolling out”. However, modern calenders have a fast-opening device, so that this procedure is not relevant.

Since the rollers are covered with material, it is important to remove it from the rollers starting with the last roll, followed by roller 4, etc. If this is not done as quickly as possible, there is a risk that the PVC-hide on the rollers burns and produces hydrogen chloride, which can cause serious damage to the roller surface.

Pick-off ratio

There is a relationship between the roller speed in the calender and the pick-off speed with regard to the appearance of the film surface. It has been found that the result is good with a speed ratio between the penultimate calender roller and the pick-off rolls of 1:1.67, with a film thickness of 250 µm. This ratio depends on the film thickness (see Table 2.3).

Table 2.3 Peel Ratio as a Function of Film Thickness

Film thickness (µm)	100	150	200	250	300	800
Pick-off ratio	1:3.02	1:2.63	1:1.89	1:1.67	1:1.53	1:1.15

Surfaces of the following rollers

The surfaces of the rollers downstream, especially in the take-off and temperature control areas, are important in order to avoid surface defects. It should be noted that the film adheres better to glossy surfaces than to matte ones. On the other hand, the better adhesion also leads to problems with wax stains or picking marks if the film partially "sticks" more strongly. The two opposite characteristics can be combined.

The first two take-off rollers should have high-gloss across their entire width, possibly pixelated, to ensure that the hot film is pulled off evenly from the last calender roller and to avoid excessive constriction. To prevent wax stains, the rollers must be constantly cleaned. This can be done by rotating rollers wrapped in absorbent paper or by stationary "rollers" wrapped in "Molton" cloth. The remaining rollers are matted in the middle area up to approx. 200 mm from the edges, with the remaining 200 mm remaining high-gloss. This ensures that the film adheres better in the area of the high-gloss edges, and that the constriction is reduced. In the middle area, the lower adhesion of the film to the rollers prevents waxing and picking marks.

Web speed in the post-calender units

The web speed in the following units, more precisely the speed at which the web is wound up, is also referred to as the recommended speed. It depends primarily on the film thickness and, in rare cases, also on the type of film produced or the film surface. In addition to the production width, the recommended speed is the measure of the throughput in the calender. As a rule, the web speed decreases slightly within the post-calender units in order to take account of the shrinkage due to cooling.

With this known standard speed and the ratio of the speed between the penultimate calender roller and the speed of the pick-off-unit, the basic setting values in the calender can be represented with known friction between the rolls. In Figure 2.26, the speed in the thin area is limited by the maximum speed of roller 5 at 60 m/min.

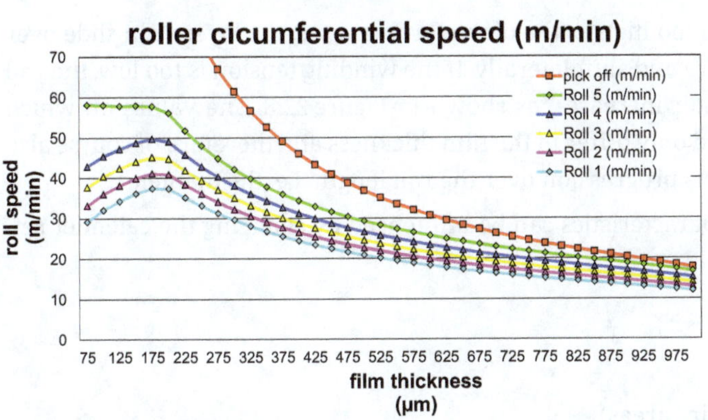

Figure 2.26 Roller peripheral speed as a function of film thickness

Winder

The last production stage of the calender line is the winder. Since the film subsequently becomes statically charged due to the constant redirection during transport over the rollers, the static electrical charge must be discharged before winding up. Otherwise, the risk of telescoping is not guaranteed even if the winding tension is adjusted correctly.

The flatness is ultimately influenced by the setting and regulation of the winding tension, especially with thin films (50–200 µm) (Figure 2.27). When winding, the film should be as cold as possible to prevent "stretching" in slightly thicker areas. It is therefore recommended to install a pair of rollers with good cooling after the edge trimming, shortly before winding. The telescopic, straight-edge winding of the jumbo reel is also determined by the course of the winding tension.

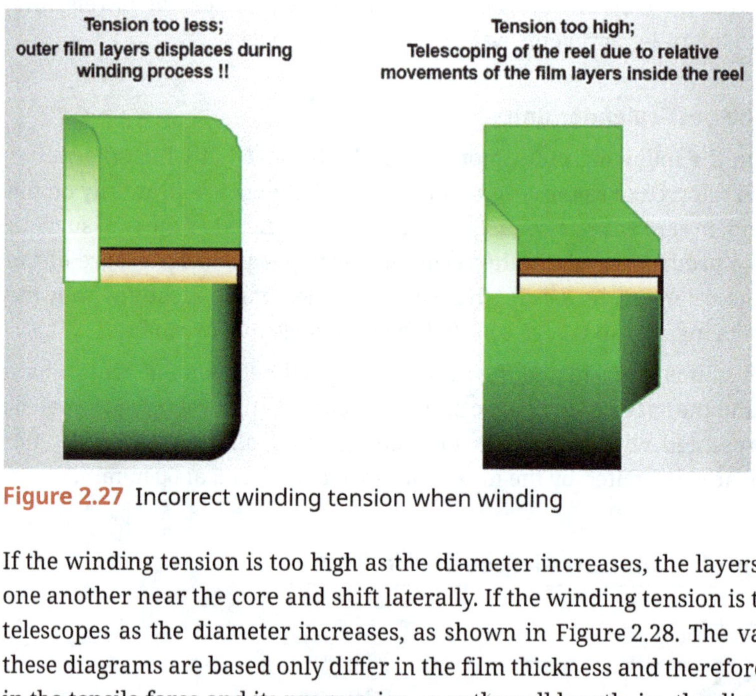

Figure 2.27 Incorrect winding tension when winding

If the winding tension is too high as the diameter increases, the layers can slide over one another near the core and shift laterally. If the winding tension is too low, the roll telescopes as the diameter increases, as shown in Figure 2.28. The values on which these diagrams are based only differ in the film thickness and therefore of course also in the tensile force and its progression over the roll length, i.e. the diameter.

The following quality characteristics can be influenced by changing the calender settings:

- Flow lines
- Friction stripes
- Burns (stripes or entire areas)

- Flatness (poor flatness can also have other reasons, and so this problem is dealt with in Section 2.2.6)
- Gas inclusions (air bubbles). These slide errors can also have causes other than the calender setting, and this is therefore explained in more detail in Section 2.2.6. This particularly includes air cracks (craters).

To resolve problems with one of the above-mentioned quality characteristics, the steps suggested in the following chapters sections should help.

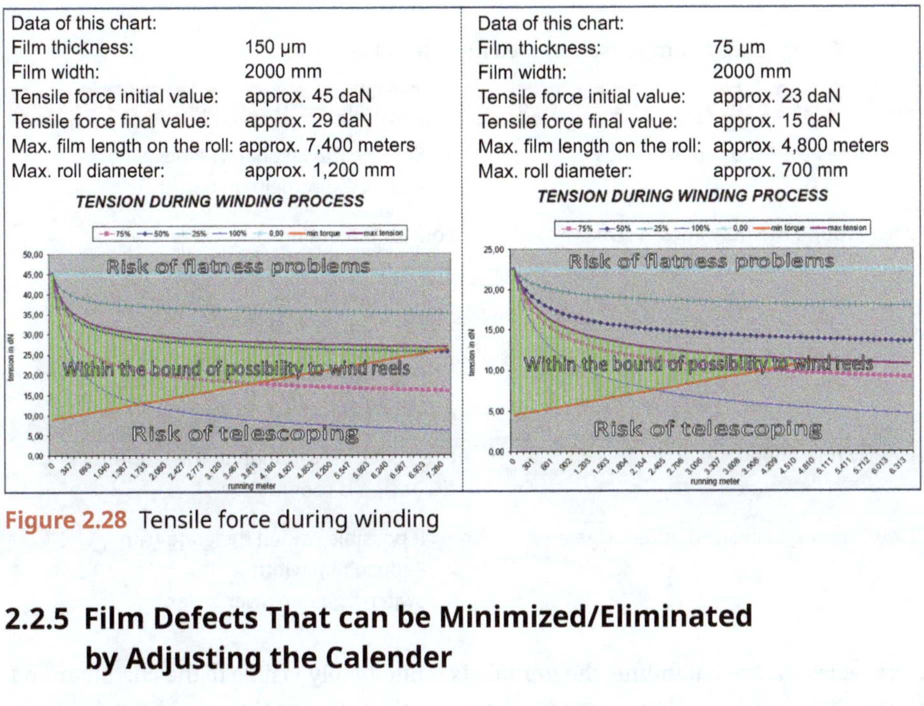

Figure 2.28 Tensile force during winding

2.2.5 Film Defects That can be Minimized/Eliminated by Adjusting the Calender

Flow lines

Flow lines arise from the rotation of the banks and are a typical sign of a calendered film. The goal can only be to reduce these flow lines to a minimum so that there are no restrictions on processing for the customer.

This is particularly important for transparent cardboard boxes (e.g. round cans), office supplies (e.g. transparent sleeves), items for printing, decoration, and advertising. When calendering, a smooth rotational movement of the bank in the last gap is therefore important.

The measures to reduce the flow lines first begin with a check of the basic settings of the driving values, as described in Section 2.2.3. Afterwards, the appearance of the flow lines must be analyzed, and then the improvement measures can be initiated (Figure 2.29). Some compensation suggestions are listed in Table 2.4.

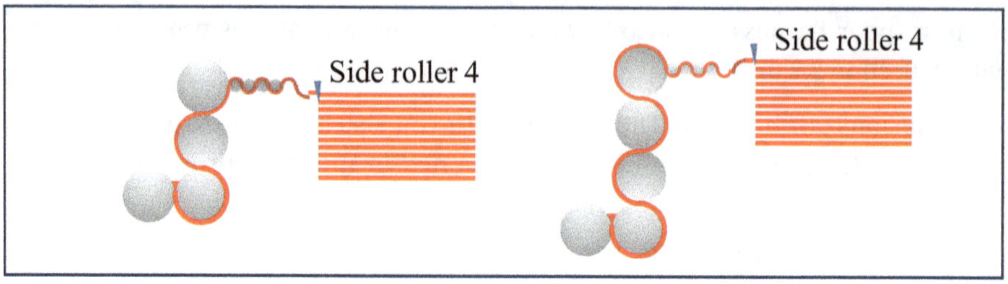

Figure 2.29 Problems with flow lines on 4- or 5-roll calenders

Table 2.4 Suggested Compensation Measures for Flow Lines

Problem	Suggested actions
Christmas-tree-shaped flow lines	1. Enlarge the last bank (especially) 2. Adjust the remaining banks
Area around the flow lines is streaky matt	Four-roll calender: 1. Heat up roller 2 2. Cool down roller 3.
	Five-roll calender: 1. Heat up roller 3 2. Cool down roller 4.
Greasy flow lines	Feed back information to the laboratory—the recipe is too "fat"
Flow lines exclusively in outer edge area	▪ If possible, widen the edge trim ▪ Reduce film width ▪ Switch to a narrower order.

For sheets that are cut in-line, the top side is immediately visible to the customer and is usually printed. It is important to note that this is always the side of the 4th calender roller. This means that films from a 4-roll calender present the better side, whereas films from a 5-roll calender present the worse side, unless the pallets are turned.

Flow lines can be "hidden" by embossing. The embossing roller can be high-gloss or have fine pixels (this also improves printability). Corona treatment also improves printability.

To make the flow lines invisible on both sides, the film can be passed through a gap formed by two glossy or pixel rollers. It is important to ensure that the rolls cannot touch each other, even if there is no film between them (automatic opening).

Friction stripes

Friction stripes (also called rubbing stripes) occur when the hide partially adheres to the rollers. The reasons for this can vary. The cause usually lies in the calender driving values that have been set, uneven adhesion on the roll surface, irregularities in the mixing (incorrect mixing), or insufficient homogenization, e.g. due to mixing times that are too short or worn mixing tools.

However, it is also possible that the plasticization in the fluxing unit (usually in the co-kneader) was insufficient because the process chamber was not completely filled.

Friction stripes can become dangerous if the entire hide sticks to the calender roll. The risk that the calender runs empty is then very high. Therefore, corrective measures must be taken immediately (Table 2.5). This work is carried out directly on the rollers, i.e. in the danger area ("line of fire"). It may therefore only be carried out by experienced and specially trained people. Whenever possible, aids must be used, as shown in Figure 2.30.

Table 2.5 Suggested Compensation Measures for Friction Stripes

Problem	Suggested actions
Shortly after starting	▪ Rub the rollers thoroughly with a dry cotton cloth (see Practical tip 1).
After the gap in the transition from one roller to the other	▪ Rub the rollers with a dry cotton cloth and Viennese lime (see Practical tip 2).
Friction stripes visible in the film	▪ Increase friction slightly ▪ Check external lubricant.
Pickoff line wobbles or moves up and down	▪ Information to the laboratory ▪ Check external lubricant.

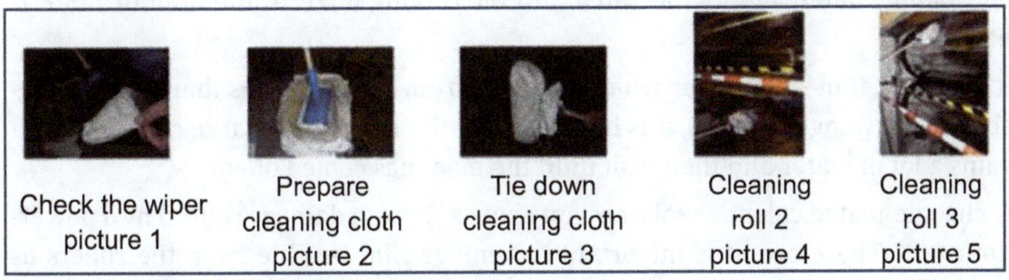

Figure 2.30 Elimination of friction stripes

This is work in the "increased danger zone"! This work involves a higher risk and therefore requires the permission of the shift manager, and can only be carried out by specially trained operators. It is also recommended to use tools with an extension to be able to do the work without getting into the danger zone in front of the rollers.

1. Always carry out work shortly after starting, before increasing to the recommended speed and the need to restrict access to the rollers comes into force:
   - Wipe all rollers dry—you can use AJAX or "greenstone"
   - Check every hour during the first shift that the hide detaches easily from the rollers
   - If a deposit is found, the rollers must be rubbed with a dry cotton cloth.
2. Wipe again if friction stripes are observed during production.
   - In more difficult cases, the cloth can be moistened with Shellsol T (be careful—note the safety data sheet) and coated with greenstone or diamond paste. Act carefully. First start in a smaller area and wait one turn to see whether the sticking becomes stronger. Only then do large-scale work. To do this, use tools such as a wiper with a handle. Finally, be sure to wipe with a dry cloth.

Burnings

Burnings (thermal degradation) can occur anywhere in the production process, including in the hot mixer. Severe thermal damage is a very serious incident due to the evolution of HCl gas, which combines with the moisture in the air to form hydrochloric acid, which attacks metals, including stainless steel. Therefore, severe burnings musts be avoided whenever possible, and the hot mixer must have an emergency outlet to immediately empty the batch into the cooling mixer if the limit of 135°C is exceeded.

If leftover PVC film—no matter where it comes from—decomposes thermally (recognizable by the pungent smell), it is best to throw the mass into a large container that contains a lot of water and then wait until the mass has cooled down.

Even chrome-plated calender rolls are not protected from damage caused by a powerful "burner". Therefore, it is important to remove film residue from the rollers as quickly as possible. If the chrome surface of the rollers is thermally damaged (recognizable by matt spots), the rollers must be changed. Polishing does not help here. Non-chrome-plated rollers can be polished in the calender, but this takes time.

Burnt material can be removed from calender rolls by the following process—but note, this is dangerous work in an increased danger zone ("line of fire").

- The material must be removed from the rollers as quickly as possible
- First operate the calender at slow speed or, if necessary, switch it off
- You must start with the last reel, then the next reels
- Personal protective equipment must be used (gloves and safety glasses)
- Spray the hide intensively with "polishing oil" or coat it with "stearin wax"
- Lift the hide at one point and blow air (compressed air) between the hide and the roller
- Once the hide is removed, polish all rollers with 1 µm diamond paste and—whenever possible—make the restart with opaque film.

Even if the calender stops due to a power failure, the hide must be removed as quickly as possible. The emergency power generator must be used for this. If it does not work or there is not one, the reels need to be spun manually. This can be done with a chain pipe wrench that matches the drive shaft diameter. To do this, turn the rollers backwards and first remove the material from the banks.

If burnings are observed, the following actions are suggested:

- First, determine the cause and source of the thermal degradation, and note the appearance of the "burnings"
- Take a sample from the "top" and "bottom" of the storage container in front of the plasticization unit and keep it ready for a laboratory test
- Manually measure the mass temperatures of the gelled material, the skin temperatures on the rolling mill and on the calender rolls, and manually measure the surface temperatures of the rolling mill and calender rolls, and compare them with the settings.

Further measures are listed in Table 2.6.

Table 2.6 Suggested Compensation Measures for Burnings

Problem	Suggested actions
Burned areas or color degradation can already be noticed in the material out of the fluxing unit	■ Immediately inform the laboratory for appropriate information, and consider the following countermeasures ■ Add some stabilizer into the dry blend before entering the fluxing unit ■ In serious cases, stop and open the calender.
Color degradation on the hide in the 2-roll mixing-mill or in the calender	■ Reduce the amount of material in the gap ■ Check stabilizer dosage.

Table 2.6 Suggested Compensation Measures for Burnings (*continued*)

Problem	Suggested actions
Burned stripes in the hide on the 2-roll-mixing-mill	▪ Adjust the roller gap filling more evenly ▪ Adjust turning bucket settings ▪ Stop adding remaining foil and check for burned areas.
Burned spots stick to the rollers	▪ Rub the affected areas of the rollers intensively but carefully with greenstone or diamond paste on a cloth ▪ If necessary, stop and open the calender and polish it.
The whole hide sticks to the rollers	▪ Stop and open the calender ▪ Read the instructions regarding the removal of burnt material from calender rolls (see above)
Intense yellowish discoloration of the hide	▪ Inform the laboratory for appropriate information ▪ Add stabilizer, and possibly also internal lubricant.
Burnings appear as short, narrow lines	▪ Immediately inform the laboratory for appropriate information, and consider the following causes/countermeasures ▪ Dry blend not homogeneous—insufficient mixing ▪ Lubricant content too low ▪ Dosage incorrect ▪ Batch not homogeneous—increase mixing time ▪ Mixing tool worn out.
Burnings appear as long, narrow streaks	▪ Increase pick-off speed (increase gap), especially with thin films

2.2.6 Film Defects That are Subject to Both Calender Settings and External Influences

Bad flatness (cross-profile)

If a flatness problem is identified, the film cross-section must be known. This can be seen from a diagram of the film cross-section. To do this, we can use a printout of the thickness gauge recordings, or the cross-profile of the film must be recorded manually. To do this, the thickness is measured every 50 mm with a micrometer across the entire width of the film, and the values are noted on a sheet of paper, preferably graph paper.

Another option is to tap the finished roll with a stick. Areas with a bright tone and a “spring back” of the rod are thicker than areas with a dull tone and no “spring back” of the rod.

Sample dimensions: Width: full width / Length: 2.500 mm

Level	Defect / acceptable Waviness	numbers of waves per meter lengthwise	width of the waves	Sample
0	Film is free of defect	none	none	
1	very light	(1 wave at edge and/or centre per meter)	Width of wave approx. 200 mm	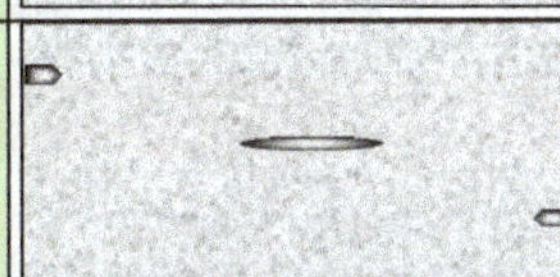
1a	slight	(2 waves at edge and/or centre per meter)	Width of wave approx. 300 mm	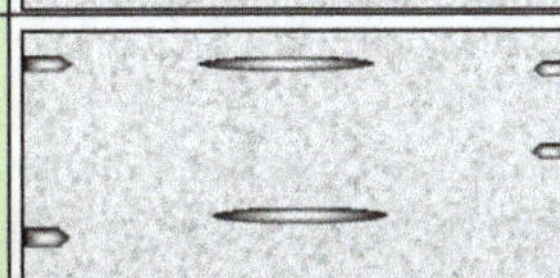
2	pronounced	(4 waves at edge and/or centre per meter)	Width of wave approx. 300 mm	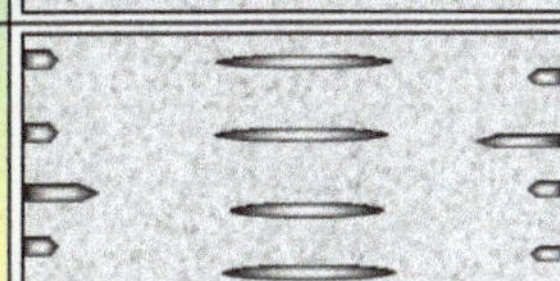
2a	significant	(5 waves at edge and/or centre per meter)	Width of wave approx. 500 mm	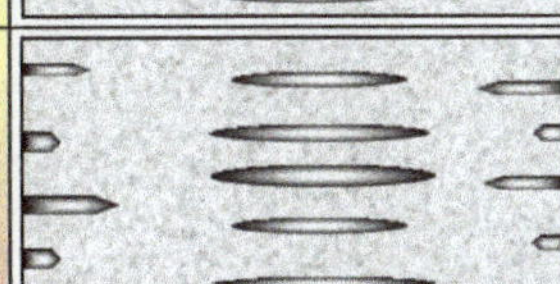
3	heavy	(6 waves at edge and/or centre per meter)	Width of wave approx. 600 mm	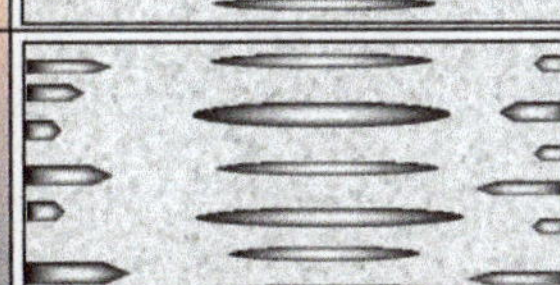
3a	extreme	(>6 waves at edge and/or centre per meter)	Width of wave approx. 600 mm	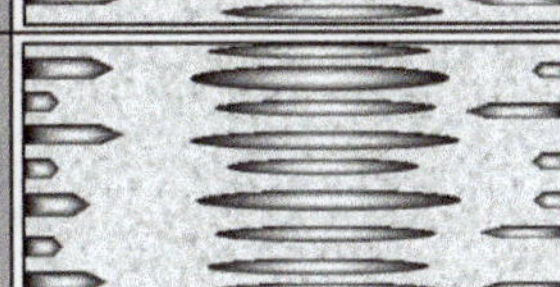

Figure 2.31 Assessment of the flatness for a pattern size with a full calender width, and a length of 2.5 m

Today it is usual that calenders are equipped with a thickness gauge. This control loop effectively helps to keep the thickness distribution across the running direction within narrow limits. However, a few principles should be observed:

- If the film is too thick in the middle, the bending pressure should first be increased, up to the maximum
- If the film is too thin in the middle, then the axis crossing adjustment should first be reduced until the penultimate roller and the last roller are parallel (value "0")
- If the film is too thick (thin) overall, then all gaps must be reduced (enlarged)
- This representation can be used to assess the flatness (Figure 2.31). It can also serve as a criterion if everyone involved agrees, thereby avoiding lengthy discussions.

If poor flatness is observed, then the actions listed in Table 2.7 are suggested.

Table 2.7 Suggested Compensation Measures for Poor Flatness

Problem	Suggested actions
The thickness profile shows different values at the edges	▪ Temporarily switch off the thickness control. ▪ Manually adjust the calender gap so that the thickness is the same on both edges.
The film thickness only increases sharply in the outermost edge area	▪ The film is too wide for this calender. If possible, reduce the width or produce a different (narrower) order

The film constricts strongly on the first and/or second pick-off roll	▪ Insert/position the spreader rollers as shown in the picture below
	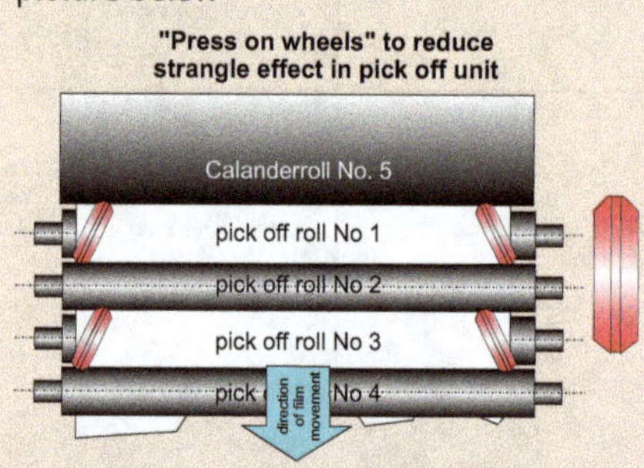
	▪ Matt the subsequent rollers after the second pick-off roll in the middle area
	▪ Electrostatic charging rods can also be used to reduce constriction of the film web. However, great caution is required:
	  ▪ The tip of the rod must be protected against contact
	  ▪ The charging rods must be switched off as soon as there is no material on the roller, and during starting-up and switching-off the line.

Problem	Suggested actions
Speed fluctuations in the calender or pick-off-unit (ampere fluctuations)	▪ Stop and open the calender ▪ Contact maintenance to find the cause and rectify it.
Speed fluctuations of one or more calender rolls after a calender-roll change	▪ Inform maintenance ▪ Check alignment of cardan shafts (possibly installed incorrectly)
The pick-off-line is not constant (it moves up and down or wobbles)	▪ Stop and open the line ▪ Polish the calender rolls with 1 µm diamond paste ▪ Polish pick-off rollers 1 and 2.

The film is too thick in the middle
Note: The film should always be approx. 1/100 thicker in the middle to support straight-edge winding
	▪ Increase bending pressure
	▪ If this is not possible (because the maximum pressure has been reached), increase axis crossing
	▪ If both the above are at a maximum, or if the bending pressure fluctuates:
	  ▪ Contact maintenance to bleed the hydraulic system or adjust the pressure in the nitrogen pressure tank
	  ▪ Adjust the axis crossing indication display.

The film is too thin in the middle
Note: If the film is too thin in the middle, there is a risk of wrinkles forming in the middle area when winding it up
	▪ Reduce axis crossing
	▪ If this is not possible, reduce the bending pressure but respect the minimum pressure
	▪ If both the above are at a minimum, or if the bending pressure fluctuates, contact technical service to:
	  ▪ Bleed the hydraulic system, or
	  ▪ Adjust the pressure in the nitrogen pressure tank.

"Piston rings" in the film become visible on the rolls	▪ Adjust the cleaning roller (wiper roll); it must be parallel to the pick-off roller ▪ Adjust the spreader rollers[a] as shown in the picture above ▪ Renew spreader rollers[a] ▪ Polish pick-off rollers 1 and 2 with diamond paste ▪ Replace the pick-off rollers with newly ground and chrome-plated rollers[a].
Helical darker spirals can be seen in the film *Note: This is only possible if the heating channels are helical*	▪ Heating channel of the first and/or second take-off roller clogged[a] (only visible in opaque films).

[a] In order to make this setting, maintenance must be informed. It may also be the case that the calender needs to be extended.

Thickness fluctuations in the running direction

In contrast to the flatness errors (thickness fluctuations in the cross-profile), thickness fluctuations in the film running direction (MDO) are not due to the setting but rather to the machine. Since the thickness measuring system always measures transversely to the web, it is possible that in the event of longitudinal fluctuations, an error in the cross-profile is "simulated". To record longitudinal fluctuations with the thickness gauge, the transverse movement must be switched off.

Table 2.8 Suggested Compensation Measures for Thickness Fluctuations in the Machine Direction (MDO)

Problem	Suggested actions
Film thickness varies in the direction of travel (MDO)	▪ Check whether the roller temperature control circulation pumps are running
	▪ Check whether the heating system pressure fluctuates or is too low ▪ To do this, the maintenance service must be informed. It may also be the case that the calender needs to be stopped and opened ▪ Check the filling level in the expansion tank ▪ Check positioner ▪ Check control valves (at 0.5 bar both must be closed) ▪ Avoid constant "opening and closing"!
Surface temperatures of the rollers (measurement is only possible with an analogue device)	▪ Check roller surface temperatures with a pyrometer ▪ (If there are fluctuations, a heating channel is blocked)

Arc running

If the film thickness is OK, but the unrolled film does not run straight, it is referred to as "arc running" (also called the "banana effect") (Figure 2.32). The possible causes can be found in Table 2.9.

Figure 2.32
Example of arc running

Table 2.9 Suggested Compensation Measures for Arc Running

Problem	For checking	Suggested actions[a]
Unrolled film runs in an arc	Does the axis-crossing of the penultimate roller have the same values on both sides (see diagram)?	If not, correct the setting to the same values on both sides[b]
	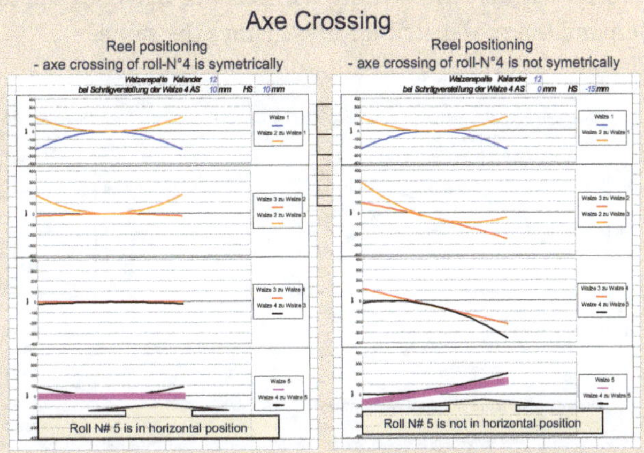	
	Measure roller temperatures on both sides (HS & DS) manually with a pyrometer (calender and all heated rolls in the post-calender train)	If the temperature difference from one side to the other is greater than 3°C, the rollers must be replaced.
	Check all rolls of the post-calender train to ensure that they are horizontal and parallel to the axis	If deviations are found (often visible through one-sided "fluttering edges"), realign the rollers.
	Is the position of the winder exactly aligned with the calender line (post-calender unit)?	If the winder is displaced, realign it.
	Are the winding axes (star/discs) horizontal and parallel to the rollers in front of the winder?	If the winding axes are twisted, realign them.

[a] Maintenance must be informed. It may also be the case that the calender needs to be stopped and opened.

[b] If the inclination adjustment is not symmetrical, then roller 5 is at an angle to the take-off rollers (see diagram for a 5-roll L-calender).

Persistent problems with flatness and thickness distribution

If flatness cannot be improved by the above-mentioned measures, this may be due to poor calender roll grinding. It is possible that the deviation in film thickness is caused by a deviation in the profile of the calender roll. To find out, several thickness measurements must be taken at different values of the axis crossing and roll bending pressure settings. Figure 2.33 provides an indication of how this can be accomplished. It turns out that the thickness difference between the theoretical thickness distribution and the real distribution is always the same.

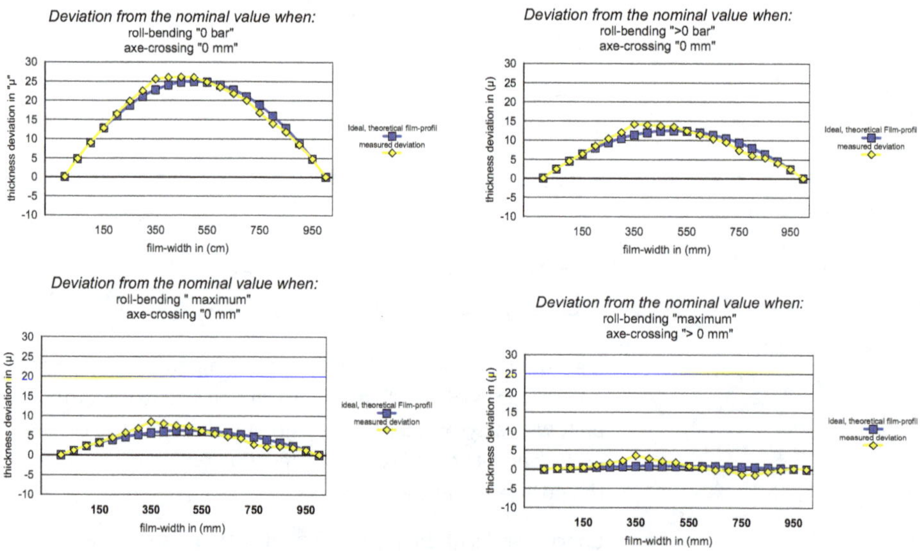

Figure 2.33 Errors in grinding uncovered by thickness measurements

2.2.7 Film Defects That are Predominantly or Exclusively due to External Influences, Such as Foreign Bodies

Modern calender lines are equipped with a surface inspection unit, which takes photos of the defect and makes information about it available immediately (usually). In this section, we will provide information on how film defects can be detected, even in older systems that do not have a film inspection unit.

This section does not cover further information about the general condition of the calender line and its basic parameters, namely:

- Polishing the calender rolls
- Cleaning interval and the cleanliness of the air intake filters
- Basic settings of the calender

- Recommended speed
- Use of production surplus
- General cleanliness of the surroundings of the calender and the calender line itself.

Regarding the latter point, it is important that the area around the production facilities is clean, and free from insects, and this point is not mentioned in the sections that follow. Specifically, it is important to keep the doors and windows closed. Ventilation should always function properly and lead to a slight excess pressure in the production rooms. To prevent insects from entering the production rooms, special light traps must be installed. However, these devices should not be positioned in the immediate vicinity of the production facilities, but rather away from them in poorly lit corners. The placement should be at an angle of 90° to light sources such as lamps, windows, etc. and the height—if possible—should be approx. 2.5 m. Regardless of whether sticky traps or blue light traps are used, regular maintenance is required. The number and type of insects and, of course, the period of control must be considered.

It is not so important to produce the film in a cleanroom, but rather in a *cleaned* room!

The following information is intended to provide guidance on what to do if there are deviations in film quality due to foreign particles. Please note:

- A claim to completeness cannot be made
- Tips, comments, and hints are the basis for improvements
- New findings should be included in lists!

2.2.7.1 Air Bubbles

The causes of air bubbles are varied, and can result from both settings and external influences. This problem will first be discussed in more detail here, as finding the cause is just as necessary as quality problems caused by external influences. To do this, you must first proceed in the same way as when searching for foreign particles. The microscope provides a very good aid for the assessment, as described in Section 2.2.2.

Air or gas bubbles that are not coupled with a foreign body arise due to the setting parameters of the calender or the plasticization unit. However, not only should the gas bubbles enclosed in the film be examined, but also the open "craters". An indication of an external influence is given when the air bubble sticks to a foreign body. This problem is also considered in Figure 2.34 for the corresponding foreign body. Table 2.10 lists actions that can be taken if an air bubble is observed

Table 2.10 Suggested Compensation Measures for Air Bubbles or Craters

Appearance	Cause	Suggested actions
Bubbles are distributed throughout the film with no apparent structure	Too much "air" in the plasticized mass	▪ Check vacuum suction ▪ Change and clean the powder container.
Bubble "chains" are present in the running direction	Air under the bank in the gap between roller 2 and roller 3[a]	▪ Increase mass temperature ▪ Reduce material in gap between roller 1 and roller 2 ▪ Increase temperature of roller 1 and roller 2 ▪ Install a hide ripper roller on roller 2.
Bubbles are along the flow lines	Hide too hot	▪ Heat down the two penultimate rollers ▪ Increase pick-off ratio.
Bubbles are in the same place after each revolution	Particle sticks on the last roller	▪ Rub the last two rollers with a dry Molton cloth[b]
Small bubbles	Last roller is too hot	▪ Reduce the temperature of the last calender roller
Larger craters	Hide on calender roller is too thick	▪ Reduce friction ▪ Reduce the gap and reduce the pick-off ratio.
Very large craters	Newly chromed rollers	▪ Stop and open the calender ▪ Heat the rollers up to 200°C for 24 h, and then allow to cool down.

[a] The feed material must already rotate in the gap between rollers 1 and 2. The formation of "tongues" must be avoided.

[b] A cloth moistened with polishing oil and diamond paste can be used first, but the final step should be wiping with a dry cloth.

2.2.7.2 Foreign Particles

Figure 2.34 serves as a guide to the procedure for determining the nature of foreign particles in the film, with individual sources of contamination being discussed below.

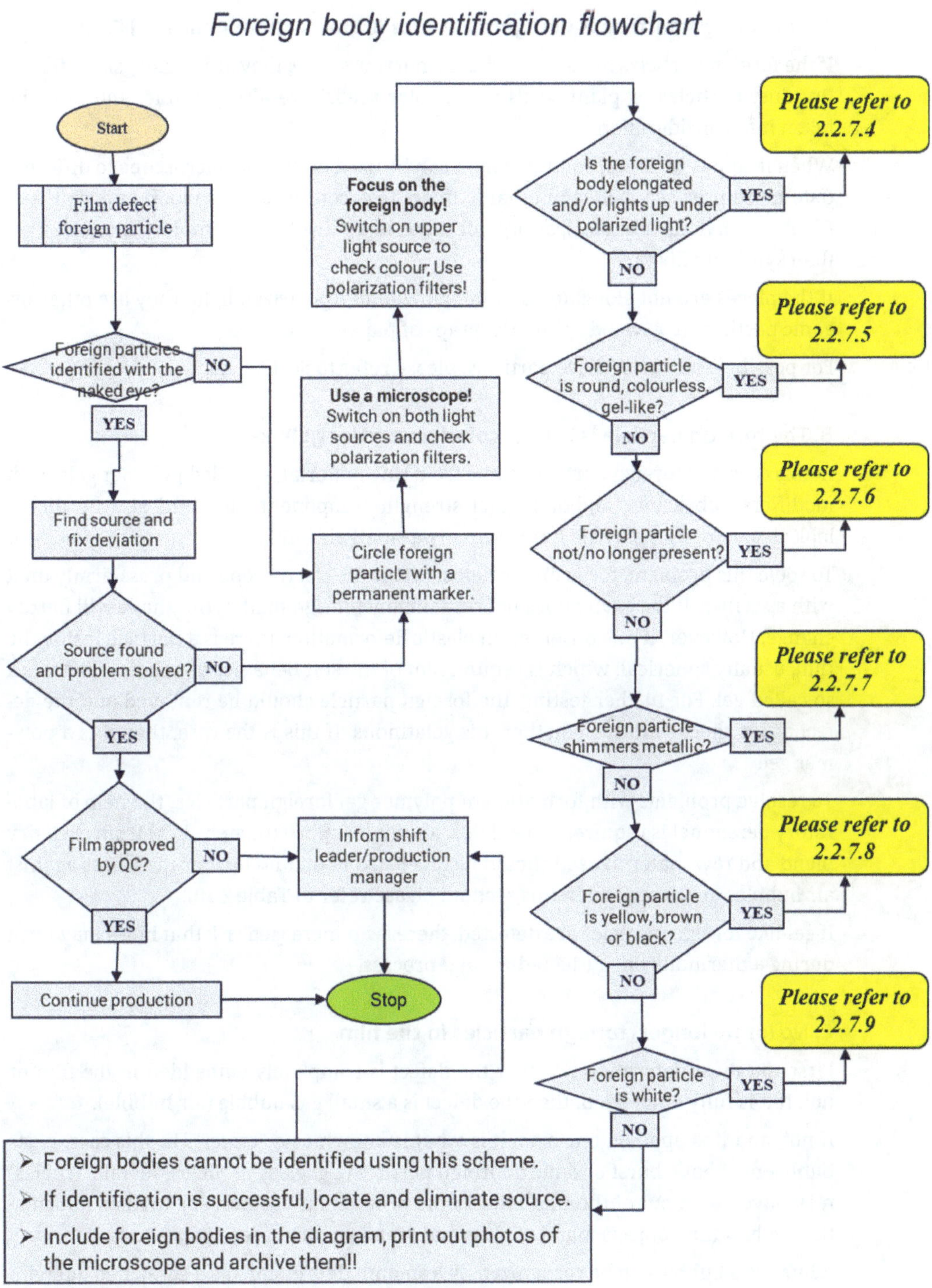

Figure 2.34 Measures for identifying foreign bodies

A. The foreign particle is elongated and/or it glows under polarized light

If the foreign particles have an elongated, narrow shape, they are usually small fibers. But dust particles or plant seeds (e.g. poplar seeds) are also possible, and must be taken into consideration.

When it comes to fibers, polarized light can be used under the microscope to differentiate more precisely between organic fibers (e.g. cotton, sheep's wool) and synthetic fibers (e.g. nylon, polyester), as organic materials shine brighter under polarized light than synthetic fibers.

If the pieces are not elongated and but glow under polarized light, they are other organic particles (e.g. wood splinters, scraps of paper).

For possible sources of these particles, please refer to Section 2.2.8.

B. The foreign particle is round, colorless and/or gel-like

In this case, the foreign particles come from raw material, so-called polymer gels from modifiers, lubricants, and/or impact strength components. It could also be that it looks like a foreign particle but is actually a small air bubble.

To solve this problem, focus on the defect under the microscope and press firmly on it with a scriber. If the cavity does not contain any foreign matter, the shape will hardly change. However, if one observes an elastic deformation, there is a particle inside the film, usually spherical, which is reminiscent of the soft behavior of gelatin, then it is a so called gel. For further testing, the foreign particle should be removed and the defect again checked to see whether it is gelatinous. If this is the case, then it is a polymer gel.

To resolve problems with formation of polymer gel foreign particles, the help of laboratory personnel is required. To aid this, a sample (approximately 3 kg) from each dry blend and raw material used should be taken. (Note that no further measures against air bubbles are described in this section: please refer to Table 2.10).

If gel-like foreign particles are detected, there is an increased risk that holes may form during a thermoforming (deep-drawing) process.

C. No (or no longer) foreign particles in the film

First you must determine whether the defect is completely embedded in the film or not. If it is fully embedded, then the defect is a small gas bubble (air bubble).

If not, and it is open on one side, it is what is known as a "crater". In this case, a gas bubble may have burst or a hard foreign particle (e.g. glass particles, metal particles) may have fallen out of the film later in the production process. For further information on handling objects made of glass and metal particles, see Figure 2.34.

A burst gas bubble can be recognized by a smooth depression (as if scratched out with a spatula). In contrast, a hard particle that falls out usually leaves a somewhat jagged shape behind. In this case, it is important to check the film for prick marks, as the hard particles may have slightly damaged the roller. In this case, the calender rolls and/or rolls in the pick-off unit must be polished.

The cause of an error due to gas bubbles lies in the production parameters. See the guidance at the start of Section 2.2.7. The possibility that "foreign particles that have fallen out" is a reason for a more detailed search, but unfortunately no possible sources can be seriously mentioned here.

D. The foreign particle has a metallic shimmer

If the particles have a metallic shimmer under the top light of the microscope (turn off the bottom light), the challenge is to figure out what type of metal they might be. To do this, we should carefully remove as many particles as possible and place them on a white sheet of paper. Then we can check whether the particles are magnetic or not. To do this, we take a permanent magnet, hold it under the paper and move it back and forth. If the particles follow the magnet, it is ferromagnetic material, such as cast iron, steel, or rust.

If the particles do not follow the magnet, the next step is to check whether the particles are very hard. To do this, we pierce the particle with a scriber. If it is very hard and the particle "pops off", it is probably a small piece of glass.

If it is hard but the needle penetrates the particle a little, it is stainless steel, aluminium, brass, bronze, or (if it is black) possibly also hardened rubber (see Section E).

The color can provide further information about the origin of the foreign particles:

- If they appear yellowish or reddish under the reflected light of the microscope, they are most likely bronze or brass
- If they are relatively soft and silvery, they are probably a light metal such as aluminium
- If they are hard and have a silvery sheen, they are probably stainless steel
- If they are very dark or brown and ferromagnetic, they are rust particles
- If they are shiny, they may be pieces that have recently erupted.

E. The foreign particle appears yellow, brown, or black

If the particles are yellowish, brown, or black, and have the size of a PVC grain, they are most likely particles from the PVC powder used.

If the particles are larger and have a sharp line, we need to expose them from the surrounding material. If, upon poking with a scriber, it splinters into many small pieces, then it is burnt PVC, oil carbon, burnt rubber seals, or something similar.

If this is not the case, or if the color is yellowish or light brown, then the material has most likely originated from something stuck to a hot surface, such as the pipes or in the hot mixer. It is also possible that these particles were already present in the raw material. To be on the safe side, we can check again whether the particles are magnetic.

F. The foreign particle appears white

In this case we must proceed in the same way as when checking the black particles. First the particles are exposed and then pierced with a scriber.

- If the particles shatter into many small particles, moisten them with a drop of water and check if the water drop has become milky. If this is the case it is aluminium oxide (Al_2O_3)
- If the particles break into many small pieces but are not water-soluble, they are shear-hardened PVC particles resulting from strong agitation in a pipe or mixer
- If the particles do not shatter into small pieces but the scriber draws a sharp line, they are rubber particles or plastic materials other than PVC, such as PP, PET, or PE.

The foreign particle remains unidentified

If the particles cannot be identified according to these recommendations, it is important to document them precisely (if possible, with a photo) and store them in the archive. If these errors occur again, it is advantageous if it is known when and with which type of film the error occurred. This makes later identification easier.

In addition, the shift supervisor must be informed to ensure that the information about the unknown error is reported to production management, the laboratory, and the planning department (logistics).

2.2.8 Some Possible Sources of Foreign Particles

The following subsections discuss the following sources of foreign particles, along with guidance on the action to take:

- Section 2.2.8.1 Organic fibers (including natural fibers, dust particles, and plant seeds)
- Section 2.2.8.2 Synthetic fibers (for example, nylon and polyester)
- Section 2.2.8.3 Other organic particles (for example, wood and paper)
- Section 2.2.8.4 Ferromagnetic metals (for example, cast iron and steel)
- Section 2.2.8.5 Rust
- Section 2.2.8.6 Glass particles
- Section 2.2.8.7 Stainless steel
- Section 2.2.8.8 Light metal (for example, aluminium)

- Section 2.2.8.9 Brass or bronze
- Section 2.2.8.10 Small black dots (size similar to a PVC grain)
- Section 2.2.8.11 Medium-sized black dots (larger than a PVC grain and/or breaking into small pieces)
- Section 2.2.8.12 Burned rubber
- Section 2.2.8.13 Unburnt rubber and non-PVC plastics (for example, PP, PET, and PE)
- Section 2.2.8.14 Shear-hardened PVC
- Section 2.2.8.15 Aluminium oxide (Al_2O_3).

2.2.8.1 Organic Fibers

Including natural fibers, dust particles, and plant seeds.

Belt of conveyor belt frayed?

- Flare the edge of the conveyor belt
- Align the band so that it runs without touching the housing
- Make sure that all support rollers run smoothly.

Parts of Molton cloth in the mixer?

- Open and check the mixer
- Check the drainpipes using the pipe magnet.

Molton cloth in the extruder hopper?

- Let the funnel run empty, and check.

Cotton strings used to tie something?

- Remove any existing cotton cord fasteners.

Strong air currents in the calender area?

- Ensure no (or low) air currents in the production rooms
- Close windows and ensure doors are "airtight"
- Remove covers in front of air source outlets
- Close doors and gates
- Switch off or clean fans
- Close compressed air leaks.

Cleaning roller cloth frayed?

- Check the cleaning rollers (rollers 1 and 2) and turn them if necessary or renew.

Clothes releasing threads?

- Ensure as little activity as possible in the production area (number of people working on the calender).

Flower seeds?

- Keep windows and door closed—especially in spring when (for example) poplar trees are in full bloom
- Check/clean outside air intake filter.

2.2.8.2 Synthetic Fibers

For example, nylon and polyester.

Big bag cut open with a blunt knife?

- Do not cut the big bag open with a knife
- Pull on a string or use scissors and remove the string
- Pull out the ticking, and make sure that the product does not come in contact with the outer big bag casing.

Raw material from PE bag included in the dry blend?

- Only use a sharp knife (trapezoidal blade—not hook blade) to open the bags
- When cutting, pay attention to the "right" side; not the "ventilation side".

Belt from the conveyor belt frayed?

- Flare the edge of the conveyor belt
- Align the band so that it runs without touching the housing
- Make sure that all support rollers run smoothly.

Strong air currents in the calender area?

- Ensure that there is no (or low) air current in the production rooms (close windows and ensure that the doors "airtight")
- Remove partial covers from source outlets

- Close doors and gates
- Turn off fans
- Close compressed air leaks.

Filter on day container and/or main silo defective?

- Check filter visually—if possible, use filters made of PVC filter fabric

Intake filter for conveying air defective?

- Check the intake filter—use PVC filter fabric if possible.

Clothes secreting threads?

- Ensure as little activity as possible in the production area (minimize the number of people on the calender)
- Wear lint-free clothing.

2.2.8.3 Other Organic Particles

For example, wood and paper.

Raw material from paper sack included in the mix?

- Inform your purchasing manager, and change packaging, if possible
- If unavoidable, use a very sharp knife with a trapezoidal blade (not a hooked blade) when cutting to open the paper bags
- When cutting, pay attention to the "right" side; do not cut on the "venting side".

Broken wooden knife?

- Do not use broken wooden knives; sand-smooth any points of breakage.

Pallets

- Remove wooden pallets from the production area.

Broomsticks/hand brushes

- Replace brooms with wooden handles—use only plastic brooms
- Check the hand brush.

Regenerated material contaminated?

- Keep regrind very clean
- Do not use regrind that has been stored for a long time
- Containers must also be very clean on the outside
- Change the pallet with the regrind.

Cleaning roller—paper layers frayed?

- Check the layers of paper on the cleaning rollers and replace them if necessary.

Corrugated cardboard blanks—covers, underlays made of corrugated cardboard or paper?

- Remove all paper and/or paper covers, underlays, etc.

2.2.8.4 Ferromagnetic Metals

For example, cast iron and steel.

Scrape grinder

- Check blades (rotor and stator blades) for chipping, and change if necessary.

Pipe magnets

- Check and clean pipe magnets in the mixer drain. If parts stick, follow "backwards".

Magnetic grates

- Check magnetic grates in extruder hoppers for adhering particles.

Planetary-gear extruder

- Examine the thrust ring
- Check planets
- Check plasticificate.

Kneader

- Check the kneader shaft for abrasion
- Check cutting blades.

Deflection rollers of conveyor belts?

- If particles are very small, check the bearings of the discharge, drive, and support rollers.

Regenerated material contaminated?

- Check the container used (e.g. pour over a magnetic grate).

No loose metal parts in the product flow (e.g. from repair work)?

- Search the system carefully and conscientiously, and remove any loose parts.

2.2.8.5 Rust

Pipe magnets

- Check pipe magnets in mixer drain and mills for adhering particles.

Magnetic grates

- Check magnetic grates in extruder or kneader hoppers for adhering particles.

Deflection rollers of conveyor belts?

- If particles are very small, check the bearings of the discharge, drive, and support rollers.

Regenerated material contaminated?

- Check the container used (e.g. pour over a magnetic grate).

Rust particles in the raw material?

- Pour the mixture and raw material components through a magnetic grate, and check.

Conveyor air cooler

- Open the cooler housing, remove the tube bundle, and check the housing with cover.

Pinch valve housing

- Check the rubber lining of the pinch valves
- Check flange connections.

Dosing screws for the daily containers

- Loosen the bearing shield and pull the worm out to the rear, and check.
- If necessary, dismantle and clean the conveyor screw (sandblast it).

2.2.8.6 Glass Particles

Glass objects

- Glass bottles and other movable glass parts should be removed from the production area
- Glass fluorescent tubes must be protected with a plastic cover.
- Windowpanes must be checked for damage.

2.2.8.7 Stainless Steel

Abrasion in dosing screws (also, dosage of small components)?

- Listen to running noises with a "stethoscope".

Abrasion from mixing tools?

- Check whether the mixing tool is rubbing on the hot-mixer container or on the cooling mixer container.

Plow blades on the 2-roll-mixing-mill made of stainless steel?

- Remove any burrs that may be present
- Check whether the plow blades have been adjusted correctly.

Stripping jaws in rolling mills and calenders made of stainless steel?

- Have the stripper jaws adjusted.

Granulated PVC film?

- Check the container used.

Metal parts in the raw material?

- Check the mixture and raw material components for shaking channels.

Other parts made of "VA" steel in the product flow?

- Search the system carefully and conscientiously, and remove any loose parts.

2.2.8.8 Light Metal

For example, B-type aluminium.

Abrasion in rotary valves?

- Check running noises in rotary valves with a stethoscope
- Open (replace) if there is reasonable suspicion.

Broken tab from the ventilation line in the silo?

- If there is reasonable suspicion, open and check the manhole cover.

Dosing screw for the daily container?

- Check the cover and bearings of both bearings of the dosing screw
- If suspected, remove the dosing screw.

Aluminium blades used in the premixing plant?

- No aluminium shovels should be used.

Loose aluminium parts on the system (e.g. as knocking rods)?

- Search the system carefully and conscientiously, and remove any loose parts.

Regenerated material contaminated?

- Check the container used.

2.2.8.9 Brass or Bronze

The following measures should guard agains brass or bronze abrasion.

Brass rolling mill knives?

- Remove any burrs that may be present
- Readjust the contact pressure, so that it is as low as possible
- Install new, sharp knives if they are "blunt"
- Knife sharpened too sharp (edges can break)
- Check material selection.

Plow blades on the 2-roll-mixing-mill made from brass?

- Remove any burrs that may be present
- Have the turning blades adjusted correctly (TD).

Stripper jaws made from brass?

- Have the stripper jaws adjusted (TD).

Granulated PVC film?

- Check the container used.

Other brass parts in the system?

- Search the system carefully and conscientiously, and remove any loose parts.

2.2.8.10 Small Black Dots

If the black dots are small (the size of a PVC grain), then the following should be considered.

Extruder temperatures >210°C

- Heat up the gelling part (temperature should not be more than 190°C in central spindle and zone 1; max. 210°C in zone 2)
- Clean the extruder with cleaning mixture
- Measure the wear of the gelling part (carry out wear calculation)
- Add more planets if still possible
- The screw conveyor may need to be replaced
- Measure the temperature of the feeding housing:
  - If it is too warm, cool it down
  - If not possible, rinse the housing (limescale deposits) with a limescale remover (observe regulations for corrosive substances).

Co-kneader

- Check cooling of the screw
- Check the temperature of the case
- Check the wear and tear situation of the teeth
- Does the storage ring have the correct diameter?
- Check the cutting blade at the exit

- Check the housing to see if there are any visible traces of grinding from the screw
- Check the heating/cooling units to avoid overheating
- Is the configuration of the teeth OK?

Heating/cooling unit (HCU) overheating

- The temperature control is not working properly.

Feed belt & swivel belt

- Clean all rollers—including the support rollers
- Clean the conveyor belt and frame.

Dry blend and dry blend components

- Examine the individual components using the shaking trough
- Make a sample with the likely defects in the laboratory on the 2-roll mill, and compare it with the defects in the film
- If possible, use and report other equivalent components.

Temperature of the supply line

- The temperature in the delivery lines (including to the silos) must not be more than 60°C during filling
  - If they are warmer, the cooler of the conveyor fan is defective
- If there is reasonable doubt, open and clean the cooler:
  - Check the housing thoroughly and clean the galvanizing, and replace if desired
  - Tube bundles must be made of stainless steel.

Feed fan suction filter

- Check filter fabric for holes and cleanliness, and change it if necessary.

2.2.8.11 Medium-sized Black Dots

If the black dots are medium-sized (larger than a PVC grain) and/or breaking into small pieces, then the following should be considered:

Shaft seal hot and cold mixer

- Remove and clean the mixing tool
- Replace the shaft seal if necessary.

Extruder thrust ring with cutting teeth

- Clean thoroughly (follow cleaning instructions)
- Remove caking between Teflon cap and central spindle
- No cavity between Teflon cap and central spindle
- Test the effectiveness of the cooling of the thrust ring
- Rinse the cavity of the thrust ring, and if necessary, clean with limescale remover (TD—follow regulations when dealing with corrosive substances).

Co-kneader

- Open the kneader and clean the storage ring and discharge screw area thoroughly
- Clean nozzle with cutting device
- Check barrel and screw temperature
- Empty and check the vacuum container
- Check the function of the vacuum unit.

Chopper (chipping roller) on the rolling mill (for cutting the stripe crosswise)

- To test produce with stripes, swivel the shredder (shredder roller)
- Clean the shredder (shredder roller) and remove it if necessary.

Limiting jaws of rolling mill and calender

- Clean under the limiting jaws.

Plow blades on the 2-roll mixing-mill

- Clean turning blades, including support beams.

Roll edges and end faces on the calender and rolling mill

- Clean roller edges and front sides
- Remove any burnt PVC and cracked oil
- Seal bearings, replace felt rings, and replace split oil seals.

Support rollers and deflection rollers of the conveyor belts

- Clean drive, deflection, and support rollers
- Clean the belt.

2.2.8.12 Burnt Rubber

Simmer ring in the hot mixer

- Remove the tool and check the cleanliness under the floor scraper
- Replace oil seal if necessary
- Mix the order on another mixer.

Mixer lid seal

- Open the lid and check the seal; replace if necessary
- Mix the order on another mixer.

Conveyor belt/swivel belt

- Clean drive, deflection, and support rollers
- Clean the belt.

Pinch valves—rubber seal

- Check the pinch valves of the PVC types used
- Replace the rubber ticking if necessary.

Delivery hoses

- Check for external damage
- Check inside for wear.

Dry blend/dry blend components

- Examine the individual components using the shaking trough
- Replace components according to plan if replacement is possible
- Make a sample with the likely defects on the laboratory rolling mill and compare it with the defects in the film.

2.2.8.13 Unburnt Rubber and Non-PVC Plastics

For example, PP, PET, and PE.

Mixer cover seal

- Check the sealing for defects and/or breakouts.

Feed conveyor/swivel conveyor

- Inspect the belt for defects and/or abrasion
- Examine the strap to see if it is touching the frame anywhere
- Check support rollers for smooth movement.

Rubber rollers

- Check rubber rollers for abrasion and/or defects.

Hoses for silo filling

- Inspect the internal part for abrasion.

Dry blend and dry blend components (raw materials)

- Switch to other components (if possible)
- Write it down and inform the laboratory
- Take a sample of the bad mixture (approx. 3 kg) and give it to the laboratory.

Examining samples—recommendations

- Sieve samples using a sieve (500 µm mesh size)
- Examine the remaining particles with a vibrating trough.

2.2.8.14 Shear-hardened PVC

Delivery lines

- Knock on it when it is "empty"
- If the lines do not sound "empty," dismantle the section and inspect for flow hardening (TS).

Hot mixer

- Check the inside of the mixer container and clean if any incrustations are found.

Mixer drainpipe

- Check the mixer drainpipe and clean if any incrustations are found.

Dry blend and dry blend components (raw material)

- Insert another funnel and watch
- Use replacement products or products from another silo in the mixing plant, and record the measures carefully
- Take a sample "top and bottom" (approx. 3 kg) from the funnel.
- Take samples of the individual components (approx. 3 kg) from the daily container or small component dosage (do not forget liquids).

2.2.8.15 Aluminium Oxide (Al_2O_3)

Dry blend and the individual raw materials, especially the COPO

- Take COPO from another silo
- Take a sample from the COPO silo used so far and sieve it very carefully in the laboratory using a fine sieve (500 µm).
- Make a rolling mill skin with the sieve residue, and check for defects under the microscope.

Storage and conveyor system

- At the next opportunity, check the inner wall of the main silo, day container, rotary valve, switches, and pipes (all parts made of aluminium that come into contact with the product) for corrosion
- Make sure that no aluminium can come into contact with COPO for any significant length of time.

2.2.9 Summary

When calendering high-quality rigid PVC films, a clean environment and a reproducible basic setting of the production system are essential. This does not just begin with the film production in the calender, but rather with the selection of raw materials, their storage, and random quality checks. For this reason, the entire production line has also been considered in this chapter, starting with filling the mixer with the individual components and ending with winding the film in the winder.

Only if the setting parameters in interaction with the individual machines, apparatus, and devices are known can the necessary search for quality defects be successful in a timely manner.

For these reasons, the quality assurance equipment and the basic settings of the interacting machines and devices have been discussed in the introduction to this chapter. It is not so important that the settings mentioned are always followed exactly in all companies; what is more important is that the basis is uniform within an individual company. Equally important is good and honest communication between the machine operators and the other departments in the company.

Of course, the above suggestions for avoiding errors must be individually adapted to the circumstances of the respective company. Some suggestions can certainly be omitted, whereas others need to be added. These notes listed can only be viewed as examples.

References for Chapter 2

[1] Kopsch, H.: *Calender Technology*, Carl Hanser Verlag, 1978

[2] Kohlert, C. [lecture], Klöckner Pentaplast

[3] *Herders Konversations-Lexicon* (3rd edition), Herder Verlag, 1905

[4] Kohlert, C. et al.: *Kalandrieren von Polymeren*, Deutscher Verlag für Grundstoffindustrie, Leipzig, 1992

[5] Krüger, E.: "Mixing dry blend", no further information available

[6] Compounding (X-Compound) [brochure], Troester, *https://www.troester.de/products/x-compound/* (accessed March 2025)

[7] Planetary Gear Extruder [brochure], Entex, *https://www.entex.de/* (accessed March 2025)

Additional Sources for Chapter 2

TRM heating mixers [brochure], PLAS MEC, *https://www.plasmec.it/trm-heating-mixer/* (accessed March 2025)

Mischeroptimierung [brochure], MTI, *https://www.mti-mixer.de/anwendungen/mischeroptimierung/* (accessed March 2025)

Heiz-/Kühlmischer [brochure], Mixaco, *https://www.mixaco.com/industriemischer/heiz-kuehlmischer/* (accessed March 2025)

3 Rheology of Polymers

Frank Kleinert

3.1 Polymers

Polymers are chemical substances made up of macromolecules. A distinction is made between synthetic polymers (plastics) and natural polymers (e.g. polysaccharides and proteins). The essential building block for plastics is crude oil, natural gas or (with renewable feedstocks), sugar, alcohol, or CO_2. It is reasonable to assume that around 4% of the petroleum produced is used to produce plastics, and that the proportion of bio-based plastics is very low (due to higher costs and a high demand for agricultural land).

Over 70% of global production is for "low-cost commodity resins", which have a favourable ratio of performance to costs, and which are primarily represented by polyethylene, polypropylene, polyvinyl chloride, and polystyrene. Polyethylene (PE) is differentiated into high-density and low-density types (HDPE and LDPE, respectively). The high density of HDPE is caused by macromolecules with weakly branched polymer chains, whereas the strongly-branched chains of LDPE leads to its low density. The three synthetic plastics consumed in the largest quantities in Germany in 2017 were polyethylene (28%), polypropylene (17%), and PVC (13%) [1]. Around one-third of the plastics produced are used for packaging, 15% for construction applications, and 15% for textiles.

The production of plastics is essentially based on polymerization processes. In each case, another monomer is added to the existing molecular chain. In addition to polymerization, there are two other reactions that are also called step growth reactions. The polymers are created from at least 2 bifunctional groups. If the macromolecule is formed without the separation of other molecules, this is called polyaddition. However, if molecules are split off during the reaction (such as water in PET production), this is called polycondensation. The properties of the resulting macromolecule are therefore determined by the monomers and the length of the macromolecule. The main polymerization reactions and the polymers resulting from them are listed in Table 3.1.

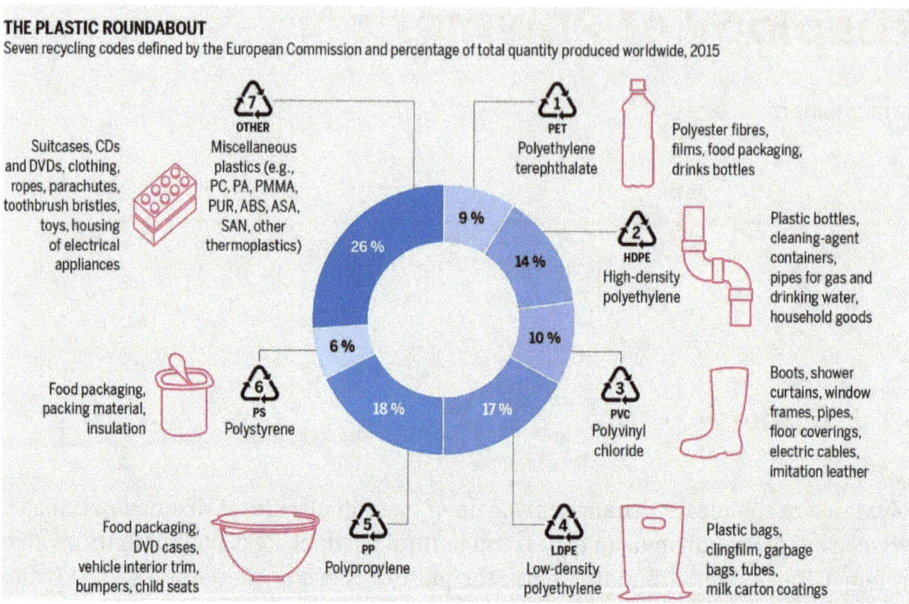

Figure 3.1 Share of different plastics with recycling codes in Germany in 2017 (Source: [1])

Table 3.1 Polymerization Reactions and the Resulting Polymers

Reaction system	Polymer	Abbreviation
Bulk polymerization	Polystyrene	PS
	Polyvinyl chloride	PVC (also known as modified PVC or M-PVC)
	Low-density polyethylene	LDPE
Suspension polymerization	Polystyrene	PS
	Polyvinyl chloride	S-PVC
	Polypropylene	PP
	High-density polyethylene	HDPE
Emulsion polymerization	Polystyrene	PS
	Polyvinyl chloride	PVC (also known as emulsion PVC or E-PVC)
	Polytetrafluoroethylene	PTFE

Reaction system	Polymer	Abbreviation
Polyaddition	Polyurethane	PURe
	Polyamide 6	PA 6
Polycondensation	Polyethylene terephthalate	PET
	Polyamide 66	PA 66

3.2 Rheology of Polymers

Rheology, the study of flow, deals with the deformation and flow behavior of matter. When it comes to plastic melts, the focus is on plasticity theory, elasticity theory, and fluid mechanics. With the "economic miracle years" in the 1950s and 1960s a new need arose for rheology as a science. Following this, the wide use of plastics and the necessary the need for standardized characterization drove the adoption of more rigorous measurement methods and mathematically based evaluation procedures based on melt rheology.

3.2.1 Rheometry

Since the 1990s, rheological measuring devices have become a standard tool used in research and development laboratories [2–6].

By comparing the beginnings of rheometry (measuring methods for determining rheological properties) with current rheometers, the rapid growth of the technology can be understood. The comprehensive description of plastics that resulted allow properties to be predicted and processing flows to be simulated, ultimately opening up new areas of application.

Various measuring devices and methods have been developed to measure the viscosity of liquids and melts. These range from flow cups (which determine the viscosity of colours using a stopwatch), to MFI measuring devices (which work similarly but have a built-in heater for polymers), to high-pressure capillary viscometers (which determine the flow time as throughput and pressure increase).

A completely different measuring principle is implemented using stirrers or drag flows. In this case, the polymer to be measured is moved with a plate, and the resistance is determined using a torsion bar, with the measuring device generating a stationary or an oscillatory shear flow.

3.2.2 The "Spaghetti Analogy" for Macromolecules

The so-called "spaghetti analogy" is a useful metaphor to explain the behavior of macromolecules. The length of the spaghetti strands corresponds to the length of the macromolecules, while the strength of the individual spaghetti strands corresponds to the strength of the main chain-forming bonds (e.g. C–C bonds in PE or PVC).

When frozen, the spaghetti forms a relatively solid "heap", due to their propensity to stick together (corresponding to the intermolecular forces in a polymer). If you heat this heap, then the individual spaghetti strands become easier to remove from the pile—a process that works better at higher temperatures, with short strands, and when butter or oil has been added. This corresponds approximately to the situation when a polymer is heated, which is easier with short chains and with the help of an internal lubricant.

Just as in spaghetti, the long macromolecules and intermolecular forces in polymers make it difficult to create linearly aligned regions with high crystallinity and high transparency. Similarly, the fact that the macromolecules are not usually of uniform length (due to the manufacturing process) means there is no true "melting point", but instead a temperature range over which the solid liquifies. However, if the temperature is too high, or the heat persists for too long, then the macromolecular chains will decompose, just as overcooked spaghetti will tend to fragment if overcooked.

3.2.3 Viscosity Measurement

Viscosity can be interpreted as a measure of the internal resistance to flow. In addition to measuring the MFI (melt flow index) values and using the flow function $\tau(\dot{\gamma})$ determining the viscosity function $\eta(\dot{\gamma})$ at different temperatures is the most commonly used description method of the rheological material behavior of plastics. By measuring the viscosity function, it is possible to describe the mathematical relationship between the shear rate and the resulting viscosity. This results in the possibility of defining the function with a few material constants, and also in modelling flow processes computationally.

Figure 3.2 shows the viscosity curves of three different types of PVC film. Despite identical PVC raw materials, the flow curves vary in their toughness. This is due to the use of around 10% fillers (green curve) or the use of a high proportion of MBS (methacrylate butadiene styrene) modifiers (red curve).

The curves of the three PVC formulations show the typical course of viscosity for polymer melts as a function of shear rate. As the shear rate increases, the Newtonian behaviour of the polymer melt changes into shear-thinning behavior. The cause of the drop in viscosity is a change in the structure of the polymer melt. This can be explained by considering the relatively linear, lightly-branched structure of the macromolecules, which orient themselves in the direction of flow and thereby cause a decrease in the entanglement point density of the partially interpenetrating molecules [6].

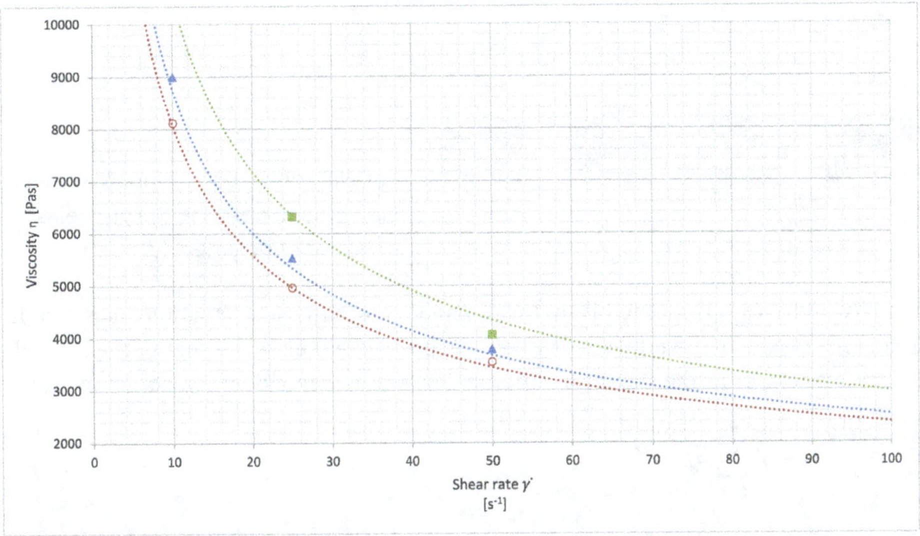

Figure 3.2 Melt viscosity vs shear rate for three PVC film formulations

The viscosity is therefore strongly dependent on the shear rate and can be described with a power function.

Viscosity function:

$$\eta\,(\dot{\gamma}) = k\dot{\gamma}^{n-1}$$

The factor k is referred to as consistency, and n as the flow index. From Figure 3.2, a different consistency factor can be derived for the three PVC types (which corresponds to the apparent zero-shear viscosity), but a similar flow index.

3.2.4 Mechanical Models to Describe the Rheological Behavior of Materials

As with the "spaghetti analogy", which describes the behavior of macromolecules in simplified terms, the rheological behavior can be illustrated by connecting Maxwell elements in parallel.

The Maxwell elements are created from the combination of springs (an elastic body) and dampers or shock absorbers (a viscous fluid). By connecting three elements in parallel, the viscoelastic behavior of polymers such as polypropylene, polyethylene, and polyvinyl chloride can be described with sufficient precision (Figure 3.3).

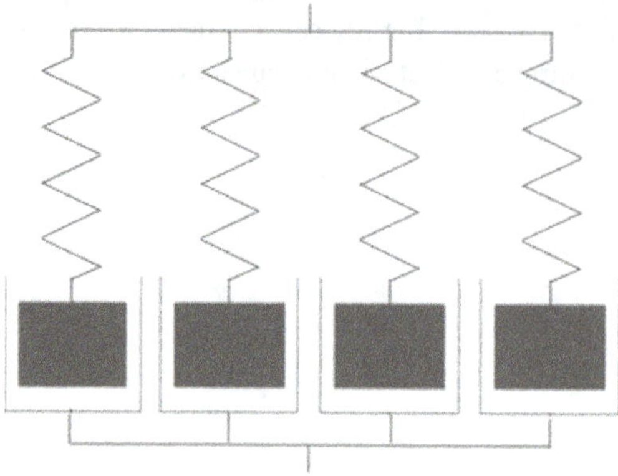

Figure 3.3
Parallel connection of four Maxwell elements to describe the viscoelastic behavior of polymers [7]

The advantage of this method is that well-known mechanical calculation rules can be used. This results in the total stress τ being expressed either as the sum of the individual stresses (τ_i), or as the shear (γ) multiplied by a time-dependent relaxation module:

$$\tau = \sum_{i=1}^{3} \tau_i = \gamma \sum_{i=1}^{3} G_i e^{-t/\lambda_i}$$

The Cox–Merz rule [8] is of practical importance for rheology. It combines the data from stationary measurements (e.g. from the high-pressure capillary viscometer) with the data from an oscillatory measurement (e.g. from a cone plate viscometer). For the linear viscoelastic region, i.e. in the region of small deformations, the following applies:

$$G*(\varpi) = \sqrt{G'^2 + G''^2} \triangleq \tau(\dot{\gamma}) = \eta(\dot{\gamma})\,\dot{\gamma}$$

On this basis, the viscosity curve of polymers can now be described over a wide range of shear rates. If the curve is calculated using three to four Maxwell elements connected in parallel, material constants are obtained that represent the rheological behavior much better than the viscosity curve, because (in addition to the dependency on viscosity from the shear rate) data is obtained on the relaxation behavior of the polymer.

3.2.5 Recipes and Raw Materials

Recipes are a well-kept secret in polymer manufacturing companies. The cost of raw materials (and therefore recipes) often accounts for more than 70% of a company's total costs. The recipes contain technical expertise, many years of production optimization, and also knowledge about recipe costs.

Simple standard formulations usually consist of three to four components, which are essentially made up of the polymer, anti-blocking additive, and the color. This means that there is direct competition with many identical providers, and so to be able to survive in this marketplace, recipes and processes must be adjusted to optimize costs to the nearest cent.

In principle, polymer films can consist of any number of individual components to support the production process and determine the properties of the end product. In practice, recipes calling for 15–20 individual components are at the extreme end of the spectrum of complexity. If multilayer films are produced, each individual layer has a recipe, which is then co-extruded or laminated with adhesion promoters to form a composite film.

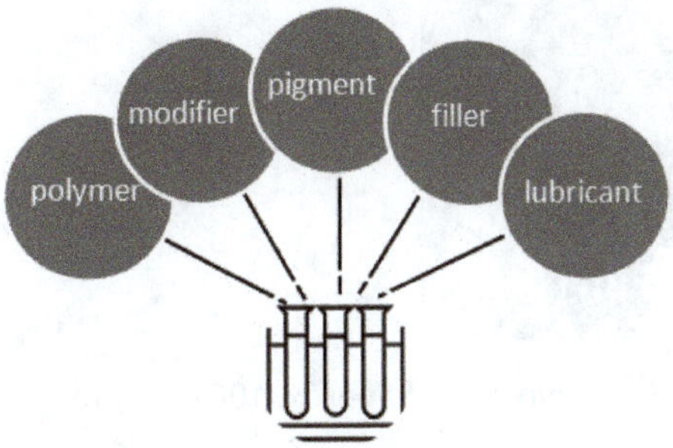

Figure 3.4
Recipe structure for polymer products

3.2.6 PVC Recipes

As an example of the complexity of polymer recipes, we discuss here recipes for PVC films. There are various types of PVC polymers, which (as described in Section 3.1) differ in the process used to manufacture them. However, they also differ in their physical properties, such as melt viscosity (for PVC this is often also the case for the *k* value measured) and in the porosity.

The *k* value correlates with the mechanical properties, such as the porosity of the PVC grain, and leads to different degrees of absorption for stabilizers, plasticizers, and lubricants (see also Figure 3.5).

3.2.6.1 Suspension PVC

About 90% of global PVC production capacity is based on the suspension process, giving suspension PVC (S-PVC). In the batch process, vinyl chloride, water, protective colloids, and organic peroxides are added to the reactor as initiators. Polymerization occurs through pressure and stirring. After polymerization, the water must be removed and the PVC dried.

By choosing the stirring conditions and the protective colloids (polyvinyl alcohol or cellulose ether), it is possible to control the grain size and the grain porosity (which is essential for additive absorption) in a targeted manner. The S-PVC then has a particle size distribution between 50–200 µm, and is available as a white powder.

S-PVC can be used to produce very transparent films that have good thermal stability. S-PVC hardly absorbs any water even under sterilization conditions, and has good electrical insulation properties. The electrical charge on the film and the phenomenon of "plate out" are disadvantages. When analyzing the coverings, one often uses polyvinyl alcohol, which is an essential component of the suspension process.

Figure 3.5 S-PVC as a free-flowing powder with a grain size of approx. 100–150 µm (Source: [9])

3.2.6.2 Bulk PVC

Bulk PVC, also known as "modified PVC" (M-PVC), is polymerized by liquefying vinyl chloride under pressure. The result is a PVC with the "least contamination" from the manufacturing process. The raw material properties are like S-PVC, and for rigid PVC films, a *k*-value of 57–60 is used. For soft PVC films, *k*-values of 65–70 are typical, although the porosity of the PVC grains is more important here.

3.2.6.3 Emulsion PVC

Emulsion PVC (E-PVC) is polymerized from vinyl chloride in water under pressure. Emulsion aids (surface-active substances) are used to achieve better wetting. The E-PVC is obtained by spray-drying, and a very fine powder with a grain size of 10–50 µm results.

Figure 3.6 SEM image of E-PVC as a fine-grained PVC with a grain size of 2–40 µm (Source: [9])

What is interesting is the significantly smaller particle size and the higher *k*-value. The small particle size leads to a higher absorption ratio of the PVC due to the larger surface area. At the same time, E-PVC leads to faster plasticization during the processing process, influences the shine of the end-product, and has an antistatic effect at doses above 10%. The clouding of the film during steam sterilization is a disadvantage.

3.3 Mathematical Methods in Recipe Design

3.3.1 Introduction

For a long time, the development of recipes was a "trial and error" process. New recipes were created through many experiments, and the effectiveness of the experiments was based on the "know-how" of the employees. This has resulted in a complex product portfolio without the use of modern production systems such as modular construction kits.

This means that many medium-sized companies produce many customer-specific products that often differ only in minor ways, but nevertheless cause very complex production changes (changeover).

3.3.2 Statistical Design of Experiments

The statistical design of experiments should be the basis for recipe development, and should be done in addition to:

- Determination of the minimum scope of the experiment
- "Fine planning" of the experiment to eliminate disturbances
- Sequential experimental planning and experimental evaluation.

Since current formulations are a compromise between performance and cost, film formulations usually consist of 5–20 individual components. This means that a statistical design of experiments is not three-dimensional, but rather 5–20-dimensional. To simplify this, you can use modular construction kits that are always kept stable, and then only change two or three components that determine the properties.

In the case of optimizing a PVC formulation for tensile impact strength, for example, three modules would be used:

- Internal and external lubricants and thermal stabilizers
- Auxiliary materials, such as E-PVC and copolymers
- S-PVC and impact modifiers.

This means that in the detailed planning, modules 1 and 2 would be unchanged, and there would only be a variation in module 3. Interactions between modules 1 and 2 would be neglected. Theoretically, this approach is effective. In practice, however, there are two possible problems:

- This simplified experimental plan does not necessarily represent the best result in relation to the target size
- This product must also be able to be reproduced permanently.

This means that the most cost-effective recipe is not necessarily the optimum for the development process, but rather the product of material costs multiplied with the production costs.

3.3.3 Optimization of Order Planning

With the increase in product diversity and the different factors influencing the operating result of production, planning has become more important in recent years. The automotive industry has played an absolutely pioneering role, having optimized its workflow from demand planning (forecast) to controlling the supplier industry (just in time) to daily production planning.

In comparison, the film industry has developed very rough planning tools, and therefore still has great potential to optimize its results. A typical behavior in medium-sized industries is to fulfil customer requests at short notice and to regularly question existing production processes. If it is a standard product that is always produced on the system, or a product that does not cause any loss, then the changeover is not significant (either for time or material). In contrast, small, irregularly scheduled orders are critical for planning and production, because having a production plan that changes at short notice also means having to make a forecast, meaning that material requirements, personnel requirements, and machine occupancy are unlikely to be optimal. At the same time, due to the additional customer requests, all other customer orders are disadvantaged, and in most cases the company's results are worsened.

References for Chapter 3

[1] *Plastic Atlas 2019*, Heinrich Böll Stiftung, 2019

[2] Ferry, J.: *Viscoelastic Properties of Polymers*, John Wiley & Sons, 1980

[3] Kulicke, W.: *Flow Behaviour of Substances and Mixtures of Substances*, Hüthig & Wepf, 1986

[4] Pahl, M. G.-M.: *Practical Rheology of Plastics and Elastomers*, VDI Publishing, 1995

[5] Laun, H.-M.: *Plastics* (40). Rubber+Rubber, 1987, p. 554 ff

[6] Gleißle, W.: *Rheology II*, Plenum New York, 1980, p. 457 ff

[7] Kleinert, F.: *Rheological Models for the Formation of Transfer Functions* [dissertation], Martin Luther University, 1999

[8] Cox, W., Merz, E.: Correlation of dynamic and steady flow viscosities, *Journal of Polymer Science*, 1958, 28: 619–622

[9] SEM images PVC, Vinnolit GmbH & Co KG, 2020

4 Mathematical Methods in Film Production

4.1 Basics of Modelling Roll Gap Flow

Christian Kohlert

The structural variants of various calendering systems for hard and soft films are described in the literature [1]. Mathematical methods have long been used to calculate calendered thickness and expected thickness, throughput, and the contact pressure on the rolls for the purpose of designing the roll construction and roll bearings.

Starting from the basic equation of mass conservation:

$$\frac{\partial \rho}{\partial t} + \frac{\partial \rho v_i}{\partial x_i} = 0 \tag{4.1}$$

and the law of conservation of momentum:

$$\rho\left(\frac{\partial v_i}{\partial t} + v_j\frac{\partial v_i}{\partial x_j}\right) = \rho f_i + \frac{\partial \tau_{ij}}{\partial x_j} \tag{4.2}$$

and the law of conservation of energy:

$$\rho c_v\left(\frac{\partial T}{\partial t} + v_i\frac{\partial T}{\partial x_i}\right) = -\frac{\partial q_1}{\partial x_1} \pm \phi \tag{4.3}$$

and considering various boundary conditions (neglect of radiation energy, wall sliding, incompressibility, inertial forces, and chemically non-reactive substances), the following formula is then obtained for the throughput:

$$Q = h^2 \int_{-1}^{+1} \left(1 - \frac{y}{h}\right) \dot{\gamma}^d \left(\frac{y}{h}\right) + 2hu_1 \tag{4.4}$$

This becomes a function:

$$\dot{\gamma} = f_1(\tau) f_2(T) \tag{4.5}$$

which is used as an engineering approximation for the formulation of the one-dimensional rheological equation, where f_1 and f_2 are monotonic functions that can either take an analytical expression or are available in tabular form for further use in numerical methods.

Because of the ignorance of previous material damage in the mixing process, exact determination of the parameters turns out to be complicated without an inline process viscometer and therefore only leads to approximate solutions. An example is determining the consistency coefficient μ at a given material temperature and the flow index n as a characteristic of the viscosity anomaly within the power law equation:

$$\tau = \mu \dot{\gamma}^n \tag{4.6}$$

After solving the initial equations described above, considering the boundary conditions of the rolling process, the pressure profile [2] is obtained over several steps as follows:

$$p(x) = \int_{-x_2}^{x} \left[k \frac{\left\{ \frac{2n+1}{n} \frac{u}{r} \left[(x^2 - x_1^2) + (R + h_0) \left(\sqrt{R^2 - x^2 - \sqrt{R^2 - x_1^2}} \right) \right] \right\}^n}{\left(r + h_0 - \sqrt{R^2 - x^2} \right)^{2n+1}} \right] \mathrm{d}x \tag{4.7}$$

where:

x_1 = exit coordinate

x_2 = entry coordinate

The calculation of the contact pressure, also known as the gap load, is essential for the roll deflection and thus later optimization of the minimum thickness tolerance.

This results in a force parallel to the *x*-axis:

$$F = b \int_{-x_2}^{+x_1} p(x)\, dx \tag{4.8}$$

which is applied as a square force (Figure 4.1).

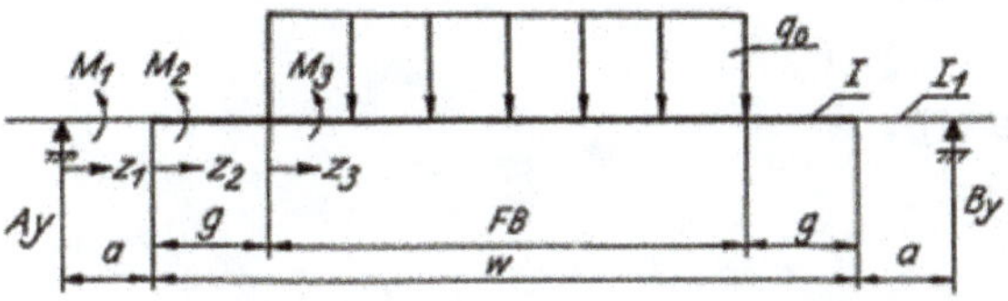

Figure 4.1 Schematic representation of the forces and quantities acting on the roller (Source: [1])

Further optimization through roll bending and axis crossing is covered below.

Roll calculations:

To calculate the camber or determine the residual error in the gap height after compensation through roll bending, axis crossing, and camber, the information in Table 4.1 is required (see also Figure 4.2).

Table 4.1 Quantities Required for Roll Calculations

Quantity	Notation	Units
Hide width on the roller	f_b	mm
Specified gap load (if known)	q	N/mm
Inclination adjustment	s_b	mm
Roll-bending force (or, if applicable, pressure in the cylinder)	R_b	N
Camber of the roller (if known)	h	mm
Roller material (steel or cast)	St/GGG	—
Roller diameter	D_a	mm
Roller body width (face of roll)	w	mm
Central drilling diameter	D_i	mm
Pin diameter (main bearing)	D_z	mm
Distance from main bearing to edge of roller body	a	mm
Distance from main bearing to roll bending bearing	b	mm

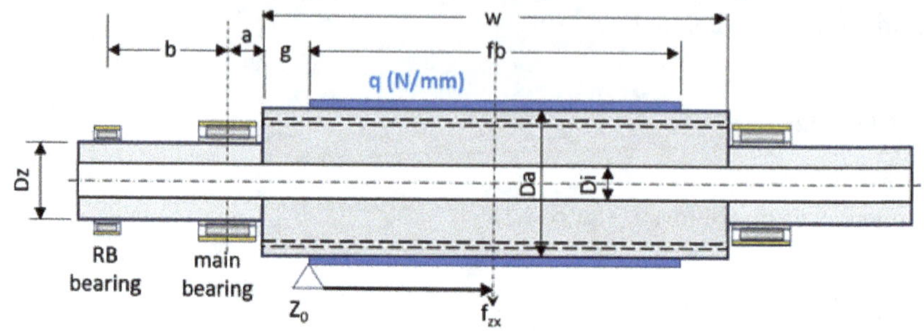

Figure 4.2 Roller geometry (Source: [1])

Calculation of the camber:

$$\alpha = \left(\frac{\pi}{180} * 20\right) + \left(\frac{7}{18} * \pi * \frac{w - f_b}{w}\right) \tag{4.9}$$

$$\beta = \left(\frac{\pi}{180} * 160\right) - \left(\frac{7}{18} * \pi * \frac{w - f_b}{w}\right) \tag{4.10}$$

$$bb_z = \frac{h}{1 - \sin\alpha}\left[\sin\left[\alpha + \left(\frac{\beta - \alpha}{w} * \left(z + \frac{w - f_b}{2}\right)\right)\right] - \sin\alpha\right] \tag{4.11}$$

Calculation of roll bending line due to gap load:

$$g = \frac{w - f_b}{2} \tag{4.12}$$

$$f_z = \frac{q f_b}{24 E J_b}\left[\begin{array}{l} \frac{z^4}{f_b} - 2z^3 - 6\,(a + g)\,z^2 + \left(6 f_b a + 6 f_b g + {f_b}^2\right)\ldots \\ (z + g + a) + 4\left(g^3 + 3ag^2 + 3a^2 g + \frac{J_b}{J_z} a^3\right) \end{array}\right] \times 1000 \tag{4.13}$$

where:

J_b = moment of inertia of the barrel

J_z = moment of inertia of the bearing journal

Gap deformation due to axis crossing:

$$s = \sqrt{\left(s_b \frac{f_b}{w}\right)^2 + {D_a}^2} - D_a \tag{4.14}$$

$$s_z = s\left(1 - 2\frac{z}{f_b}\right)^2 \tag{4.15}$$

Gap deformation due to roll bending:

$$rb_z = R_b \left(\frac{b}{2EJ_b}\right)\left[z^2 - zf_b - a\left(f_b + \frac{J_b}{J_z}a\right)\right] \tag{4.16}$$

References for Section 4.1

[1] Kögler, M.: Beitrag zur Optimierung einer Kalanderlinie [Contribution to the optimization of a calender line] [Diploma Thesis], Merseburg, TH, Leuna-Merseburg, 1986

[2] Kohlert, C. et al.: *Kalandrieren von Polymeren*, Deutscher Verlag für Grundstoffindustrie, Leipzig, 1992

4.2 Mathematical Models and Software Complexes

Tamara Chistyakova, Andrey Polosin

4.2.1 Introduction

Modern production of polymer films is a multi-stage project characterized by many parameters of raw materials, equipment, technological regimes, products (more than 100), and complex cause-and-effect relationships between them (about 800). Its main distinguishing features are continuity, large capacity (from 1000 kg/h), high energy intensity (specific energy consumption over 2500 kJ/kg), variety of product range, diversity, and flexibility of the systems.

The main steps of film production are producing a homogeneous viscous polymer mass (extrudate), forming the extrudate into a film, and cooling the film to fix its structure [1]. The hardware for these stages respectively consists of dosing, mixer-plasticizers of the extrusion type (single-screw, co-kneader, and twin-screw) [2, 3], the calender [4], and multi-roll coolers.

Each unit has a variable configuration that allows the number, type, and arrangement of structural elements (screw elements of extruders, calenders, and cooling rolls) to be changed. The diversity of the production range is due to a variety of film recipes depending on the intended use, a wide range of thicknesses (0.025–1.2 mm), widths (100–3000 mm), and degrees of shrinkage (0–60%), as well as a large selection of colors.

The need to release a wide range of products to fulfil customer orders leads to frequent reconfiguration of production for new types of films. During reconfiguration, not only does the technological regime change, but configuration of the main units is also required.

Waste is created in the manufacture of products—the edges of the film are cut to provide the specific web width, and changeover rolls are needed when the configuration changes. To save resources, these are crushed and returned to production as secondary raw materials [5]. Therefore, the production of films involves systems with recyclate, which have a negative impact on product quality and can, for example, lead to color deviations from the standard. To reduce the color deviation to an acceptable value according to the regulations, it is necessary to correct flow rates of dyes in the extrudate production phase.

In general, maintaining a certain product quality requires setting values of about 30 control measures, considering their different influence on the quality indicators of semi-finished products in the main production stages [6].

The following characteristics of production data significantly complicate the adoption of correct management decisions:

- Large number of measured parameter values (more than 6 million values per month of production)
- High accumulation rate (the sensor sampling interval is sometimes 10–30 s)
- Variety of sources (a system for generating production tasks, a "supervisory control and data acquisition" (SCADA) system, a thickness gauge, a video surface control system, a spectrophotometer, etc.) [7].

The lack of time and professional experience to analyze this data results in no more than 5% of the data being incorporated into product quality management decision-making. In addition, the incompleteness of information due to limited control over the quality of the extrudate (material homogeneity and thermal state) and important customer properties of the film (e.g. thickness variation and degree of shrinkage) contributes to making erroneous management decisions. This reduces productivity and increases the mass of disposable waste (defective film sections that are cut out and discarded).

An analysis of digitalization trends in high-tech industrial production has shown that a rational solution to the problem of increasing the efficiency of decision-making in product quality management is the development of problem-oriented computer systems. Such systems implement a single approach that allows the integration of individual digital solutions based on mathematical modeling and artificial intelligence, as well as data processing and decision support for production management, taking into account semi-finished products and finished products. To predict product quality, methods for obtaining production data, including regression analyses and machine learning methods (e.g. artificial neural networks, boosting decision trees), are used [8–12].

Mathematical models that describe physical processes in the main stages of production act as "virtual sensors" that make it possible to determine the values of process state parameters and quality indicators of semi-finished products and finished products, the possibilities of automatic control of which in of production are limited. Mathematical models of extrusion [13–44] and calendering [45–48] of polymers described in the literature are systems of equations based on the conservation laws of physics, the properties of substances (mass, momentum, energy) and the laws of rheology, and which usually consider some of the key features of the simulated processes:

- The phase transition of the polymer from a solid state to a viscous state (during extrusion)
- Abnormally viscous behaviour of the melt
- Leaks of melt through the gaps (radial gap between the body and the screw thread, and side and intermediate roll gaps of the screw engagement) and through axial slots in the screw thread (during extrusion)
- Heat exchange of the polymer with the structural elements of the unit (body and screw of the extruder, and calender rolls)
- Dissipation of mechanical energy due to the work of viscous friction forces.

When modelling extrusion, the melting mechanisms observed in extruders of various types are described:

- Film melting (melting a continuous densified solid phase—a plug surrounded by a melt) in single-screw extruders
- Dispersed melting (melting of the solid phase dispersed in the melt in the form of individual particles) into individual and twin-screw extruders.

The models make it possible to calculate the distributions of the state parameters of the polymer phases (velocity, pressure, and temperature), the length of the melting zone (during extrusion), productivity, electricity consumption depending on the geometric parameters of the aggregates, the parameters of the properties of the polymers, and the regime parameters of the processes.

However, the scope of these models is mainly limited to the study of the regularities of the processes of heat and mass transfer of polymers in extruders and calenders of certain types and configurations, in certain functional zones of extruders (solid polymer transport, melts, and melt transport), and gaps between the rollers of calenders or in channels of separate screw elements.

Simulation results provide information necessary for equipment design, but not for extrudate and film quality control. Models are typically implemented in general-purpose or problem-oriented software modelling environments (e.g. ANSYS Polyflow [49], Ludovic [50]), which are poorly focused on solving control problems, have a difficult user interface for production personnel, and relatively high purchase and maintenance costs. The models created do not allow a comprehensive quantitative assessment of the quality of polymer materials, and for the productivity and energy consumption of units, depending on the control measures for different types of polymers.

A systematic approach to combining the positive properties of the developed mathematical models and methods for calculating the properties of extrusion and calendering has not yet been implemented. The application of this approach enables the development of uniform methods and algorithms for structural synthesis, parametric tuning, and numerical analysis of physically based mathematical models to control hardware-flexible, multi-range production of polymer films according to resource and energy-saving criteria.

Therefore, the development of mathematical models and software systems is relevant, because they make it possible to solve complex problems, specifically evaluating the quality indicators of extrudate and film that are not controlled in production when information is incomplete, and to provide flexible multi-components in a way that conserves resources and controls energy use.

The following sections contains a description of the physical basics, mathematical models of the main phases of the production of polymer films, and problem-oriented software systems. These will help senior production personnel to determine the values of control measures in the phases that affect the specified customer properties of the film, while at the same time meeting the productivity and energy consumption requirements of the production system for different film types and system configurations. The use of the created mathematical models and software systems as an effective digital tool to support management decision-making will therefore help to improve the customer properties of polymer films, and increase productivity and efficiency in the use of production resources (raw materials and energy).

4.2.2 Characterization of the Production of Polymer Films as a Control Object

The production of polymer films as a control object is characterized by a set of vectors consisting of input parameters X, variables V, disturbing influences F, and output parameters Y, as shown in Figure 4.3 (using the example of the extrusion calender production of colored polymer films).

Each of these vectors contains the corresponding parameters of the s-th main production stage (s = 1, extrudate production; s = 2, film production; s = 3, film cooling). The input parameters of the s-th stage are the characteristics of the corresponding unit M_{EQs} (extruder, calender, cooling device) and the parameters of the properties of the polymer in the corresponding aggregate state P_{Ms} (e.g. the polymer heats up in the extruder and melts during the transition from solid to viscous condition in which it is during calendering).

The characteristics of the extruder $M_{EQ1} = \{T_{extrud}, D, L\}$ is determined by its type T_{extrud}, and by the diameter D and length L of the screw(s). The extruder types differ in the number of screws n_{scr} and the type of movement of the screws. The screw of a single-screw extruder only rotates, whereas the screw of an oscillating extruder simultaneously performs a rotational and axial reciprocating movement with amplitude S_0. The screws of a twin-screw extruder, which mesh with one another, rotate in the same direction (extruder with one-way rotation of the screws) or towards each other (counter-rotating extruder).

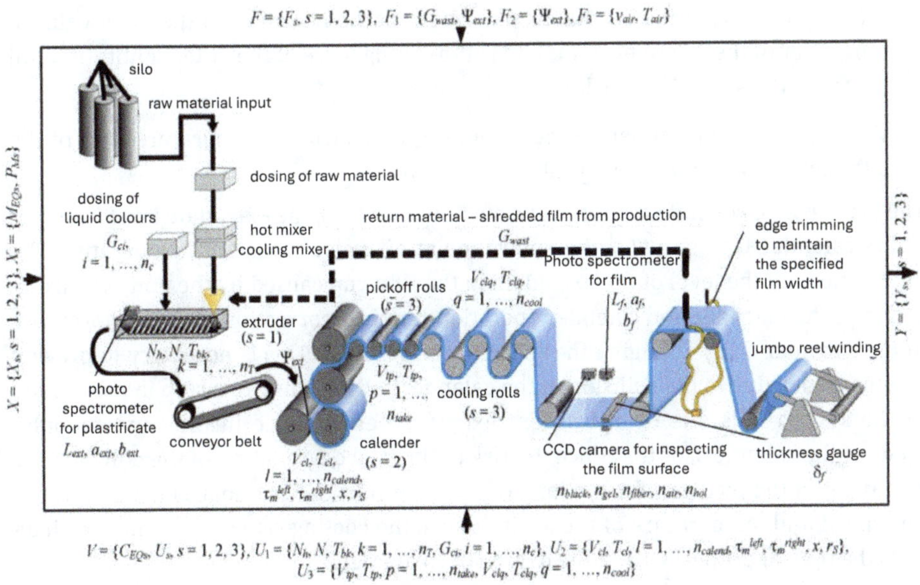

Figure 4.3 Description of the production of polymer films as a control object (Source: C. Kohlert [lecture], Klöckner Pentaplast)

The characteristics of the calender $M_{EQ2} = \{n_{calend}, D_c, L_c\}$ is determined by the number n_{calend}, and by the diameter D_c and the length L_c of its rolls.

The cooling characteristics $M_{EQ3} = \{n_{take}, D_t, L_t, n_{cool}, D_{cl}, L_{cl}\}$ is determined by the numbers n_{take} and n_{cool}, the diameters D_t and D_{cl}, and the lengths L_t and L_{cl} of the pick-off and cooling rolls.

The variable parameters of the s-th stage are the unit configuration C_{EQs} and the control actions U_s. The control actions in the stages are:

- N_h, N = revolutions per minute of the screws of the hopper and the extruder [L/min]
- T_{bk} = temperature of the k-th thermal zone of the extruder body [°C]
- n_T = number of thermal zones
- G_{ci} = flow rate of the i-th liquid dye fed to the extruder [kg/h]
- n_c = number of colors pigments
- V_{cl}, T_{cl} = peripheral speed [m/min] and surface temperature [°C] of the l-th calender roll
- τ_m^{left}, τ_m^{right} = operating times of the left and right electric motors lifting/lowering the external calibration roller of the calender to control the film thickness $\delta_{f,s}$

- x, r_s = spatial offset (horizontal displacement of the bearings) of the inner calibration roller of the calender [mm] and roll-bending force of the outer calibration roll [N] to set the same film thickness
- V_{tp}, V_{clq}, T_{tp}, T_{clq} = peripheral speeds [m/min] and surface temperatures [°C] of the p-th pick-off and q-th cooling rolls.

The disturbances include the return waste flow rate G_{wast}, the extrudate level (reserve) in the calender feed gap Ψ_{ext}, the environment air velocity v_{air} and temperature T_{air}. Fluctuations in the level of the extrudate in the first gap caused by the non-synchronization of the extruder and calender operation requires correction of the performance of the extruder G_1 by changing the speed of the screw N. This is necessary to prevent an emergency due to the collision of the calender rolls, leading to a stop in production (if the system fails due to insufficient extruder performance in the "undersupply" mode of the calender), or to reduce the risk of thermal degradation of the polymer due to staying in the feed gap for too long (up to 5 min) (if the calender is fed too much). The speed and temperature of the air influence the cooling rate of the film, which affects the optical properties and the degree of shrinkage.

The production output vector Y contains the output vectors of the main stages:

$$\begin{aligned}
&Y_1 = \{G_1, E_1, Q_1\}, Q_1 = \{\varphi_{ext}, \gamma_{av}, I_d, L_{ext}, a_{ext}, b_{ext}\},\\
&Y_2 = \{G_2, E_2, Q_2\}, Q_2 = \{\delta_f, D_{\overline{\delta}}, n_{black}, n_{gel}, n_{fiber}, n_{air}, n_{hole}, L_f, a_f, b_f\},\\
&Y_3 = \{T_f, Q_3\}, Q_3 = \{S_{ff}, S_{fc}\}.
\end{aligned} \tag{4.17}$$

where:

G_1, G_2 = productivity of the extruder and calender [kg/h]

E_1, E_2 = specific energy consumption of extruder and calender [J/kg]

Q_s = vector of material quality indicators at the s-th level

φ_{ext} = proportion of the solid phase in the extrudate

γ_{av} = average degree of mixing of the extrudate (average shear stress accumulated by the polymer melt during its stay in the extruder)

I_d = index of thermal destruction of the extrudate (characterizes the degree of irreversible changes associated with the destruction of polymer macromolecule chains during thermal exposure to it in the extruder) [%]

L_{ext}, a_{ext}, b_{ext} = color coordinates of the extrudate in the CIE-Lab model

δ_f = distribution of the film thickness over its width w_f [μm]

$D_{\overline{\delta}}$ = limit variation in film thickness [μm]

n_{black}, n_{gel}, n_{fiber}, n_{air}, n_{hole} = number of surface defects of various types (black dots, gels, inclusions of unmelted polymer fibers, air bubbles, holes) on a given film surface

L_f, a_f, b_f = color coordinates of the film, which differ from the color coordinates of the extrudate due to the thermal action on the extrudate during its formation on the calender

T_f = film temperature distribution along the length of the cooler [°C]

S_{fl}, S_{fc} = degree of shrinkage of the film in the longitudinal and transverse directions [%]

The main operating modes of the production system are reconfiguration to a new film type $T_{film} = \{K_f, Q_f^0\}$, productivity G_0 and film production under the influence of disturbances F. The film type is determined by the component composition K_f (including the type of film-forming polymer T_{polym}) and the following quality requirements:

$$Q_f{}^0 = \{\delta_{f0}, D_\delta{}^{max}, n_{black}{}^{max}, n_{gel}{}^{max}, n_{fiber}{}^{max}, n_{air}{}^{max}, n_{hole}{}^{max}, L_f^*, a_f^*, b_f^*, \Delta E_f{}^{max}, T_{f0}, S_{fl0}, S_{fc0}\}. \quad (4.18)$$

where:

δ_{f0} = specified thickness [μm]

$D_\delta{}^{max}$ = maximum allowable thickness deviation [μm]

$n_{black}{}^{max}$, $n_{gel}{}^{max}$, $n_{fiber}{}^{max}$, $n_{air}{}^{max}$, $n_{hole}{}^{max}$ = maximum permissible number of surface defects of the corresponding types

L_f^*, a_f^*, b_f^* = reference color coordinates

$\Delta E_f{}^{max}$ = maximum permissible color deviation from the standard

T_{f0} = reference temperature profile of cooling [°C]

S_{fl0}, S_{fc0} = specified degree of shrink in the corresponding direction [%]

The values of the parameters of its physical properties in each phase of PMs depend on the type of polymer T_{polym}.

The control problem when reconfiguring production for a new task $Y_0 = \{T_{film}, G_0\}$ is as follows:

For given input parameter "X", determine the configuration of the units "C_{EQs}". Likewise, the permissible values of control measures in the main phases of production $[U_s{}^{min}; U_s{}^{max}]$; $U_s \subset [U_s{}^{min}; U_s{}^{max}]$. This is intended to ensure the required extrudate quality ($\varphi_{ext} \leq \varphi_{ext}{}^{max}$, $\gamma_{av} \geq \gamma^{min}$, $I_d \leq I_d{}^{max}$), as well as the required film quality ($|\delta_{f0} - \delta_f(0)| \leq \Delta s^{max}$, $|\delta_{f0} - \delta_f(w_f)| \leq \Delta\delta^{max}$, $\Delta\delta \leq \Delta\delta^{max}$, $\Delta E_f \leq \Delta E_f{}^{max}$, $|T_{f0} - T_f| \leq \Delta T^{max}$), even while limiting power and energy consumption.

$$G_2 \geq G_0 \quad (4.19)$$

and

$$E_1 + E_2 \leq E^{max}$$

where:

U_s^{min}, U_s^{max} = vectors of routine threshold values of control actions in the *s*-th phase

φ_{ext}^{max}, γ^{min}, I_d^{max} = maximum permissible values of the corresponding extrudate quality indicators, depending on the restrictions on the number of surface defects of the film (non-melting, gels, black dots)

$\delta_f(0)$, $\delta_{f(wf)}$ = film thicknesses at the edges of the film web [µm]

$\Delta\delta^{max}$, ΔT^{max} = maximum permissible deviations of the layer thickness and temperature, chosen depending on the measurement errors of these initial parameters [µm, °C]

ΔE_f = color deviation of the film from the standard

E^{max} = maximum allowable energy consumption of the production system [J/kg]

Limitations on the properties of material homogeneity (φ_{ext}, γ_{av}) and thermal state (I_d) of the extrudate arise from the suitability of the extrudate for processing on a calender, to exclude film scrap resulting from poor preparation of the extrudate. For example, an irrational extruder configuration and/or extrusion procedure leads to violation of agitation conditions, melting, mixing of the polymer, and formation of dead zones. Increasing the residence time in the extruder at high temperature contributes to the development of thermal degradation in the polymer, which results from the destruction of its macromolecule chains and a strong decrease in molecular weight. The accumulation of degradation products in the extrudate leads to the appearance of black dots on the film surface, the number of which exceeds the specified limit ($n_{black} > n_{black}^{max}$).

The restriction of the temperature profile T_f for cooling the film determines its correspondence with the reference temperature profile T_{f0}, which determines the cooling law, which in turn determines the required optical properties (transparency and gloss) and the degree of shrinkage in the longitudinal direction for a given type of film.

The problem of production control under the influence of disturbances *F* is as follows: For a given configuration of the production system $C_{Ps} = \{M_{EQs}, C_{EQs}, s = 1, 2, 3\}$ and the characteristics of the produced film $H_{film} = \{T_{film}, P_{Ms}, s = 1, 2, 3\}$ the rational values of the control actions in the main stages of production $U_S^0 = [(U_s^*)^{min}; (U_s^*)^{max}]$ which provide the minimum limit change in the thickness of the film under the influence of disturbances $F_s D_g\left(U_S^0\right) = \min_{\text{var } U_s}\left[D_g\left(U_s\right)\right]$, where ΔE_{ext}, ΔE_{ext}^{max} is the current and maximum permissible deviation of the extrudate color from the standard, ΔS^{max} is the maximum permissible deviation of the shrinkage value of the film [%] and ΔG^{max} is the maximum permissible deviation of the extruder throughput from the calender throughput, selected from the condition of synchronous operation of these units, when the requirements for the quality of the extrudate (ΔE_{ext} and ΔE_{ext}^{max}), film quality (n_{black}, n_{black}^{max}, n_{gel}, n_{gel}^{max}, n_{fiber}, n_{fiber}^{max}, n_{air}, n_{air}^{max}, n_{hole}, n_{hole}^{max}, $|S_{fl0} - S_{fl}|$, ΔS^{max}, $|S_{fc0} - S_{fc}|$, and ΔS^{max}), and extruder performance ($|G_2 - G_1|$ and ΔG^{max}) are met.

The need to control the color of the film according to the color of the extrudate is due to the requirement to reduce the losses of the film. These losses occur due to the large delay in the response of the control system to the color deviation from the standard when changing the dye consumption, according to the film color deviation [51]. The values of the color coordinates L_{ext}, a_{ext}, and b_{ext} are taken as the reference color coordinates of the extrudate L^*_{ext}, a^*_{ext}, and b^*_{ext}, which ensure that the values of the color coordinates of the film correspond to the reference values $\Delta E_f \leq \Delta E_f^{max}$ every time production is changed to a new color type T_{film}.

4.2.3 Mathematical Models of the Main Stages of Polymer Film Production

To solve the control problems posed, mathematical models have been developed that allow calculation of the initial parameters of the main production stages depending on the control measures, and which are tailored to various characteristics and configurations of the units of the main stages and varieties of processed polymers:

$$Y_S = \Phi_S \left(X_S, C_{EQS}, U_S, A_S\right), s = 1, 2, 3 \tag{4.20}$$

where:

A_s = vector of adjustable coefficients of the equations of the mathematical model for calculating the output parameters of the s-th stage

Mathematical modelling of extrusion using a co-kneader as an example

The extrusion modelling method is based on a cellular approach to mathematically describing processes in sectional units. This makes it possible to consider the variability of the configuration of the screw of the fluxing unit, which is provided by changing the installation order of kneader elements of different types along the axis of the screw. A screw extruder of each type has a specific set of standard screw elements, which are in geometric different parameters and can be used in different functional areas of the screw-extruder [52–54].

For example, the screws of oscillating extruders (co-kneaders) consist mainly of elements with continuous thread (for creating pressure in the transport zone of the polymer melt), transport elements (which have a single thread with an axial slot and are used to transport solid polymer particles), and kneading elements (which have a two-start thread with three axial slots and are used in the melting zone). A special type of screw element is the element with a throttle ring. This element, like the mixing element (kneading element), has a two-way thread that is interrupted by three axial slots. However, this element is approximately 50% longer because there is a smooth (non-threaded) area in its center that mates with the throttle ring.

The throttle ring is inserted into the co-kneader housing. This creates a narrow radial gap between the throttle ring and the smooth section (core) of this special screw element. This gap reduces the flow rate of the melted polymer. Therefore, an element with a confining ring is used to form a region of a densified melt for the purpose of its subsequent degassing (mainly to remove air, which contributes to the appearance of air bubbles in the polymer film).

The efficiency of this element can be adjusted by changing the value of the radial clearance between the ring and the core (replacing this ring with a ring of a different diameter). To degas the melt, a special vacuum unit is used, which is connected to a vent in the extruder housing, and which creates a vacuum in the degassing zone. Typical elements of the screw of the co-kneader are shown in Figure 4.4.

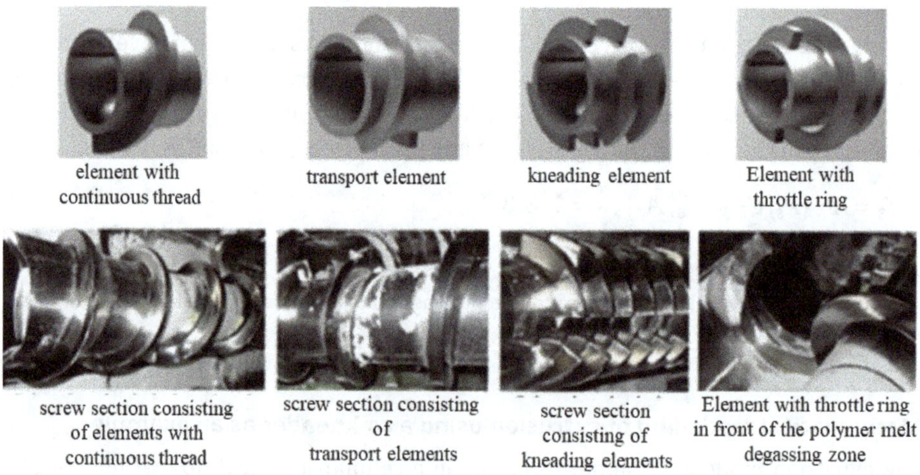

Figure 4.4 Typical elements of an oscillating co-kneader screw (Source: Buss-AG, Klöckner Pentaplast)

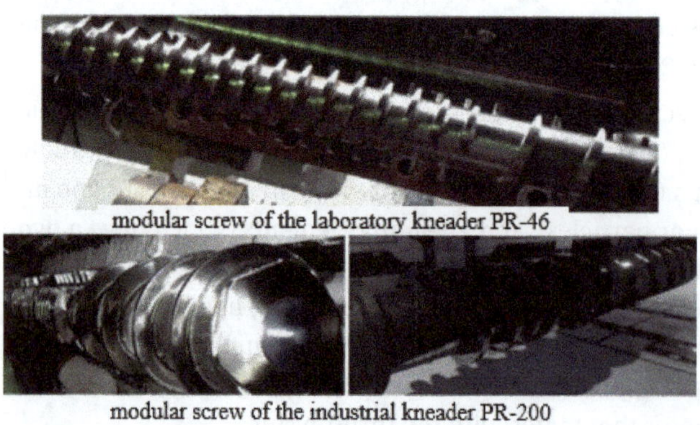

Figure 4.5 Modular screw laboratory and industrial co-kneader PR200 (Source: Buss-AG, Klöckner Pentaplast)

By arranging similar elements in sections and combining sections of different types, it is possible to adjust the co-kneader to process different types of polymers when switching production to a new type of film. Modular screws of a laboratory and industrial oscillating extruder are shown in Figure 4.5.

A library of mathematical models has been developed for the melting and flow of polymers in the channels of typical screw elements for single-screw and twin-screw extruders, as well as co-kneaders [55–59].

During modelling, the following general assumptions were made (for all types of screw extruders):

- The principle of reverse movement of the casing and the screw element is applied
- The curvature of the channel of the screw element is insignificant (a flat flow channel model is used)
- The process is stationary in time
- The movement of the polymer is stationary along the housing wall length
- The thermophysical properties of the polymer phases are independent of temperature and pressure
- The melt is an incompressible inelastic fluid, the flow of which is characterized by the absence of radial flow, inertial and mass forces, and sliding along the surface of the casing

Heat transfer in the melt is characterized by the following conditions:

- The convective mechanism of heat transfer predominates along the length of the channel
- The temperature profile is formed completely along the flow channel width.

The polymer melting mechanism depends on the type of extruder. In single-screw extruders, the compacted, coherent solid phase (plug) of the polymer is melted, which is formed by high pressure in a continuous screw channel with gradually decreasing depth. Therefore, the system is modelled on a solid plug on the back wall of the canal, with:

- A melt film between the plug and the heated inner surface of the barrel
- An area of circulating melt near the pushing throttle ring of the channel, which is created by the flow of the plasticized material.

In co-kneaders and twin-screw extruders, a continuous solid phase of the polymer is not formed. This is due to the friction of the screw in the co-kneader between the kneading teeth in the housing, and in a twin-screw extruder due to the threads of the counter-screw. For this purpose, a two-phase system is modelled, consisting of a melt (continuous phase) and individual solid particles suspended in the melt (disperse phase). The melting and flow model in the screw element channel is a system of equations:

- The mass balance equation
- The balance equations for viscous friction and pressure forces
- The heat balance equation for the solid phase and the polymer melt
- The interfacial material balance

Note that the interfacial heat balance equation includes (and is contained by) kinematic and temperature boundary conditions.

The melting and flow model in the channel of the screw element of a co-kneader can therefore be described using the following sets of equations [60]:

(1) Equations for calculating the radius R_s^j and the temperature T_s^j of solid particles:

$$\rho_s \bar{v}_{\mathrm{mz}}^j \left(\mathrm{d}R_s^j/\mathrm{d}z_1\right) = -\xi_{\mathrm{melt}}^j,\ z^{j-1} < z_1 \le z^{j-1} + Z_{\mathrm{e}}^j \tag{4.21}$$

$$\varphi_s^j \rho_s c_s \bar{v}_{\mathrm{mz}}^j \left(\mathrm{d}T_s^j/\mathrm{d}z_1\right) = s_{\mathrm{sm}}^j \alpha_s^j \left(T_{\mathrm{melt}} - T_s^j\right) - s_{\mathrm{sm}}^j \xi_{\mathrm{melt}}^j c_s \left(T_{\mathrm{melt}} - T_s^j\right) \tag{4.22}$$

(2) Equations for calculating the speeds of translational v_{mz}^j and circulation currents v_{mx}^j, pressure P_{m}^j and temperature T_{m}^j of the melt:

$$\left(1 - \varphi_s^j\right) \int_0^{H'} v_{\mathrm{mx}}^j \,\mathrm{d}y_1 = \dot{Q}_\delta^j + h_{\mathrm{f}}^j \dot{Q}_{\mathrm{c}}^j,\ \left(1 - \varphi_s^j\right) W^j \int_0^{H'} v_{\mathrm{mz}}^j \,\mathrm{d}y_1 = Q_{\mathrm{m}}^j \tag{4.23}$$

$$\partial P_{\mathrm{m}}^j/\partial x_1 = \partial \left[\eta^j \left(\partial v_{\mathrm{mx}}^j/\partial y_1\right)\right]/\partial y_1 \tag{4.24}$$

$$\partial P_{\mathrm{m}}^j/\partial z_1 = \partial \left[\eta^j \left(\partial v_{\mathrm{mz}}^j/\partial y_1\right)\right]/\partial y_1 0 < y_1 < H^j$$

$$\begin{aligned}&\rho_{\mathrm{m}} c_{\mathrm{Pm}} Q_{\mathrm{m}}^j \left(\mathrm{d}T_{\mathrm{m}}^j/\mathrm{d}z_1\right) = W^j \left(q_{\mathrm{bk}} - q_{\mathrm{sct}}\right) - H^j W^j \left[s_{\mathrm{sm}}^j \alpha_{\mathrm{m}}^j \left(T_{\mathrm{m}}^j - T_{\mathrm{melt}}\right) + \right.\\ &\left. s_{\mathrm{sm}}^j \xi_{\mathrm{melt}}^j c_{\mathrm{Pm}} \left(T_{\mathrm{m}}^j - T_{\mathrm{melt}}\right)\right] + W^j F_\varphi^j \left\{V_{\mathrm{bx}} \tau_{\mathrm{mbx}}^j + V_{\mathrm{bz}} \tau_{\mathrm{mbz}}^j - \right.\\ &\left.\left[\left(\dot{Q}_\delta^j + h_{\mathrm{f}}^j \dot{Q}_{\mathrm{c}}^j\right)\left(\partial P_{\mathrm{m}}^j/\partial x_1\right) + Q_{\mathrm{m}}^j \left(\partial P_{\mathrm{m}}^j/\partial z_1\right)/W^j\right]/\left(1 - \varphi_s^j\right)\right\},\\ &z^{j-1} < z_1 \le z^{j-1} + Z_{\mathrm{e}}^j\end{aligned} \tag{4.25}$$

(3) Rheological model for calculating the polymer viscosity η^j:

$$\eta' = \mu' \left[\left(\partial v_{\mathrm{mx}}'/\partial y_1\right)^2 + \left(\partial v_{\mathrm{m=}}'/\partial y_1\right)^2\right]^{(n-1)/2} \tag{4.26}$$

$$\mu^j = \mu_0\left(1 - \varphi_z^j/\varphi_{\max}\right) - \beta\left\{\left(1 - h_{\mathrm{pt}}\right)\mathrm{e}^{-C_1\left(T_\mathrm{m}^j - T_\mathrm{r}\right)/\left[C_2 + \left(T_\mathrm{m}^j - T_\mathrm{r}\right)\right]} + h_{\mathrm{pt}}\mathrm{e}^{-b\left(T_\mathrm{m}^j - T_\mathrm{r}\right)}\right\}$$

(4) Equation to calculate the flow rate of the polymer through the channel of the screw element:

$$Q_\mathrm{E} = z_\mathrm{f}^j Q^j - \left(Q_\delta^j + h_\mathrm{f}^j Q_\mathrm{c}^j\right) \tag{4.27}$$

(5) Equation for calculating the specific melting rate of a polymer ξ_{melt}^j:

$$\alpha_\mathrm{m}^j\left(T_\mathrm{m}^j - T_{\mathrm{meit}}\right) = \alpha_s^j\left(T_{\mathrm{meit}} - T_s^j\right) + \xi_{\mathrm{meit}}^j\left[\left(1 - h_\mathrm{p}\right)c_s\left(T_{\mathrm{meit}} - T_s^j\right) + h_\mathrm{p} r\right] \tag{4.28}$$

(6) Boundary conditions at the entry into the channel of the screw element:

$$z_1 = z^{j-1} : R_s^j = R_{\mathrm{sout}}^{j-1}, \varphi_s^j = \varphi_{\mathrm{sout}}^{j-1}, T_s^j = T_{\mathrm{sout}}^{j-1}, P_\mathrm{m}^j = P_{\mathrm{mat}}^{j-1}, T_\mathrm{m}^j = T_{\mathrm{mout}}^{j-1} \tag{4.29}$$

$$j = 2, \ldots, N_\mathrm{e}$$

(7) Boundary conditions at the bottom and cover of the screw element channel:

$$y_1 = 0 v_{\mathrm{mx}}^j = 0 v_{\mathrm{mz}}^j = 0 H^j \tag{4.30}$$

$$y_1 = H^j$$

$$\begin{aligned} v_{\mathrm{mz}}^j = V_{\mathrm{bz}} = V_{\mathrm{rot}}\sin\varphi^j + V_{\mathrm{osc}}\cos\varphi^j = \\ \pi\left(N/60\right)\left[-D\sin\varphi^j + S_0\sin\left(\Phi_{\mathrm{ox}}\right)\cos\varphi^j\right] \end{aligned} \tag{4.31}$$

$$\begin{aligned} v_{\mathrm{mz}}^j = V_{\mathrm{bz}} = V_{\mathrm{rot}}\cos\varphi^j + V_{\mathrm{osc}}\sin\varphi^j = \\ \pi(N/60)\left[D\cos\varphi^j + S_0\sin\left(\Phi_{\mathrm{ox}}\right)\sin\varphi^j\right] \end{aligned}$$

where:

j = number of the kneading element

ρ_s, ρ_m, = density of the solid phase and the polymer melt [kg/m^3]

$\bar{v}_{\mathrm{mz}}^j$ = mean forward flow velocity (along the channel depth) [m/s]

x_1, y_1, z_1 = coordinates along the width, depth, and length of the screw channel [m]

z^{j-1} = coordinate of the input to the channel of the j-th element [m]

W^j, H^j, Z_e^j = width, depth and length of the element channel [m]

φ_s^j, φ_s^0 = current and initial volume fraction of solid particles

c_s, c_{Pm} = average specific heat capacity of solid phase and polymer melt [J/(kg·°C)]

s_{sm}^{j} = specific area of the interface [m²/m³]

α_s^j = heat transfer coefficient from the interface to the particle [W/(m²·°C)]

T_{melt} = melting temperature of the polymer [°C]

$\dot{Q}_\delta^j, \dot{Q}_c^j$ = leakage rates of melt through the radial gap (the gap between the threads and the inner surface of the body) and the axial slots in the thread [m²/s]

f^j = coefficient for adapting the model to the cutting type of the element (if the cutting is continuous, then $h_f^j = 0$ and there are no leaks through the slots. If the cutting is interrupted by cuts, then $h_f^j = 1$ and leaks occur through the slots)

Q_m^j = throughput of the translational flow of the melt [m³/s]

q_{bk}, q_{scr} = heat flux densities from the k-th thermal zone of the body to the melt and from the melt to the screw [W/m²]

α_m^j = heat transfer coefficient from the melt to the interface [W/(m²·°C)]

F_φ^j = correction that considers the increase in energy dissipation in the two-phase "solid–melt" system

V_{bx}, V_{bz} = components of the speed of movement of the body (with reverse movement) along the width and length of the screw channel [m/s]

$\tau_{mbx}^j, \tau_{mbz}^j$ = shear stresses acting near the inner surface of the body in the circulation and translational flow of the melt [Pa]

μ^j = consistency coefficient [Pa·s^n]

n = flow index

μ^0 = consistency coefficient at reference temperature T_r [Pa·s^n]

φ_{max} = volume fraction of solid particles with close packing

$ß, C_1, C_2, b$ = coefficients indicating the degree of influence that characterize the viscosity of solid particles ($ß$) and temperature (C_1, C_2, b)

h_{pt} = coefficient for fitting the model to the class of polymer and the temperature range of the process (if the polymer is amorphous and the temperature of its melt does not exceed the glass transition temperature by 100°C, then $h_{pt} = 0$ and the dependence of viscosity on temperature is described by the Williams–Landel–Ferry equation; if the polymer is amorphous with a melt temperature exceeding the glass transition temperature by more than 100°C, or a crystalline polymer, then $h_{pt} = 1$)

Q_E = flow rate of the polymer through the extruder, corresponding to productivity [m³/s]

z_f^j = number of cuts of the element

Q_δ^j, Q_c^j = leakage rates due to radial clearance and axial slots [m³/s]

h_p = coefficient for fitting the model to the polymer class (if the polymer is amorphous, then h_p = 0; if the polymer is crystalline, then $h_p = 1$)

r = specific heat of fusion [J/kg]

d_s = average equivalent diameter of polymer particles [mm]

T_0, P_0 = temperature and pressure of the polymer at the inlet to the screw channel [°C, Pa]

$R_{sout}^{j-1}, \varphi_{sout}^{j-1}, T_{sout}^{j-1}, p_{mout}^{j-1}, T_{mout}^{j-1}$ = state parameters of the solid phase (radius, volume fraction, temperature) and the melt (pressure, temperature) at the exit of the channel ($j-1$)-th element

N_e = number of screw elements

V_{rot}, V_{osc} = screw rotation and oscillation speeds [m/s]

φ^j = cutting inclination angle [rad]

Φ_{osc} = screw oscillation phase (state parameters calculated at different values of the oscillation phase are averaged over the operating cycle of the extruder, which corresponds to the full stroke of the screw) [rad]

In addition to general assumptions, the construction of the mathematical model described by Equation 4.21–Equation 4.31 assumes that the two-phase "melt–solid particle" system is monodisperse. Solid particles are non-deformable balls, friction and impacts between them are insignificant, and there is no squeezing and sticking. Therefore, a decrease in the volume fraction of solid particles is a consequence of a decrease in their radius during melting. Solid particles do not move forward or rotate relative to the melt. The heat given off by the melt to the solid phase is distributed evenly among all solid particles, so that the melting rate is the same at all points on the surface of each particle. The interaction of solid particles with the channel walls is negligible.

The heat flux densities in Equation 4.25 are calculated according to the Newton–Richmann cooling law for different thermal operating modes of the extruder:

$$q_{bk} = \alpha_{bk}\left(T_{bk} - T_{m}^{j}\right), q_{scr} = (1 - h_{scr})\,\alpha_{scr}\left(T_{m}^{j} - T_{scr}\right) \tag{4.32}$$

where:

α_{bk} = heat transfer coefficient from the k-th thermal zone of the body to the melt [W/(m^2·°C]

h_{scr} = coefficient for fitting the model to the conditions of heat transfer to the screw (if the screw is cooled, then $h_{scr} = 0$; if the screw is thermoneutral, then $h_{scr} = 1$)

α_{scr} = heat transfer coefficient from the melt to the screw at the temperature T_{scr} [W/(m^2·°C)]

The structure of the proposed model is adapted to the type of screw element, the polymer class, the temperature range of the process, and the thermal mode of the extruder. The melting and flow model in the screw channel is formed by arranging the melting and flow models in the channels of individual kneader elements according to a predetermined screw configuration. This configuration determines the sequence of elements of different types in the screw. The layout of the models is carried out using the coupling conditions of screw elements (Equation 4.30). The shaped melt and flow model in the screw channel is combined with the flow model in the melding zone of the extruder head [61, 62].

To solve the nonlinear equations of the constructed model of the extrusion process, a calculation method was developed that includes external and internal iterative cycles. The outer loop looks for the operating point of the extruder. In the inner loop, the flow rate is determined by the channel of each screw element Q_j, which (in the presence of leaks) provides the flow rate indicated in the outer loop through the extruder Q_E. With each new approximation for the flow rate, the flow rates in the channel and the leakages (e.g. in the case of a co-kneader the flow rates Q_j, Q_δ^j, Q_c^j) are recalculated one after the other for each screw element. This is also done for the distribution of the parameters of the state of the solid phase and the melting polymer along the length of the channel.

To calculate the state parameters, the balance equations are solved numerically [63]. For example, to solve Equation 4.21 and Equation 4.22 considering Equation 4.28, the Runge–Kutta method of the fourth order of accuracy is used. To solve Equation 4.23–Equation 4.25 considering the rheological model (Equation 4.26), a calculation scheme based on the theory of plane asymmetric flows under non-isothermal conditions is used [64].

If the screw has an element with a throttle ring, the calculation scheme is changed. Since this element is installed before the degassing zone, the melt pressure at the exit of its channel equals zero. The pressure of the melt at the entrance to the channel of the element with a throttle ring is calculated according to the Poiseuille equation depending on the throughput of the element, the viscosity of the melt, and the flow rate of the melt. The throughput of the element is mainly determined by the geometric parameters of the radial gap between the throttle ring and the element core. If the pressure at the outlet of the channel of the element located upstream of the element with a restrictor ring exceeds the pressure at the inlet of the channel of this element, a sign of incomplete filling of the channel with polymer is set for the previous section of the screw.

The previous section of the screw is understood as a set of elements of the screw connected in series, ending with an element with a throttle ring. The degree of filling of the channel of each element is determined as a relative volume flow:

$$\psi_j = Q_j / Q_\mathrm{d} \tag{4.33}$$

where:

Q_d = flow rate of the forced polymer flow in the channel of the screw element with the smallest total cross-sectional area of the channels (element with the minimum throughput) [m³/s]

Then the calculations are repeated starting from the first element of the previous section of the screw. In this case, the effective width of the part of the channel completely filled with polymer ($\Psi^j W^j$) is used in Equation 4.23 and Equation 4.25 instead of the full width of the channel. The external circuit continues until the condition of coupling between the screw and extruder's output die is met, that is, the melt pressure $|P_\mathrm{m}^j\Big|_{j=N_\mathrm{a}}$ at the exit of the screw channel is equal to the melt pressure P_die at the entrance to the transport zone of extruder's output die.

A disc with larger holes (like a large spinneret) is usually used as kneader output die, since the extruder, in conjunction with a calender, takes on the role of a fluxing unit for producing the plastic and does not deform the material into a film.

The flow model in the outlet die of the kneader allows the calculation of the pressure P_{die} depending on the geometric parameters of the output zone, the melt viscosity, and the throughput Q_E. The throughput Q_E determines the productivity of the extruder via the coupling state of the screw and outlet die of the kneader:

$$G_1 = 3{,}600\ \rho_m Q_E \tag{4.34}$$

The specific energy consumption of an extruder is calculated as the ratio of electricity consumption to productivity.

The solids content of the extrudate is determined by the solids content of the polymer at the end of the melting zone. In the normal process sequence, the condition for the end of the melting zone is met, which consists in the disappearance of the solid phase (zeroing of the solid phase proportion) within the screw channel. If this condition is not met before leaving the channel, then there is an extrudate defect, which is associated with the presence of non-melting and leads to a corresponding film defect.

Depending on the residence time, key figures characterizing the material homogeneity and the degree of thermal destruction of the extrudate are calculated:

$$\gamma_{av} = \frac{\tau_{av}}{Z}\sum_{j=1}^{j-N_e}\left\{\int_{z^{l-1}}^{z^{l-1}+Z_e^l}\left[\frac{1}{H^j W^j}\int_0^{w^l}\int_0^{H^l}\sqrt{\left(\frac{\partial v_{mx}^j}{\partial y_1}\right)^2+(n_{sov}-1)\left(\frac{\partial v_{mz}^j}{\partial x_1}\right)^2+\left(\frac{\partial v_{mz}^j}{\partial y_1}\right)^2}\right.\right.$$
$$\left.\left. dy_1 dx_1\, dy_1 dx_1 \right| dz_1\right\} \tag{4.35}$$

$$I_d = \left(\frac{\tau_{av}}{\tau_d}\right)\exp\left[\frac{E_d\,(T_{ext}-T_d)}{R\,(T_{ext}+273)\,(T_d+273)}\right]\times 100 \tag{4.36}$$

where:

τ_{av} = average residence time of the polymer in the extruder [s]

Z = length of the screw channel [m]

τ_d, T_d = time and temperature corresponding to the onset of polymer color change (e.g. the appearance of a yellow color in polyvinyl chloride samples) due to thermal decomposition (during testing in a thermostat) [s, °C]

E_d = activation energy of the destruction process [J/mol]

T_{ext} = is the temperature of the extrudate [°C]

R = the universal gas constant [8.31 J/(mol·K)]

The variable modular design of the extruder and the complex flow structure in it (in particular, the presence of leaks) do not allow calculation of the average residence time as the ratio of the volume of the screw channel to the volumetric productivity of the extruder.

Approaches for estimating the residence time of polymers in various types of extruders are described in the literature [65–71]. They are based on the use of both individual typical hydrodynamic models (e.g. a single-parameter diffusion model) and complex combined models of the flow structure. These were created based on typical hydrodynamic models (ideal mixing and displacement models, cell models), including the introduction of circulating flows. Typical models describe extruder zones with different mixing mechanisms and degrees. Recycled materials simulate backflows, especially leaks caused by tool resistances.

A cellular approach to modelling hydrodynamics in aggregates with a complex structure was applied [72] to create a mathematical model that makes it possible to estimate the residence time considering the complex hydrodynamic situation in the extruder. The hydrodynamic model of the extruder is formed as a set of interconnected zones that describe the structure of flows in screw sections with different mechanisms and degrees of mixing. Each zone is one of the typical hydrodynamic models describing the flow structure in the corresponding screw section.

Leaks are accounted for by including reprocessing (recycling) between cells. The resulting system of material balance equations is solved using numerical methods. This makes it possible to calculate the temporal distribution of the tracer concentration at the extruder outlet given the tracer concentration at the inlet modelled by the δ function. The average residence time in the extruder is calculated as the first-order start time of this distribution.

The color coordinates of the extrudate are calculated using nonlinear multivariate regression models obtained by processing the results of active production experiments using the least-squares method [73].

For example, if three color pigments are used to form the color (the maximum number of color pigments used in production at the same time), then the structure of the empirical model for calculating the brightness of the extrudate is as follows:

$$L_{\text{ext}} = a_{10} + a_{11}k_1 + a_{12}k_2 + a_{13}k_3 + a_{14}k_1k_2 + a_{15}k_1k_3 + a_{16}k_2k_3 + a_{17}k_1^2 + a_{18}k_2^2 + a_{19}k_3^2 \tag{4.37}$$

where:

a_{1n}, $n = 0....9$ are the empirical coefficients depending on the extruder, polymer, and dye types

$k_i = G_{ci}/G_1$ is the share of the consumption of the i-th dye in the productivity

Models for calculating color coordinates a_{ext} and b_{ext} have a similar structure. The color deviation of the extrudate from the reference is calculated using the CIE-Lab model:

$$\Delta E_{\text{ext}} = \sqrt{\left(L_{\text{ext}}^* - L_{\text{ext}}\right)^2 + \left(a_{\text{ext}}^* - a_{\text{ext}}\right)^2 + \left(b_{\text{ext}}^* - b_{\text{ext}}\right)^2} \tag{4.38}$$

The use of a mathematical extrusion model for preliminary calculation of the extrudate quality indicators Q_1, depending on the characteristics of the extruder, the polymer, and the control interventions on the process, makes it possible to reduce film scrap due to poor plasticization in the extruder.

Mathematical model for calendering

The calender model allows the calculation of productivity, specific energy consumption of the calender and film quality indicators (thickness, limit thickness variation, color coordinates). The basic element of the model is the system of equations for the material balance, the force balance, and the heat balance of the polymer melt flow in a separate nip of the calender, closed by kinematic and temperature boundary conditions [74]:

$$\int_{-\delta_m^{l,l+1}10^{-3}}^{\delta_m^{l,l+1}10^{-3}} v_{\mathrm{mz}}^{l,l+1} \mathrm{d}y_2 = \dot{Q}_c, \mathrm{d}P_m^{l,l+1}/\mathrm{d}z_2 = \partial\left[\eta^{l,l+1}\left(\partial v_{\mathrm{mz}}^{l,l+1}/\partial y_2\right)\right]/\partial y_2 - \tag{4.39}$$

$$\delta_{\mathrm{m}}^{l,l+1}10^{-3} < y_2 < \delta_{\mathrm{m}}^{l,l+1}10^{-3}, \quad z_{\mathrm{in}}^{l,l+1} < z_2 < z_{\mathrm{out}}^{l,l+1}$$

$$\rho_{\mathrm{m}} c_{\mathrm{Pm}}\left(\partial T_{\mathrm{m}}^{l,l+1}/\partial\tau\right) = \lambda_{\mathrm{m}}\left(\partial^2 T_{\mathrm{m}}^{l,l+1}/\partial y_2^2\right) + \eta^{l,l+1}\left(\partial v_{\mathrm{mz}}^{l,l+1}/\partial y_2\right)^2, \tag{4.40}$$

$$0<\tau\leq\tau^{l,l+1}$$

$$\begin{aligned} &y_2 = -\delta_{\mathrm{m}}^{l,l+1}10^{-3}: v_{\mathrm{mz}}^{l,l+1} = V_{\mathrm{cl}}/60, T_{\mathrm{m}}^{l,l+1} = T_{\mathrm{cl}} \\ &y_2 = -\delta_{\mathrm{m}}^{l,l+1}10^{-3}: v_{\mathrm{mz}}^{l,l+1} = V_{\mathrm{cl+1}}/60, T_{\mathrm{m}}^{l,l+1} = T_{\mathrm{cl+1}} \\ &\tau = 0: T_{\mathrm{m}}^{l,l+1} = T_{\mathrm{m,in}}^{l,l+1} \end{aligned} \tag{4.41}$$

where:

$\delta_{\mathrm{m}}^{l,l+1}$ = half the thickness of the polymer melt layer in the gap, which depends on the longitudinal coordinate z^2 corresponding to the calendering direction [mm]

$v_{\mathrm{mz}}^{l,l+1}$ = melt flow rate [m/s]

y_2 = coordinate along the thickness of the enamel layer [m]

$\dot{Q}_{\mathrm{c}}$ = specific flow rate (flow rate per unit of flow width), defined depending on the thickness of the melt layer in the exit section from the gap and the speed of the rolls [m^2/s]

$P_{\mathrm{m}}^{l,l+1}, T_{\mathrm{m}}^{l,l+1}$ = pressure and the temperature of the melt [Pa, °C]

$\eta^{l,l+1}$ = viscosity of the melt, calculated according to rheological equations like Equation 4.26, without cross-flow and solid particles in the melt [Pa·s]

$z_{\mathrm{in}}^{l,l+1}, z_{\mathrm{out}}^{l,l+1}$ = coordinates of the melt inlet and outlet sections [m]

λ_{m} = thermal conductivity of the melt [W/(m·°C)]

$\tau^{l,l+1}$ = residence time of the melt in the gap depending on the gap length and the flow speed

V_{cl}, V_{cl+1}, T_{cl}, T_{cl+1} = speeds and temperatures of the l-th and (l + 1)-th rolls forming the gap [m/min, °C]

$T_{\text{m,in}}^{l,l+1}$ = temperature of the melt at the entrance to the gap [°C]

When constructing the model expressed by Equation 4.39–Equation 4.41, assumptions were made about:

- The smallness of the curvature of the center line of the gap
- A slow change in the size of the gap in the direction of flow
- The flow being stationary
- The incompressibility and non-elasticity of the melt
- The predominantly shearing nature of their current
- The constancy of the pressure in the gap cross-section
- The smallness of the inertial and mass forces compared to the forces of viscous friction
- The smallness of heat transfer through thermal conductivity in the longitudinal direction compared to convective heat transfer acting in the same direction and heat transfer through thermal conductivity in the transverse direction
- The adhesion of the melt to the deforming surfaces of the rolls.

During the numerical integration of the system of Equation 4.39–Equation 4.41, the solution of the hydrodynamic equation group (Equation 4.39) and the heat balance equation (Equation 4.40) are separated at each of the successive integration steps along the longitudinal coordinate.

In the first stage, the flow velocity is calculated using an approximation for the temperature taken as an extrapolation from one step ahead along the flow axis. In the second step, the temperature in the same cross-section is calculated in a stricter approximation by integrating the heat balance equation using the finite difference method [63]. Calculations are performed at a given specific flow rate through the gap, which is determined depending on the relative caliber of the molten material k_c according to the following formula:

$$\dot{Q}_c = k_c \delta_{\text{m0}}^{l,l+1} 10^{-3} \left(V_{cl} + V_{cl+1}\right) /60 \tag{4.42}$$

where:

$\delta_{\text{m0}}^{l,l+1}$ = half the minimum thickness of the melt layer in the gap [mm]

The value of the relative caliber, which determines the thickness of the PVC layer in the exit section of the gap, is obtained according to the method of successive approximation $P_{\mathrm{m}}^{l,l+1}\Big|_{z_2-z_{\mathrm{tn}}^{l,l+1}}=0$ from the condition that the melt pressure is at the exit from the gap [45]. The calculated value of the specific consumption $\dot{Q}_{\mathrm{c}}$ can be used to determine the performance G_2 of the calender.

An important criterion for the perfection of the calendering process is the consistency of the thickness (same thickness) of the resulting film. The reason for the difference in film thickness is the deflection of the calender rolls along them length that occurs under the action of spacing forces in the material to be formed.

The distance force per unit length of the roller is determined from the distribution of the melt pressure in the gap:

$$F_{\varepsilon}^{l,l+1}=\int_{z_{\mathrm{tn}}^{l,l+1}}^{z_{\mathrm{ext}}^{l,l+1}} P_{\mathrm{m}}^{l,l+1}\mathrm{d}z^2 \tag{4.43}$$

To compensate for the deflection of the reels, traditional, widely used methods are used, namely camber or roll bending of the rolls forming the calender gap [75].

Therefore, the system of Equation 4.39–Equation 4.41 is supplemented by models that are built based on methods of analytical geometry and the resistance theory of materials, and which describe the dependence of the $\delta_{\mathrm{f}}^{\mathrm{b}}, \delta_{\mathrm{f}}^{\mathrm{cross}}, \delta_{\mathrm{f}}^{\mathrm{cb}}$ thickness on the parameters of the corresponding methods, with the aim of eliminating differences in layer thickness [76]:

$$\delta_{\mathrm{f}}^{\mathrm{b}}=d_{\mathrm{b}}\times 10^3/(1-\sin\alpha_{\mathrm{b}})\left(\sin\left\{\alpha_{\mathrm{b}}+\frac{\pi-2\alpha_{\mathrm{b}}}{L_{\mathrm{c}}\left[x_2+{}^1\!/_2\left(L_{\mathrm{c}}-w_{\mathrm{f}}\right)\right]}\right\}-\sin\alpha_{\mathrm{b}}\right) \tag{4.44}$$

$$\delta_{\mathrm{f}}^{\mathrm{cross}}=\left[\sqrt{D_{\mathrm{c}}^2+(w_{\mathrm{f}}x/L_{\mathrm{c}})^2}-D_{\mathrm{c}}\right](1-2x_2/w_{\mathrm{f}})^2\times 10^3 \tag{4.45}$$

$$\delta_{\mathrm{f}}^{\mathrm{cb}}=\frac{{}^1\!/_2 w_{\mathrm{f}} r_{\mathrm{s}}}{E_{\mathrm{c}} I_{\mathrm{cb}}\left[x_2{}^2-L_{\mathrm{c}}x_2-a_{\mathrm{n}}\left(L_{\mathrm{c}}+\frac{a_{\mathrm{n}} I_{\mathrm{cb}}}{I_{\mathrm{cn}}}\right)\right]\times 10^3} \tag{4.46}$$

where:

d_{b} = maximum increase in diameter in the center of the bale during crowning [mm]

α_{b} = camber angle [rad]

x_2 = coordinate along the width of the film wf from its edge of the film [mm]

E_{c} = elastic modulus of the roll material [MPa]

$I_{\mathrm{cb}}, I_{\mathrm{cn}}$ = polar moments of inertia of the roller sections and their journal [mm^4]

a_{n} = length of the roll neck section from the center of the bearing to the roll barrel [mm]

The models expressed by Equation 4.44–Equation 4.46 are created assuming that:

- The geometric parameters of each roll are constant and do not change when heated
- The distance from the center of the roller face to the edge of the film is the same on the left and right
- The thickness profile corresponds to the profile of the roll gap, in the area that is occupied by the PVC-hide, and has the same width as the film.

When modelling the roll bending, the calculation scheme of the beam on two hinged supports is adopted (the beam is considered statically determined).

The models expressed by Equation 4.44–Equation 4.46 enable the calculation of the PVC-hide thickness profiles (thickness distribution across the web width) δ_f through the camber, axis crossing, and roll-bending depending on the geometric parameters of the calender rolls, the parameters of the mechanical properties of the roll materials, the PVC-hide width, and the control actions x and r_s.

The boundary layer thickness difference D_δ is calculated from the difference between the maximum value $\delta_{f,max}$ and minimum value $\delta_{f,min}$ in the thickness profile: $D_\delta = \delta_{f,max} - \delta_{f,min}$. According to empirical models, the structure of which is like that of Equation 4.38, the color coordinates of the film (L_f, a_f, b_f) are calculated, depending on which color deviation of the film ΔE_f is estimated to be acceptable [73].

Mathematical model of polymer film cooling

The film cooling model makes it possible to calculate the temperature profile of the film over the length of the cooling unit. The rate of decrease in the temperature of the film in different parts of the cooling zone largely determines the important customer properties of the film, the degree of shrink rate, and transparency.

Intensive cooling of the film on the first rollers of the pick-off unit and slow cooling on subsequent rollers not only increases the transparency, but also the degree of shrink rate. The problem therefore arises of finding control interventions during cooling that ensure that the temperature profile of the film cooling corresponds to a standard. This must guarantee a specified film quality for different configurations and geometric parameters of cooling devices and types of films produced. This problem is solved using a mathematical model of cooling based on the theory of convective heat transfer.

The following assumptions were made when constructing the model:

- The process is stable over time
- The film is cooled due to the thermal conductivity in the polymer
- The heat transfer from the film to the outer surface of the roller takes place through an intermediate layer of air, and the heat transfer from the film to the surrounding air takes place through convection and radiation

- The thickness of the film δ_f is constant
- The temperature of the film T_f changes slightly along its width compared to the difference along its length
- The density ρ_f and thermal conductivity of the film λ_f do not depend on the temperature.

The model contains the film heat balance equation and boundary conditions at the bottom and top surfaces of the film as well as at the inlet and outlet of the cooling device [77]:

$$\rho_f c_f (v_f/60)(\partial T_f/\partial z_3) \times 10^{-3} = \lambda_f \left[\partial^2 T_f/\partial r_f^2 + r_f^{-1}(\partial T_f/\partial r_f) + \partial^2 T_f/\partial z_3^2\right] \quad (4.47)$$

$$0 < r_f < \delta_f \times 10^{-3}, 0 < z_3 < Z_{cool}$$

$$r_f = 0\!: -\lambda_f \left(\frac{\partial T_f}{\partial r_f}\right) \times 10^3 = (1 - h_{cont})\, K_{cool}\left(T_f - T_{clq}\right) + h_{cont} \times \\ \left\{\alpha_{air}(T_f - T_{air}) + \sigma_0 \varepsilon_f \left[(T_f + 273)^4 - (T_{air} + 273)^4\right]\right\} \quad (4.48)$$

$$r_f = \delta_f \times 10^{-3}\!: \; -\lambda_f \left(\frac{\partial T_f}{\partial r_f}\right) \times 10^3 = \\ \alpha_{air}(T_f - T_{air}) + \sigma_0 \varepsilon_f \left[(T_f + 273)^4 - (T_{air} + 273)^4\right] \quad (4.49)$$

$$z_3 = 0\!: T_f = T_{f0} \\ z_3 = Z_{cool}\!: \partial T_f/\partial z_3 = 0 \quad (4.50)$$

where:

c_f = average specific heat capacity of the film, whose temperature dependence in the operating temperature range is satisfactorily described by the cubic regression model [J/(kg·°C)]

v_f = speed of film movement in the cooling zone, depending on the peripheral speeds of the rollers of the cooling device [m/min]

z_3 = coordinate along the length of the cooling device Z_{cool} [mm]

r_f = coordinate over the film thickness [mm]

h_{cont} = coefficient for adapting the model to the type of medium with which the film comes into contact

K_{cool} = heat transfer coefficient from the film to the outer surface of the roll, calculated depending on the heat transfer coefficient from the film to the air in the gap between the film and the roll surface α_f, the thermal conductivity of the air, and the heat transfer coefficient from air to the roll surface α_{roll}, considering the thickness of the air layer [W/(m^2·°C]

α_{air} = heat transfer coefficient from the film to the ambient air at the temperature T_{air} [W/(m^2·°C]

σ_0 = Stefan–Boltzmann constant [5.67 × 10^{-8} W/(m^2·K^4)]

ε_f = degree of blackness of the film

T_{f0} = temperature of the film at the inlet of the cooler [°C]

The coefficient h_{cont} allows the model to be tuned to describe the cooling of both the film surface in contact with the rollers (h_{cont} = 0) and the film surface in contact with air (located between the rollers; h_{cont} = 1). The second boundary condition from Equation 4.50 is because the speed of the film temperature change at the last cooling roll is equal to zero.

The numerical solution of the nonlinear boundary value problem for the convective heat transfer equation (Equation 4.47) is carried out using the finite difference method (Runge's five-point difference scheme) and the method of half-dividing the grid steps by spatial coordinates. It is a two-dimensional homogeneous equation in second-order partial derivatives of elliptic type to check the convergence of the solution and to calculate the temperature profile of the film with an error not exceeding the maximum allowable error [63].

The model expressed by Equation 4.47–Equation 4.50 is tuned to the geometric properties of the cooler, the type of film (by adjusting the parameters of thermophysical properties and the heat transfer coefficient), and allows calculation of the temperature distribution of the film along the length of the cooling unit (temper unit). This depends on the control interventions at the appropriate stage, namely peripheral speeds and temperatures on the surfaces of the cooling unit's rolls.

Depending on the control interventions—the peripheral speeds of the cooling unit rolls—the shrink rates of the film in the longitudinal and transverse directions (S_{fl} and S_{fc}) are calculated. For this, linear, multifactorial regression models and functional (mechanical) models are used. Linear, multifactorial regression models generate the result from the statistical processing of data from active experiments. Functional models consider only the elastic properties of the film (the Mooney–Rivlin model) or viscoelastic properties (the generalized Maxwell model or the relaxation spectrum model, rheological model with three links) [78, 79].

Elastic body models are used to describe polymers that have pronounced elastic properties and little or no viscous properties. A sufficiently large selection of materials with elastic properties and the simplicity of the mathematical apparatus justifies the use of these models.

More complete are the models that consider both the elastic and viscous properties of polymers as well as the effect of strain intensity on the strains accumulated in the polymer. These models are applicable to different types of materials and configurations of cooling devices. However, they often have a fairly complex mathematical apparatus and require complex viscometrical experiments to determine the values of the equation coefficients depending on the type of material.

The Mooney–Rivlin model makes it possible to calculate the shrinkage rate of the finished film in the longitudinal direction S_{fl} as the sum of the shrinkage rate of the cooled film in all sections of the inter-reel space of the cooling device. The degree of film shrinkage $S_l^{u,u+1}$ in a separate section between the u-m and (u + 1)-m rollers of the cooling device are calculated by solving the following nonlinear algebraic equation by the iteration method [79]:

$$b_3\left(S_l^{u,u+1}\right)^6 + b_1\left(S_l^{u,u+1}\right)^4 + \left(b_2 - b_3 - \sigma^{u,u+1}\right)\left(S_l^{u,u+1}\right)^3 - b_3\left(S_l^{u,u+1}\right)^2 - b_1 = 0 \quad (4.51)$$

$$\sigma^{u,u+1} = 2^{n+1}\frac{\delta_f^u w_f^u}{\delta_f^{u+1} w_f^{u+1}}\left[\frac{nV^u}{60L^{u,u+1}(1-n)}\right]^n\left[1-\left(\frac{V^u}{V^{u+1}}\right)^{1-n}\right]\mu^{u,u+1} \quad (4.52)$$

where:

b_1, b_2, b_3 = coefficients of the Mooney–Rivlin model equation *d* expending on the film type [Pa]

$\sigma^{u,u+1}$ = tensile strength of the film [Pa]

δ_f^u, w_f^u = thickness and width of the film on the (u + 1)-th reel [μm, mm]

V^u, V^{u+1} = peripheral speeds of adjacent rolls [m/min]

$L^{u,u+1}$ = length of the film section in the space between the rollers [m]

$\mu^{u,u+1}$ = consistency coefficient calculated as a function of the temperature of the film in this area and found with the cooling model [Pa·s^n]

According to the relaxation spectrum model, the degree of film shrinkage in the longitudinal direction is calculated as follows [79]:

$$S_{fl} = \left(\frac{1}{\alpha_s} - 1\right) \times 100 \quad (4.53)$$

$$\alpha_s = \sqrt{\left(\sigma_{nr} + \sqrt{\sigma_{nr}^2 + 4E_f^2}\right)/2E_f} \quad (4.54)$$

where:

α_s = recovery factor

σ_{nr} = sum of the normal stresses that did not have time to relax, accumulated in all sections of the gap of the cooling unit (calculated depending on the circumferential speeds of the rolls and the temperature of the film, considering the relaxation) [Pa]

E_f = elastic modulus of the film [Pa]

According to the rheological three-link model, the degree of shrink rate of the film in the longitudinal direction is calculated as follows [79]:

$$S_{fl} = \frac{\varepsilon_{ve}}{1+\varepsilon} \times 100 \quad (4.55)$$

$$\varepsilon_{ve} + \frac{\mu_{ve}}{E_f}\dot{\varepsilon}_{ve} - 3^{\frac{n-1}{2}}\frac{\mu_v}{E_f}|\dot{\varepsilon}| - |\dot{\varepsilon}_{ve}|^{n-1}(\dot{\varepsilon} - \dot{\varepsilon}_{ve}) = 0 \quad (4.56)$$

where:

ε_{ve} = highly elastic component of the tensile elongation of the film ε

$\mu_{ve} = f_1(T_f)$ = Newtonian viscosity [Pa·s]

$\dot{\varepsilon}_{ve}, \dot{\varepsilon}$ = tensile strength rates of the film [1/s]

$\mu_v = f_2(T_f)$ = consistency coefficient [Pa·s^n]

Algorithm for modelling the main stages of polymer film manufacturing

The developed mathematical models of the extrusion process (Equation 4.21–Equation 4.38), calendering (Equation 4.39–Equation 4.46), and cooling (Equation 4.47–Equation 4.56) in the production of polymer films consider the variability of the configurations of the aggregates used (extruders, calenders, and multi-roll cooling/tempering units), and are tailored to different film types. They make it possible to carry out a structural synthesis (configuration formation) of the unit at each stage of production, and to select the values of control measures on it that ensure the specified quality of the plastificate and film. At the same time, the requirements for productivity are met and the energy consumption of the process for a new task and production of products under the influence of disruptions is considered.

The adequacy of mathematical models of the processes of extrusion, calendering, and film cooling has been evaluated (according to Fisher's criterion via standard deviation and coefficient of determination) for various types of colored and uncolored films (based on rigid PVC, polypropylene, and low-density polyethylene) in factories in Russia and Germany [55–60, 72–74, 76–79].

The algorithm for modelling the main stages of film production, adapted to film type T_{film} and system performance G_0, is shown in Figure 4.6.

For the generated production modelling task Y_0, the algorithm allows the selection of hardware for each main production stage. The characteristic M_{EQs} and the configuration of the corresponding unit C_{EQs} (extruder, calender, and cooling device) form the control ranges of control actions on the units of production, $[U_s^{min}; U_s^{max}]$ synthesize mathematical models of the main production stages and conFigure them with coefficients A_s depending on the geometric properties of the aggregates M_{EQs}, C_{EQs} and the type of processed polymers T_{polym}.

The structural synthesis of the mathematical model of each main production stage consists in establishing the structure and layout of models of the hydrodynamic and thermal processes that exist in the elements of the single calender unit (kneader elements and outlet disk of the extruder, gaps between the rolls of the calender, the film on the rollers, and between the rollers of the cooling device). When synthesizing the model of the phase of plastics production, the extruder type T_{extrud} is additionally considered, which influences the mechanism of polymer melting depending on the type of extruder, the model of melting and flow of the polymer in the channel of the screw element of the extruder is selected (e.g. for a co-kneader (Equation 4.21–Equation 4.36), which describes dispersed melting).

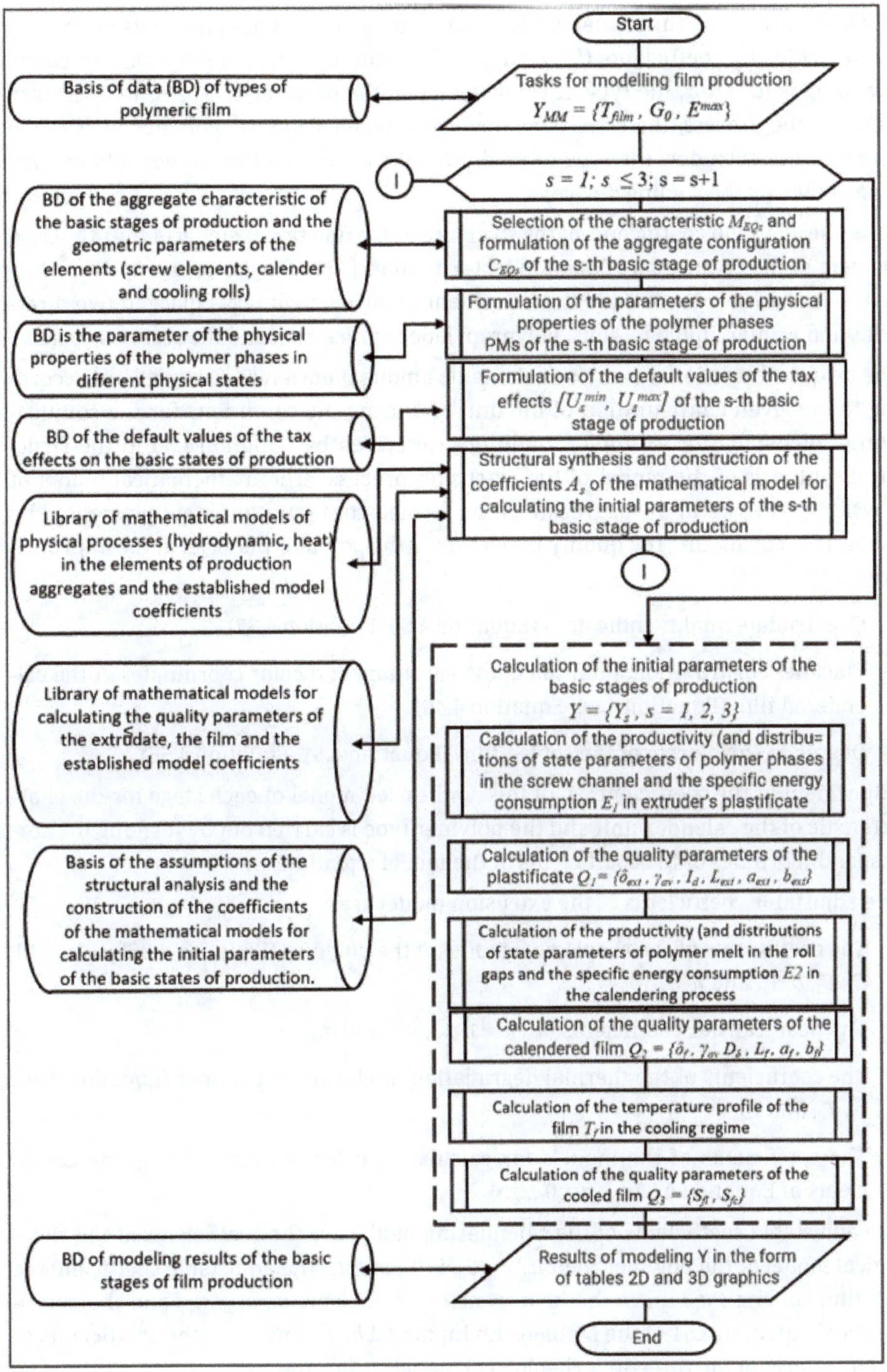

Figure 4.6 Algorithm for modelling the main stages of polymer film production

The structure of the functional model of the unit element is adapted to its properties by model-fitting coefficients (h_f^j, z_f^j, h_{pt}, h_p, h_{scr}, and h_{cont}), which are assigned corresponding values (e.g. the type of screw element, the polymer class and temperature range of the process, the conditions for heat transfer from the polymer melt to the screw in the extruder, the type of medium with which the film comes into contact when entering the cooling device).

Thus, the adaption coefficients of the structure of the functional model of the *j*-th screw element of the co-kneader (Equation 4.21–Equation 4.36) take the values $h_f^j = h_{pt} = h_p = 1$, $h_f^j = 2$, and $h_{scr} = 0$, provided the movement of the element takes place in two directions and is interrupted by slots, polypropylene is processed, and the screw is cooled.

The layout of functional models of elements among themselves is carried out according to the given configuration of the unit and using the conditions for the conjugation of elements (for example, conditions described by Equation 4.30 in the structural synthesis of the model of the extrusion process). The mathematical model of physical processes in the aggregate of the formed configuration is supplemented by models for calculating the quality indicators of the polymer material at the appropriate stage:

- Q_1 extrudate quality indicators (Equation 4.35–Equation 4.37)
- Thickness distribution, limit thickness variation, and color coordinates of the calendered film (Equation 4.44–Equation 4.46)
- Degree of shrink rate of the cooled film (Equation 4.51–Equation 4.56).

Adjustment of the coefficients A_s of the synthesized model of each stage for the characteristic of the calender unit and the polymer type is carried out by forming the corresponding values of the coefficients of the model equations.

The adjustable coefficients of the extrusion model are:

- The coefficients of the rheological model of the polymer (Equation 4.26) μ_0, φ_{max}, β, C_1, C_2, b, T_r, and n
- The heat transfer coefficients α_{bk}, $k = 1, \dots, nT$, and α_{scr}
- The coefficients of the thermal degradation model of the polymer (Equation 4.36) τ_d, T_d, and E_d
- The coefficients of the models for calculating color coordinates (e.g. the coefficients of Equation 4.37a_{1n}, $n = 0,\dots, 9$).

The adjustable coefficients of the calendering model are the coefficients of the rheological model of the polymer melt μ_0, C_1, C_2, b, T_r, and n. The adjustable coefficients of the film cooling model are the heat transfer coefficients α_f, α_{roll}, α_{air} and the coefficients of Equation 4.51 of the Mooney–Rivlin model b_1, b_2, and b_3, or the coefficients of Equation 4.56 of the three-link rheological model μ_{ve}, μ_v, and n.

The synthesized and tuned models of the production stages are used to calculate the distributions of the polymer state parameters in the channels of the aggregates:

$$S_1 = \left\{R_s^j, T_s^j, v_{mx}^j, v_{mz}^j, P_m^j, T_m^j, \eta^j, j = 1, \ldots, N_c\right\} \tag{4.57}$$

$$S_2 = \left\{v_{mz}^{l,l-1}, P_m^{l,l+1}, T_m^{l,l+1}, \eta^{l,l+1}, l = 1, \ldots, n_{calend} - 1\right\}$$

$$S_3 = \{T_f\}$$

Based on the state parameters of productivity G_1 and G_2 and the specific energy consumption E_1 and E_2 of the units at the stages of preparation and melding of the plastificate, the quality indicators of the plastificate and film Q_s (s = 1, 2, 3) are calculated. As a result, the dependencies of the output parameters of the production stages Y_s (s = 1, 2, 3) and the control interventions U_s (s = 1, 2, 3), which vary in the generated control ranges, are formed and visualized (in tabular and graphical form).

The analysis of these dependencies makes it possible to determine the values of control interventions that provide the required quality of the plastificate and the film while respecting the constraints on productivity and energy consumption for the chosen hardware design to produce a given film.

4.2.4 Software Complexes for Modelling the Main Production Stages

To automate the calculation of the initial parameters of the main steps of film production, software packages have been developed to simulate the steps of plastificate preparation, plastificate forming into a film and film cooling [80–102]. A typical functional structure of the software package for modelling the s-th stage is shown in Figure 4.7.

The software package contains a subsystem that:

- Selects the hardware design of the subunit of the calender line
- Synthesizes a functional mathematical model of physical processes occurring in the entirety of the formed configuration
- Calculates the performance, power consumption of the unit and quality indicators of the polymer material in the section
- Contains an information subsystem (a database of the characteristics of the hardware and technological design of the section) as an interface for the senior production staff.

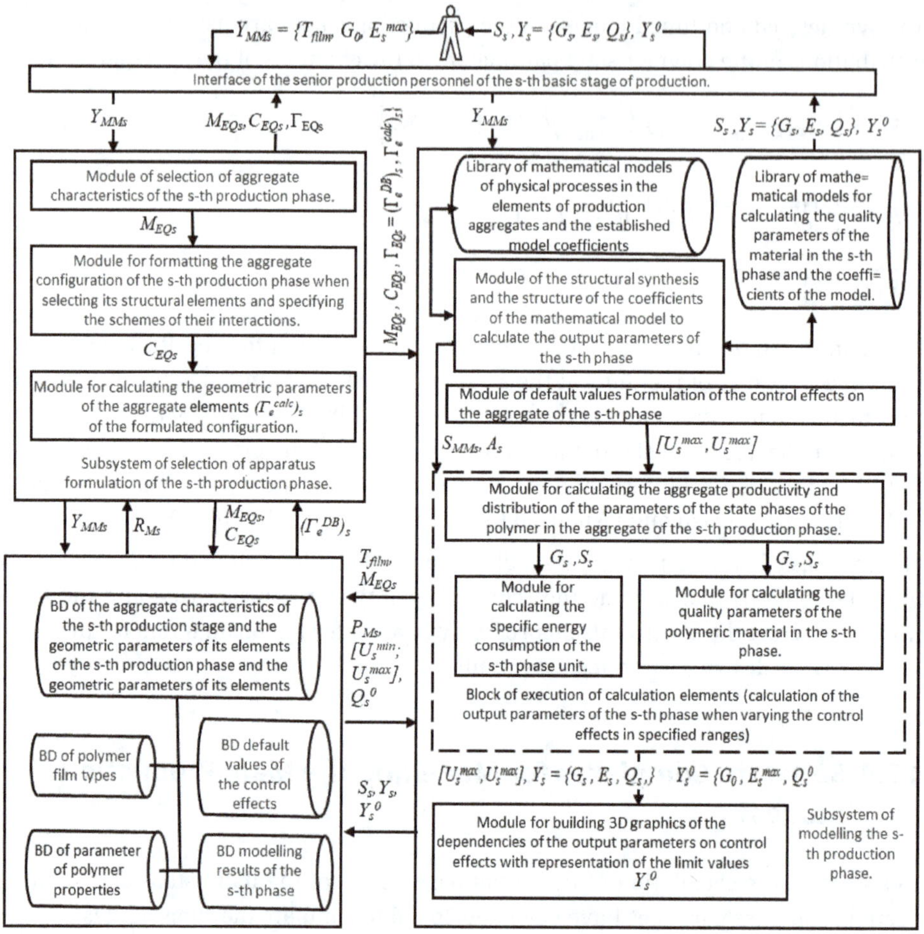

Figure 4.7 Functional structure of the program complex for modelling the *s*-th section

The stage properties database contains relational databases of the classes of the subunit of the calender line and the geometric parameters of its structural elements (screw elements and extruder meld heads, calender rolls, and cooling rolls), film types, parameters of the physical properties of polymer phases, planned areas of control actions, and simulation results. The database allows reconfiguration of the software package to account for various characteristics and configurations of the process equipment of the sections and the types of films produced, and save the output parameters of the sections calculated for this.

The unit feature selection model creates (from the database of features) a set of unit mark RMs, each of which meets the requirements of the Y_{MMs}-level modelling task.

The rules for selecting the device properties are built according to the production scheme of the type "IF condition, THEN consequence", where the condition consists of statements about the film type T_{film}, the required productivity G_0, and the maximum permissible specific energy consumption E_s^{max}. The application of this rule leads directly to a recommended decision, e.g. which brand of unit M_{EQs} to select.

For the selected characteristic of the unit, its configuration C_{EQs} is formed by selecting and arranging (from the database of equipment characteristics) the structural elements of the unit with geometric parameters corresponding to this characteristic. In this case, the number of arranged elements is controlled. For example, when forming the configuration of the extruder screw, the total length of the assembled screw elements must be equal to the length of the selected extruder screw L. When forming the calender configuration, the number of arranged rolls must be equal to the number of calender rolls n_{calend}.

Depending on the characteristic M_{EQ} and C_{EQ} configuration of the unit, a set of values of the geometric parameters of each element of the unit is formed from the database of $\left(\Gamma_e^{DB}\right)_s$ device properties.

So, the geometric parameters of the screw elements of single-screw extruders and co-kneaders are the length, pitch and number of threads, the entry and exit depth of the channel, the radial gap between the inner surface of the body and the threads, and the number and width of axial slots in the thread. The calender rolls have the geometric parameters D_c, L_c, d_b, and a_n. These properties are used to calculate the geometric parameters $\left(\Gamma_e^{calc}\right)_s$:

- Parameters of the channels of the extruder screw elements φ^j, W^j, H^j, and Z_e^j
- Calender roll parameters α_b, l_{cb}, and l_{cn}
- Lengths of the film sections that are in contact with the rollers of the cooling device
- The specific properties of the aggregate M_{EQs}, C_{EQs}, and Γ_{EQs} are transferred to the modelling subsystem.

The simulation subsystem makes it possible to synthesize the structure of and adjust the coefficients A_s of the mathematical model for processing a polymer of a given type T_{polym} in a modular unit with shaped geometric properties.

For this purpose, libraries of mathematical models of physical processes occurring in the elements of the unit and models for calculating the quality indicators of the polymer material at the stage, as well as production rules for the synthesis and adaptation of the mathematical model of the stage, are used. Based on these rules, the selection, structural adjustment, and layout of mathematical models of physical processes in the aggregate elements are carried out. The coefficients of the synthesized functional model of the aggregate are adjusted to its characteristics and type of polymer.

Model libraries enable the use of software systems to model the main stages of production of a variety of films on extrusion calender lines of various configurations. Depending on the type of film T_{film} and the brand of the M_{EQs} aggregate, the values of the parameters of the polymer properties P_{Ms} (from the database of polymers) and the regulatory restrictions of control measures U_s^{min} and U_s^{max} (from the database of regulatory areas of control measures) are formed.

With the synthesized and coordinated model, the state parameters S_s, productivity G_s, energy consumption E_s and quality indicators Q_s are calculated depending on the control actions U_s. The simulation results are visualized in the form of 2D graphics, 3D graphics of the distributions of the state parameters of the polymer in the aggregate, and 3D graphics of the dependencies of the output parameters (productivity, energy consumption, and quality indicators) on control interventions. The 3D diagrams of these dependencies show the limit values of the output parameters Y_s^0, including the maximum permissible (set) values of the quality indicators.

$$Q_s^0 \begin{pmatrix} Q_1^0 = \{\varphi_{\text{ext}}^{\max}, \gamma^{\min}, I_{\text{d}}^{\max}, L_{\text{ext}}^*, a_{\text{ext}}^*, b_{\text{ext}}^*\}, \\ Q_2^0 = \{\delta_{\text{f0}}^{\max}, D_{\delta}^{\max}, L_{\text{f}}^*, a_{\text{f}}^*, b_{\text{f}}^*\} \\ Q_3^0 = \{S_{\text{f0}}, S_{\text{f}\subset 0}\} \end{pmatrix} \tag{4.58}$$

These values are generated from the film database depending on the film type T_{film}. Parameters S_s, Y_s, and Y_s^0 defined for a given film type, and the characteristics and configuration of the unit, are stored in the database of simulation results of the *s*-th phase.

Thus, the software package generates the data necessary to determine the unit's configuration of the selected features and the values of the control actions on the unit. These data provide the specified indicators of the quality of the polymer material while meeting the requirements for productivity and energy consumption of the device.

4.2.5 Examples of Modelling the Main Phases of the Polymer Film Production

Testing of the software systems was carried out according to the data of extrusion calender production of flat films based on rigid PVC in factories in Russia and Germany.

Examples of the results of modelling the extrusion phase are shown in Figure 4.8.

The task for modelling the sections includes the following requirements:

- T_{polym} = PVC
- n_{black}^{max} = 1 piece/10 m^2
- n_{gel}^{max} = 15 pieces/10 m^2 (transparent film for packaging pharmaceuticals)
- G_0 = 950 kg/h
- E_1^{max} = 360 kJ/kg.

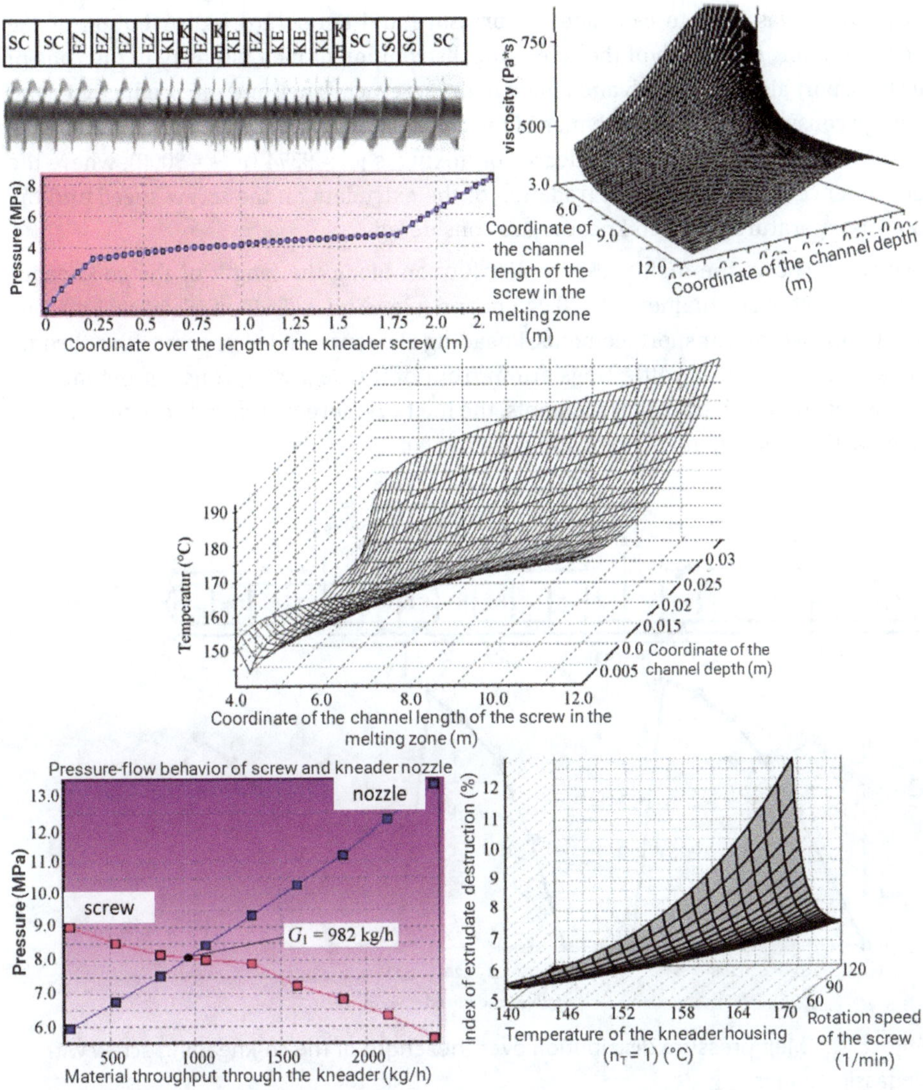

Figure 4.8 Simulation results of the extrudate preparation phase

A co-kneader PR-200 (D = 0.2 m, L = 2.28 m, S_0 = 0.03 m) was selected that meets the requirements of the task. It is constructed with a C_{scr} screw, which consists of elements with continuous thread (SC), transport elements (EZ) and kneading elements (KE).

A mathematical model of the extrusion process was created, which considers several models of dispersed melting of the polymer, mixing, and heating of the melt in the screw channel, and shaping of the melt in the outlet die. Likewise, the melt leakage through the radial gap and the crossing of the slots of the screw elements, as well as the heat exchange with a cooled screw and a heated body, are considered.

The model was used to calculate the pressure and temperature distribution of the polymer along the length of the screw, the distribution of melt viscosity in the zone of its transport along the depth and length of the screw channel, and the productivity and energy consumption of the polymer in the co-kneader (G_1 = 982 kg/h and E_1 = 288 kJ/kg at N = 1.67 L/min). The average degree of mixing is γ_{av} = 9374 (γ^{min} = 3000), where the dependence of the thermal degradation of the extrudate on the screw speed and the body temperature of the co-kneader are considered.

Figure 4.9 shows the melt pressure distribution along the length of the co-kneader screw, which has a diameter D = 0.046 m and a length L = 0.540 m (S_0 = 0.007 m). The screw consists of transport elements, kneading elements, and elements with a throttle ring. Elements with throttle rings are in front of the degassing zones of the melt. In screw sections with transport elements, the melt pressure increases more than in sections with mixing elements (kneading elements).

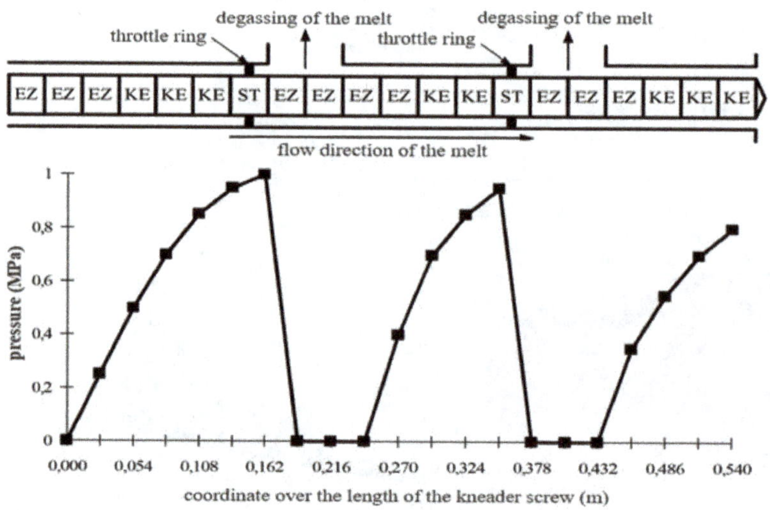

Figure 4.9 Melt pressure distribution over the length of the co-kneading screw with degassing zones

This is because each mixing element (kneading element) has three axial slots in the thread, which contribute to an increase in the leakage current through them (compared to the leakage flow through the transport element with only one slot). In degassing zones, the excess pressure drops to zero to remove volatile components from the melt. Maintaining zero pressure also prevents the melt from escaping through the vent in the housing.

A calculation example for the dependence of the limit change in thickness of the film (T_{polym} = PVC) on the size of the axis crossing of the penultimate roll of the calender and the bending pressure of the last calender roll for an L-shaped four-roll calender (D_c = 670 mm, L_c = 2500 mm) is shown in Figure 4.10. Based on the three-dimensional model of film thickness variation, the operator evaluates the change direction and the number of steps to change the control actions x and r_s to achieve the minimum thickness variation.

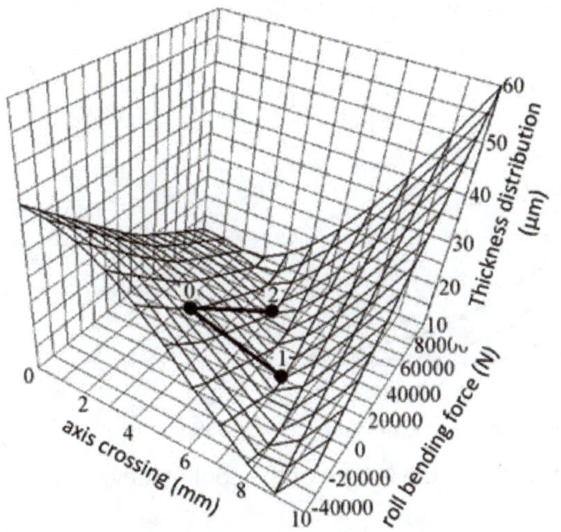

Figure 4.10 Dependence of the limiting thickness variation of the film on control interventions on the calender with display of the current (0) and best (1, 2) values of the control interventions

For example, the initial thickness difference $D_{\delta 0}$ = 28 µm (at x_0 = 4 mm, r_s^0 = 0 N) can be reduced to D_δ = 10 µm in two ways: $(x^{opt})_1$ = 7 mm, $(r_s^{opt})_1$ = 20,000 N; $(x^{opt})_2$ = 5 mm, $(r_s^{opt})_2$ = 60,000 N.

Initial (at the current values of the control actions x_0 = 6 mm, r_s^0 = 0 N) and calculated thickness variation profiles (at the best values of the control actions x^{opt} = 8 mm, r_s^{opt} = 20,000 N) result in deviations of the film thickness δ_f from the specified value δ_{f0} = 400 µm over the film width w_f = 1800 mm, and are shown in Figure 4.11. With the best values of the control actions, the limit thickness variation of the films decreases more than twice (from 28 µm to 12 µm).

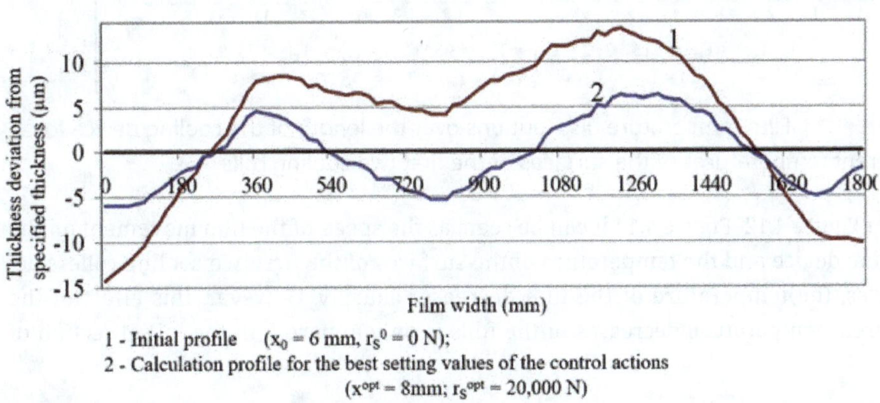

Figure 4.11 Distribution of film thickness deviations from a given thickness across the film width

Examples of the results of modelling the film cooling phase are shown in Figure 4.12–Figure 4.14. The simulation was carried out with the following input data: T_{polym} = PVC; δ_f = 300 µm; w_f = 1450 mm; ρ_f = 1390 kg/m³; λ_f = 0.17 W/(m·°C); v_f = 10 m/min; T_{f0} = 178°C. The configuration of the multi-roll cooler is shown in Figure 4.15.

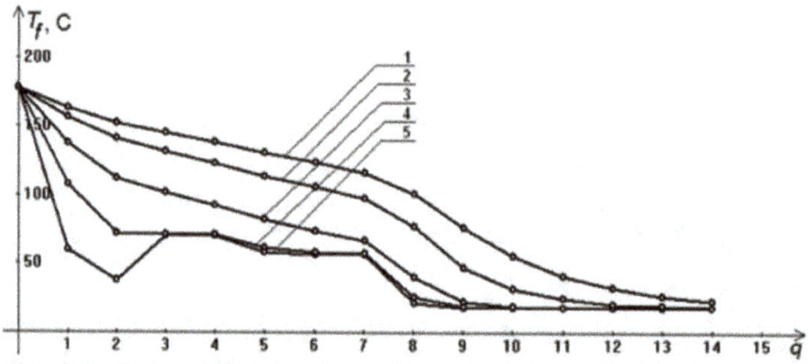

1 - v_f = 60 m/min, 2 - v_f = 40 m/min, 3 - v_f = 20 m/min, 4 - v_f = 10 m/min, 5 - v_f = 5 m/min

Figure 4.12 Film temperature distributions over the length of the cooling device for different film speeds

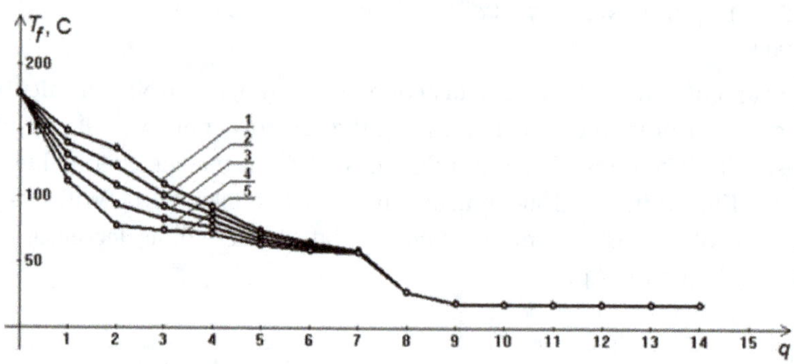

1 - $T_{cl1} = T_{cl2}$ = 115 °C, 2 - $T_{cl1} = T_{cl2}$ = 95 °C, 3 - $T_{cl1} = T_{cl2}$ = 75 °C,
4 - $T_{cl1} = T_{cl2}$ = 55°C, 5 - $T_{cl1} = T_{cl2}$ = 35 °C

Figure 4.13 Film temperature distributions over the length of the cooling device for different temperatures of the surfaces of the first two cooling rolls

From Figure 4.12–Figure 4.14 it can be seen, as the speed of the film movement on the cooling device and the temperature of the surfaces of the first two cooling rollers decreases, the temperature of the film decreases sharply. However, this effect of the different temperature decreases of the foils is only apparent in the initial section of the pick-off rolls (the first 6–7 rolls). The average temperature drops over the film thickness from bottom side CR3 to top side CR4 over the entire length of the cooling section (red to blue) is an average of 5°C.

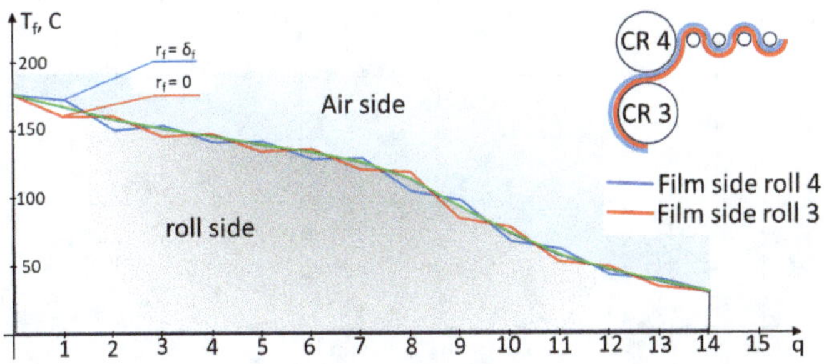

Figure 4.14 Temperature distributions of the contact side (contact with the rollers, $r_f = 0$) and air side ($r_f = \delta_f$) of the film surfaces along the length of the cooling unit (calculated above, schematic below)

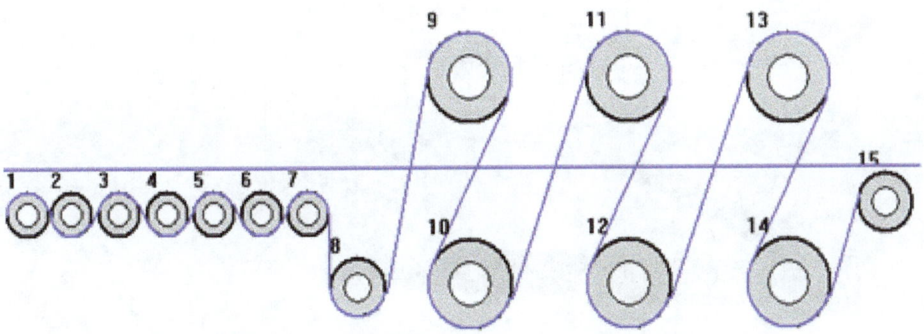

Figure 4.15 Configuration of the succession of a calender line

Examples of the results of modelling the thermal shrinkage properties of films are shown in Figure 4.16 and Figure 4.17, with designations being:

- "Length shrinkage" is the amount of longitudinal shrink
- "Width shrinkage" is the amount of transverse shrink
- "Roll number" is the number of the roller of the cooling unit
- "Temperature" is the surface temperature of the first roller of the cooling unit
- "Linear speed" is the peripheral speed of the first roller of the cooling unit.

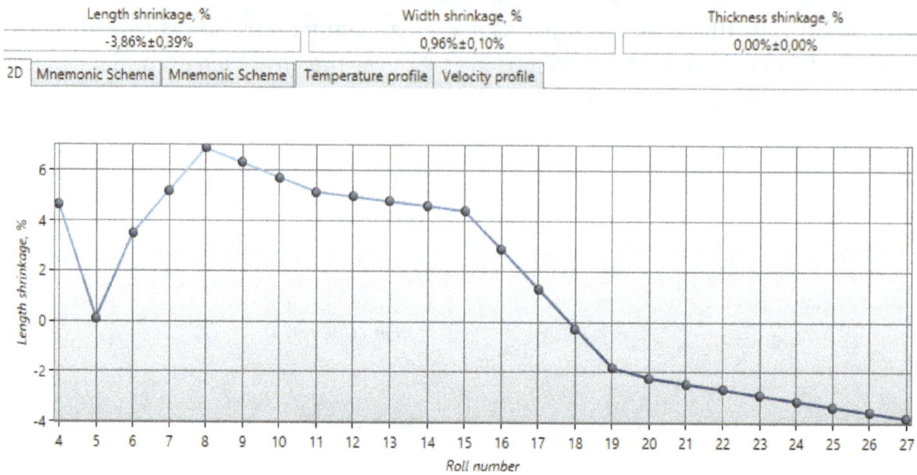

Figure 4.16 The calculated longitudinal and transverse shrinkage of the film and the distribution of the longitudinal shrinkage over the length of the cooling device

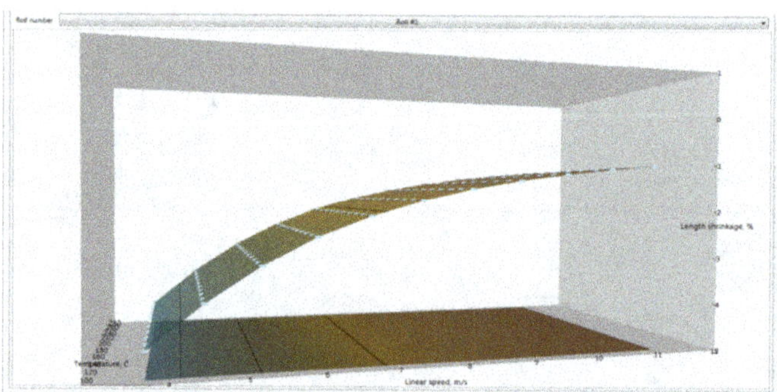

Figure 4.17 Dependence of the degree of longitudinal shrinkage calculated by the Mooney–Rivlin model on control interventions (circumferential speed and surface temperature of the first roller of the cooling device)

An example of calculating the dependence of the color deviation on the cost proportions of the color pigments fed into the extruder is shown in Figure 4.18. A 3D color variation model is shown here to help the operator select the direction of change and the number of steps to change the color pigment consumption of each color to achieve the reference film color.

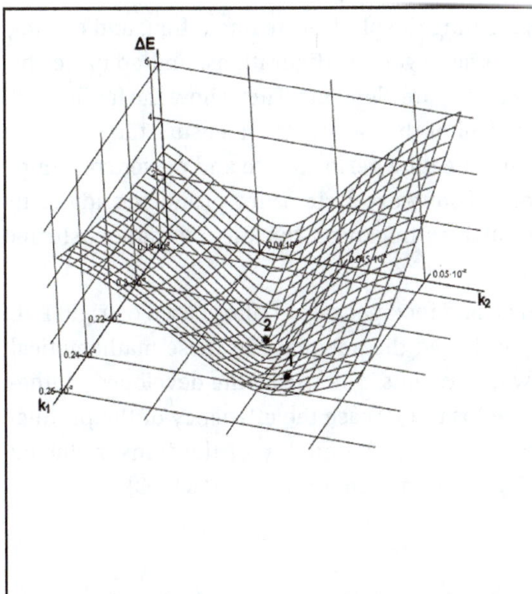

Polymer type	PVC
Film color	orange
Set data	
L_f^*	74.66
a_f^*	32.32
b_f^*	48.01
G_0	950 kg/h
Ongoing data	
L_{ext}	73.97
a_{ext}	33.06
b_{ext}	48.11
ΔE_{ext}	1.02
k_{11}	$0,25*10^{-2}$
k_{21}	$0.047*10^{-2}$
k_{12}	$0.22*10^{-2}$
k_{22}	$0.044*10^{-2}$
Control Interventions	
G_{c1}	2.1 kg/h
G_{c2}	0.42 kg/h

Figure 4.18 Dependence of the color deviation of the film on the control interventions on the extruder, with display of the current (1) and rational (2) values of the control interventions

4.2.6 Summary

In this section, mathematical models and problem-oriented software systems have been described that were developed for the digitalization of the industrial production of polymer films, and which are characterized by:

- Large outputs and energy consumption
- A wide product range
- Frequent reconfiguration to produce new types
- Different films
- Flexibility in hardware design
- Many parameters and complex cause-and-effect relationships between them.

Mathematical models and software systems help production managers solve complex problems of resource-saving management when switching production to a new type of film, enabling productivity and film production to be increased under the influence of disruptions. Mathematical models describe the physical processes in the main stages (production of the plastificate, forming the plastificate into a film, and cooling the film), consider the variability of the aggregate configurations, and adapt to the types and geometric parameters of the aggregate elements. They allow the implementation of parametric and structural control of the aggregates to ensure the specified quality of the plastificate and films while meeting performance and power consumption requirements. Models are the core of software systems for producing advice to manage customer properties of films such as thickness, thickness variation, shrink level, and color coordinates.

Tests after production of pharmaceutical and food packaging films based on rigid PVC in factories in Russia and Germany confirmed the adequacy of these mathematical models and the performance of software systems. The use of the developed mathematical models and software systems helps to increase the efficiency of the production of multi-range polymer films by improving the quality of the films, reducing waste and the time required to reconFigure the system for new tasks [103].

References for Section 4.2

[1] U. W. Gedde, M. S. Hedenqvist, M. Hakkarainen, F. Nilsson, The processing of polymeric materials, *Applied Polymer Science*, Springer: Cham, 2021, pp. 453–487

[2] J. L. White, A. F. Dean, C. C. Case, D. H. Wilson, *Polymer Mixing: Technology and Engineering*, Ed. by J. L. White et al., Munich: Carl Hanser Verlag, 2001, 241 pp

[3] K. Kohlgrüber, *Co-Rotating Twin-Screw Extruders: Fundamentals*, 2nd ed, Munich: Carl Hanser Verlag, 2020, 432 pp

[4] D. V. Rosato, *Calendering, Extruding Plastics*, Boston: Springer, 1998, pp. 418–440

[5] H. Martens, D. Goldmann. *Recyclingtechnik. Fachbuch für Lehre und Praxis*, Wiesbaden: Springer Fachmedien, 2016, 556 pp

[6] H. Gärtner, A. Schnabel, C. Kohlert, Control systems for calenders—from mixer to take-off, *Kunststoffe Plast Europe*, 1997, vol. 87, no. 6, pp. 24–27

[7] T. B. Chistyakova, F. Kleinert, M. A. Teterin, Big data analysis in film production, in: *Cyber-Physical Systems: Advances in Design & Modelling*, Ed. by A. G. Kravets, A. A. Bolshakov, M. V. Shcherbakov, Cham: Springer Nature Switzerland, 2020, pp. 229–236 (*Studies in Systems, Decision and Control*; vol. 259)

[8] C. Kohlert, S. Steinmeier, M. Kohlert, Mathematical methods in plastics processing, Математические методы в технике и технологиях: сборник трудов международной научной конференции. В 12 томах. Том 8, Санкт-Петербург : Изд-во Политехн. ун-та, 2017, с. 14–21

[9] M. Kohlert, A. König, Advanced polymeric film production data analysis and process optimization by clustering and classification methods, *Frontiers in Artificial Intelligence and Applications*, 2012, vol. 243, pp. 1953–1961

[10] M. Kohlert, T. B. Chistyakova, Advanced process data analysis and on-line evaluation for computer aided monitoring in polymer film industry, Известия Санкт-Петербургского государственного технологического института (технического университета), 2015, no. 29, c. 83–88

[11] M. Kohlert, O. Hissmann, Applied Industry 4.0 in the polymer film industry, Proceedings of the 16th TAPPI European PLACE Conference, Basel, 2017, pp. 183–190

[12] T. Chistyakova, M. Teterin, A. Razygraev, C. Kohlert, Intellectual analysis system of big industrial data for quality management of polymer films, Advances in Neural Networks, ISNN 2016: 13th International Symposium on Neural Networks Proceedings, Cham: Springer, 2016, pp. 565–572, (Lecture Notes in Computer Science. Theoretical Computer Science and General Issues; vol. 9719)

[13] Z. Tadmor, E. Broyer, Solids conveying in screw extruders. Part II: Non isothermal model, *Polymer Engineering and Science*, 1972, vol. 12, no. 5, pp. 378–386

[14] Z. Tadmor, I. J. Duvdevani, I. Klein, Melting in plasticating extruders. Theory and experiments, *Polymer Engineering and Science*, 1967, vol. 7, no. 3, pp. 198–217

[15] I. Rlein, Z. Tadmor, The simulation of the plasticating screw extrusion process with a computer programmed theoretical model, Polymer Engineering and Science, 1969, vol. 9, no. 1, pp. 11–21

[16] D. R. Hinrichs, L. U. Lilleleht, A modified melting model for plastifying extruders, *Polymer Engineering and Science*, 1970, vol. 10, no. 5, pp. 268–278

[17] J. R. Vermeulen, pp. G. Scargo, W. J. Beek, The melting of a crystalline polymer in screw extruder, *Chemical Engineering Science*, 1971, vol. 26, no. 9, pp. 1457–1465

[18] R. C. Donovan, A theoretical melting model for plasticating extruders, *Polymer Engineering and Science*, 1971, vol. 11, no. 3, pp. 247–257

[19] I. R. Edmondson, R. T. Fenner, Melting of thermoplastics in single screw extruders, *Polymer*, 1975, vol. 16, no. 1, pp. 49–56

[20] J. Shapiro, A. L. Halmos, J. R. A. Pearson, Melting in single screw extruders. Part 1: The mathematical model, *Polymer*, 1976, vol. 17, no. 10, pp. 905–918

[21] J. T. Lindt, A dynamic melting model for a single-screw extruder, *Polymer Engineering and Science*, 1976, vol. 16, no. 4, pp. 284–291

[22] A. L. Halmos, J. R. A. Pearson, R. Trottnow, Melting in single screw extruders. Part 3: Solutions for a power law temperature-dependent viscous melt, *Polymer*, 1978, vol. 19, no. 10, pp. 1199–1216

[23] N. I. Basov, I. N. Volodin, Yu. V. Kazankov, V. E. Pervushin, Hydrodynamics and heat exchange in melting in the screw channel of an extruder, *Theoretical Foundations of Chemical Engineering*, 1983, vol. 17, no. 1, pp. 61–67

[24] В. В. Скачков, Р. В. Торнер, Ю. В. Стунгур, С. В. Реутов, Моделирование и оптимизация экструзии полимеров, Ленинград: Химия, 1984, 152 с

[25] B. Elbirli, J. T. Lindt, S. R. Gottgetreu, S. M. Baba, Mathematical modelling of melting of polymers in a single-screw extruder, *Polymer Engineering and Science*, 1984, vol. 24, no. 12, pp. 988–999

[26] V. P. Pervadchuk, N. M. Trufanova, V. I. Yankov, Mathematical model and numerical analysis of heat transfer processes associated with the melting of polymers in plasticating extruders, *Journal of Engineering Physics*, 1985, vol. 48, no. 1, pp. 60–64

[27] J. T. Lindt, B. Elbirli, Effect of the cross-channel flow on the melting performance of a single-screw extruder, *Polymer Engineering and Science*, 1985, vol. 25, no. 7, pp. 412–418

[28] J. T. Lindt, Mathematical modelling of melting of polymers in a single screw extruder, *Polymer Engineering and Science*, 1985, vol. 25, no. 10, pp. 585–588

[29] K. Y. Lee, C. D. Han, Analysis of the performance of plasticating single screw extruders with a new concept of solid-bed deformation, *Polymer Engineering and Science*, 1990, vol. 30, no. 11, pp. 665–676

[30] J. L. White, Z. Chen, Simulation of non-isothermal flow in modular co-rotating twin screw extrusion, *Polymer Engineering and Science*, 1994, vol. 34, no. 3, pp. 229–237

[31] F. Thibault, P. A. Tanguy, D. Blouin, Numerical model for single screw extrusion with Poly (Vinyl Chloride) (PVC) resins, *Polymer Engineering and Science*, 1994, vol. 34, no. 18, pp. 1377–1386

[32] A. M. Voskresenskii, V. B. Voítsekhovskii, F. F. Kougiya, Modelling the processing of polymers with an arbitrary anomalous viscosity in single-screw extruders, *Chemical and Petroleum Engineering*, 1995, vol. 31, no. 1, pp. 21–27

[33] R. V. Chiruvella, Y. Jaluria, V. Sernas, Extrusion of non-Newtonian fluids in a single screw extruder with pressure back flow, Polymer Engineering and Science, 1996, vol. 36, no. 3, pp. 358–367

[34] О. И. Скульский, Численное моделирование одночервячных экструдеров, Пластические массы, 1997, no. 8, с. 39–44

[35] M.-Y. Lyu, J. L. White, Simulation of non-isothermal flow in a modular Buss kneader and comparison with experiment, *International Polymer Processing*, 1997, vol. 12, no. 2, pp. 104–109

[36] L. Zhu, K. A. Narh, X. Geng, Modelling of particle-dispersed melting mechanism and its application in co-rotating twin-screw extrusion, *Journal of Polymer Science. Part B: Polymer Physics*, 2001, vol. 39, no. 20, pp. 2461–2468

[37] B. Vergnes, G. Souveton, M. L. Delacour, A. Ainser, Experimental and theoretical study of polymer melting in a co-rotating twin-screw extruder, *International Polymer Processing*, 2001, vol. 16, no. 4, pp. 351–362

[38] M. Mehranpour, H. Nazokdast, B. Dabir, Computational study of velocity field in the KE element of a modular ko-kneader with CFD method, *International Polymer Processing*, 2003, vol. 18, no. 4, pp. 330–337

[39] K. Wilczyński, J. L. White, Melting model for intermeshing counter-rotating twin screw extruders, *Polymer Engineering and Science*, 2003, vol. 43, no. 10, pp. 1715–1726

[40] A. M. Voskresenskii, I. I. Kudin, S. E. Shakhovets, Computer simulation of the service of worm-gear machines for elastomer processing, Каучук и резина, 2006, no. 1, pp. 44–48

[41] K. Wilczyński, A. Nastaj, A. Lewandowski, K. J. Wilczyński, Multipurpose computer model for screw processing of plastics, *Polymer-Plastics Technology and Materials*, 2012, vol. 51, no. 6, pp. 626–633

[42] C. Teixeira, A. Gaspar-Cunha, J. A. Covas, Flow, and heat transfer along the length of a co-rotating twin screw extruder, *Polymer-Plastics Technology and Materials*, 2012, vol. 51, no. 15, pp. 1567–1577

[43] L. Sardo, B. Vergnes, R. Valette, Numerical modelling of the non-isothermal flow of a non- Newtonian polymer in a co-kneader, *International Polymer Processing*, 2017, vol. 32, no. 4, pp. 425–433

[44] B. Monchatre, C. Raveyre, C. Carrot, Influence of the melt viscosity and operating conditions on the degree of filling, pressure, temperature, and residence time in a co kneader, *Polymer Engineering and Science*, 2018, vol. 58, no. 2, pp. 133–141

[45] К. Колерт, А. М. Воскресенский, В. Н. Красовский, Интенсификация процессов каландрования полимеров, Э.-О. Регер, Ленинград: Химия, 1991, 224 с

[46] А. М. Воскресенский, А. А. Галкин, К. Колерт, Модели каландрования, основанные на методе итераций распределения эффективной вязкости, Известия Санкт-Петербургского государственного технологического института (технического университета), 2007, no. 2, с. 54–57

[47] R. Magnier, J.-F. Agassant, P. Bastin, Experiments and modelling of calender processing for shear thinning thermoplastics between counter rotating rolls with differential velocities, *International Polymer Processing*, 2013, vol. 28, no. 4, pp. 437–446

[48] А. М. Воскресенский, Г. В. Сыкалов, А. А. Пантелеев, Компьютерная модель каландрования листовых термопластов с высоким размерным качеством, Известия Санкт- Петербургского государственного технологического института (технического университета), 2014, no. 25, с. 56-61

[49] A. Lewandowski, K. J. Wilczyński, A. Nastaj, K. Wilczyński, A composite model for an intermeshing counter-rotating twin-screw extruder and its experimental verification, *Polymer Engineering and Science*, 2015, vol. 55, no. 12, pp. 2838–2848

[50] B. Vergnes, G. Della Valle, L. Delamare, A global computer software for polymer flows in corotating twin screw extruders, *Polymer Engineering and Science*, 1998, vol. 38, no. 11, pp. 1781–1792

[51] Т. Б. Чистякова, А. С. Разыграев, А. Н. Полосин, К. Колерт, Гибридная система управления цветом в экструзионном и экструзионно-каландровом производстве полимерных материалов, Вестник Саратовского государственного технического университета, 2012, no. 1, вып. 2, с. 194–201

[52] C. I. Chung, *Extrusion of Polymers: Theory and Practice*, Munich: Carl Hanser Verlag, 2000, 369 pp

[53] U. Berghaus, E. Bürkle, H. Potente et al., *Screw Extrusion: Science and Technology*, ed. by J. L. White, H. Potente, Munich: Carl Hanser Verlag, 2003, 444 pp

[54] C. Rauwendaal, *Polymer Extrusion*, 5th ed, Munich: Carl Hanser Verlag, 2014, 950 p

[55] А. Н. Полосин, Т. Б. Чистякова, Математическая модель одношнековой экструзии для управления качеством пластиката в многоассортиментных производствах полимерных пленок, Системы управления и информационные технологии, 2009, no. 2, с. 87–92

[56] А. Н. Полосин, Т. Б. Чистякова, Математические модели осциллирующего движения и плавления полимеров для проектирования и управления экструдерами, Системы управления и информационные технологии, 2006, no. 4, с. 30–36

[57] А. Н. Полосин, Т. Б. Чистякова, Математическая модель осциллирующего смесителя для исследования энергопотребления и качества полимерных материалов, Химическая промышленность сегодня, 2011, no. 7, с. 35–48

[58] А. Н. Полосин, Т. Б. Чистякова, Методика автоматизированного синтеза математических моделей для управления осциллирующими и двухшнековыми экструдерами в многоассортиментных производствах полимерных материалов, Системы управления и информационные технологии, 2009, no. 4.1, с. 174–179

[59] A. N. Polosin, T. B. Chistyakova, Mathematical modelling for resource and energy saving control of extruders in multi-assortment productions of polymeric films, *Journal of Physics: Conference Series*, 2018, vol. 1015, iss. "Mathematical simulation and data processing", 032104

[60] T. B. Chistyakova, A. N. Polosin, Mathematical models and software package for extrusion control at flexible multi-assortment productions of polymeric materials, *Bulletin of the South Ural State University*, Series: Mathematical Modelling, Programming & Computer Software, 2019, vol. 12, no. 4, pp. 5–28 (in Russian)

[61] Т. Б. Чистякова, А. Н. Полосин, Методы и технологии синтеза математических моделей процессов экструзии в гибких производствах полимерных материалов, Вестник Саратовского государственного технического университета, 2011, no. 4, вып. 4, с. 170–180

[62] A. N. Polosin, T. B. Chistyakova, The mathematical models and program complex for synthesis of reciprocating extruders with adjustable configurations, *Journal of Physics: Conference Series*, 2019, vol. 1202, 012007

[63] Т. Б. Чистякова, А. Н. Полосин, Л. В. Гольцева, Математическое моделирование химико-технологических объектов с распределенными параметрами, Санкт-Петербург: Профессия, 2010, 239 с

[64] А. Н. Полосин, Технология моделирования неизотермического течения расплавов полимеров в одношнековых экструдерах на базе теории плоских потоков, Информационные технологии моделирования и управления, 2007, no. 3, с. 376–386

[65] D. B. Todd, Residence time distribution in twin-screw extruders, *Polymer Engineering and Science*, 1975, vol. 15, no. 6, pp. 437–443

[66] J. P. Puaux, G. Bozga, A. Ainser, Residence time distribution in a co-rotating twin-screw extruder, *Chemical Engineering Science*, 2000, vol. 55, no. 9, pp. 1641–1651

[67] S. Hoppe, C. Detrez, F. Pla, Modelling of a co-kneader for the manufacturing of composite materials having absorbent properties at ultra-high-frequency waves. Part 1: Modelling of flow from residence time distribution investigations, *Polymer Engineering and Science*, 2002, vol. 42, no. 4, pp. 771–780

[68] B. Vergnes, Calculation of average residence time in a ko-kneader, *International Polymer Processing*, 2011, vol. 26, no. 5, pp. 587–589

[69] M. Brito-Bazan, L. Fradette, P. A. Tanguy, Experimental flow visualization and residence time distributions in a co-kneader, *International Polymer Processing*, 2012, vol. 27, no. 4, pp. 414–426

[70] B. Monchatre, C. Raveyre, C. Carrot, Residence time distributions in a co-kneader: A chemical engineering approach, *Polymer Engineering and Science*, 2015, vol. 55, no. 6, pp. 1237–1245

[71] M.-Y. Lyu, J. L. White, Residence time distributions and basic studies of flow and melting in a modular Buss kneader, *Polymer Engineering and Science*, 1998, vol. 38, no. 9, pp. 1366–1377

[72] Т. Б. Чистякова, А. Н. Полосин, Система имитационного моделирования для управления экструдерами в гибких производствах полимерных материалов, Автоматизация в промышленности, 2010, no. 7, с. 35–40

[73] А. Н. Полосин, Т. Б. Чистякова, А. С. Разыграев, К. Колерт, Математическое обеспечение гибридной системы управления цветом экструдированных и каландрированных полимерных материалов, Вестник Саратовского государственного технического университета, 2012, no. 1, вып. 2, с. 241–249

[74] Е. Б. Назарова, Т. Б. Чистякова, М. А. Сокунова, Программный комплекс моделирования неизотермического течения полимерного материала в зазоре каландра, Известия ОрелГТУ. Серия: Информационные системы и технологии, 2005, no. 2, с. 123–127

[75] Т. Б. Чистякова, В. Ю. Плонский, Адаптивная система управления качеством готового продукта на примере процесса каландрования, Приборы и системы управления, 1999, no. 3, с. 6–10

[76] Т. Б. Чистякова, С. И. Куликов, А. Н. Полосин, К. Колерт, ПО для управления толщиной каландрованных тонких материалов, Автоматизация в промышленности, 2007, no. 1, с. 8–11

[77] Т. Б. Чистякова, Н. А. Сергеев, К. Колерт, Математическая модель системы охлаждения каландровой линии для управления качеством полимерного материала, Химическая промышленность, 2005, Т. 82, no. 2, с. 72–80

[78] Т. Б. Чистякова, А. Б. Авербух, К. Колерт, Математическое моделирование процессов усадки/растяжения полимерной пленки для управления каландровой линией, Химическая промышленность, 2005, Т. 82, no. 10, с. 498–507

[79] Т. Б. Чистякова, А. М. Аразтаганова, К. Колерт, Математические модели и программные комплексы для исследования и управления качеством термоусадочных полимерных пленок, Известия Санкт- Петербургского государственного технологического института (технического университета), 2018, no. 46, с. 123–127

[80] T. B. Chistyakova, A. N. Polosin, Computer modelling system of industrial extruders with adjustable configuration for polymeric film quality control, Proceedings of 2017 IEEE II International Conference on Control in Technical Systems (CTS), St. Petersburg: Saint Petersburg Electrotechnical University "LETI", 2017, pp. 47–50

[81] A. N. Polosin, T. B. Chistyakova, Hybrid computer modelling system of extrusion processes for quality control of multi-assortment polymeric films, *Journal of Physics: Conference Series*, 2020, vol. 1479, 012077

[82] T. B. Chistyakova, A. N. Polosin, Computer modelling system for energy- and resource-saving control of multi-assortment polymeric film production, in: *Cyber-Physical Systems: Design and Application for Industry 4.0*, Ed. by A. G. Kravets, A. A. Bolshakov, M. V. Shcherbakov, Cham: Springer Nature Switzerland, 2021, pp. 119–131, (Studies in Systems, Decision and Control; vol. 342)

[83] M. M. Fozilov, T. B. Chistyakova, A. N. Polosin, Decision support system for resource-saving control of extruders at multi-assortment productions of polymeric films, *Journal of Physics: Conference Series*, 2021, vol. 1864, 012081

[84] Т. Б. Чистякова, В. Ю. Плонский, В. Н. Красовский, А. М. Воскресенский, К. Колерт, Свидетельство об официальной регистрации программы для ЭВМ no. 2000611254 Российская Федерация. Система управления червячным осциллирующим смесителем: no. 2000611096: заявл. 12.10.2000: опубл. 05.12.2000

[85] Т. Б. Чистякова, А. Н. Полосин, П. В. Голубев, С. А. Баранов, Свидетельство об официальной регистрации программы для ЭВМ no. 2002611911 Российская Федерация. Программный комплекс математического моделирования процесса плавления полимеров для проектирования осциллирующих экструдеров: no. 2002611659: заявл. 18.09.2002: опубл. 12.11.2002,

[86] Т. Б. Чистякова, А. Н. Полосин, В. В. Попов, Свидетельство об официальной регистрации программы для ЭВМ no. 2006610988 Российская Федерация. Система моделирования процесса одношнековой экструзии полимерных материалов: no. 2006610153: заявл. 23.01.2006: опубл. 16.03.2006

[87] Т. Б. Чистякова, А. Н. Полосин, К. Колерт, Р. В. Антипин, Д. В. Серебряков, Свидетельство об официальной регистрации программы для ЭВМ no. 2006610989 Российская Федерация. Программный комплекс для дистанционного проектирования модульных осциллирующих экструдеров на базе математической модели гидродинамики: no. 2006610154: заявл. 23.01.2006: опубл. 16.03.2006

[88] Т. Б. Чистякова, А. Н. Полосин, М. А. Букина, Свидетельство об официальной регистрации программы для ЭВМ no. 2007613435 Российская Федерация. Система моделирования гидродинамики двухшнековых экструдеров: no. 2007612350: заявл. 15.06.2007: опубл. 15.08.2007

[89] Т. Б. Чистякова, А. Н. Полосин, К. Колерт, И. Ю. Попов, С. В. Защиринский, Свидетельство о государственной регистрации программы для ЭВМ no. 2018618421 Российская Федерация. Конструктор виртуальных моделей экструдеров с перенастраиваемой конфигурацией шнеков: no. 2018615476: заявл. 28.05.2018: опубл. 12.07.2018

[90] Т. Б. Чистякова, К. Колерт, Е. Б. Назарова, М. А. Сокунова, Свидетельство об официальной регистрации программы для ЭВМ no. 2006610992 Российская Федерация. Математическая модель неизотермического течения полимерного материала в зазоре каландра для управления качеством пленки: no. 2006610157: заявл. 23.01.2006: опубл. 16.03.2006

[91] Т. Б. Чистякова, С. И. Куликов, Д. С. Басенко, С. Е. Никитин, И. В. Ковалев, А. Ю. Наумов, С. А. Баранов, Свидетельство об официальной регистрации программы для ЭВМ no. 2006610991 Российская Федерация. Программный комплекс для управления толщиной полимерной пленки: no. 2006610156: заявл. 23.01.2006: опубл. 16.03.2006

[92] Т. Б. Чистякова, К. Колерт, Н. А. Сергеев, Свидетельство об официальной регистрации программы для ЭВМ no. 2003611870 Российская Федерация. Система геометрического и математического моделирования процессов охлаждения каландровых линий: no. 2003611328: заявл. 18.06.2003: опубл. 12.08.2003,

[93] Т. Б. Чистякова, А. Б. Авербух, В. А. Козырь, Программное обеспечение модуля исследования усадки полимерной пленки при управлении каландровым производством, Известия ОрелГТУ. Серия: Информационные системы и технологии, 2005, no. 2, с. 20–30

[93] T. B. Chistyakova, A. M. Araztaganova, C. Kohlert, Computer system for thermal shrinkage polymer films obtaining processes control, Вестник технологического университета, 2016, Т. 19, no. 17, с. 101–104

[94] Т. Б. Чистякова, А. Б. Авербух, А. М.Аразтаганова, А. М. Воскресенский, К. Колерт, Е. Б. Назарова, Свидетельство о государственной регистрации программы для ЭВМ no. 2014662554 Российская Федерация. Программный комплекс для управления процессом усадки полимерных пленок на базе библиотеки математических моделей: no. 2014660430: заявл. 15.10.2014: опубл. 03.12.2014

[95] Т. Б. Чистякова, А. С. Разыграев, А. Н. Полосин, К. Колерт, Программный комплекс для управления цветом тонких жестких полимерных материалов, Автоматизация в промышленности, 2012, no. 7, с. 12–18

[96] C. Kohlert, B. Schmidt, A. Schnabel, F. Michels, M. Schaaf, A. Razigraev, T. Zistjakova, Patent DE no. 102011113543. Gefärbte polymere Formkörper, Verfahren und Vorrichtung zur Herstellung der Formkörper: no. 102011113543.3: Anmeldetag 15.09.2011: Veröff. 06.06.2013

[97] C. Kohlert, B.Schmidt, A. Schnabel, F.Michels, A. Razigraev, T. Chistyakova, M. Schaaf, Patent EP no. 2756029. Coloured Polymeric moulded Bodies, and Method and Device for Producing the moulded Bodies: no. 12762532.5: filing date 07.09.2012: publ. 23.07.2014

[98] C. Kohlert, B. Schmidt, A. Schnabel, F. Michels, A. Razigraev, T. Chistyakova, M. Schaaf, Patent US no. 2014/0234608. Coloured Polymeric moulded Bodies, and Method and Device for Producing the moulded Bodies: no. 14/343939: filing date 07.09.2012; publ. 21.08.2014,

[99] C. Kohlert, B. Schmidt, A. Schnabel, F. Michels, A. Razigraev, T. Chistyakova, M. Schaaf, Patent CN no. 103890051 Coloured Polymeric moulded Bodies, and Method and Device for Producing the Moulded Bodies: no. 201280044742.3: filing date 07.09.2012: grant date 25.11.2015,

[100] К. Колерт, Б. Шмидт, А. Шнабель, Ф. Михельс, Патент no. 2609174 Российская Федерация. Окрашенное полимерное формованное изделие, способ и устройство для получения формованного изделия: no. 2014114845: заявл. 07.09.2012: опубл. 30.01.2017, Разыгра

4.3 Complex Modelling of Calendering Technology Using VR Glasses

Christian Kohlert

The aim is to theoretically calculate the film thickness distribution and film shrinkage of a given calender line for the purpose of optimizing production and training operators. For this purpose, all programs from Section 4.2 are summarized for the complex calculation of the expected quality parameters of film thickness and film thickness distribution, as well as longitudinal shrinkage. The following instructions refer to the use of a 4-roll calender and must of course be adapted accordingly for a 3- or 5-roll calender.

Entering the values to calculate the film thickness and its cross-profile

- Input axis crossing roll 3 and roll bending roll 4 hollow (crowned)
- Input *V* (speed) and *T* (temperatures) of all reels
- Input all necessary calender details (diameter, position, roll data, etc.)
- Input of material values (film type, viscosities, thermodynamic parameters, etc.)
- Store all data input from above under code for recall
- After all required x–y data and material data (see list above) have been entered into the VR glasses or read in via a code, the program can be started.

  The operator then enters the new, desired process data into the VR glasses according to the target specification. This new data must include all speeds and roll temperatures, as well as the pre-loading of roll 2, the axis crossing of roll 3, and the roll bending pressure of roll 4.

  After this data has been transferred to the VR glasses program, the calculation can be started with "data transfer". This then calculates the expected film thickness and creates a suggestion for the optimal setting of the skew adjustment and bending pressure.

Later possible expansions are:

- Transverse shrinkage
- Inclusion of the throughput through gelling unit Buss kneader and/or planetary roller extruder
- Display of film quality values such as film defects using a CCD camera, film color, and gloss using measuring sensors and finding the cause of deviations (data mining).

Relevant images are shown in Figure 4.19–Figure 4.28.

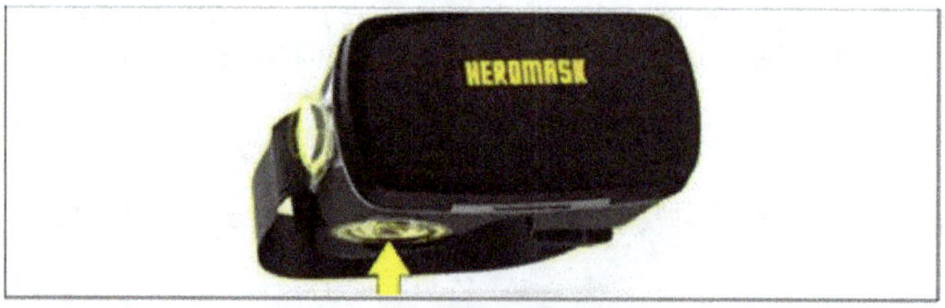

Figure 4.19 Putting on the glasses and starting the program

Figure 4.20 Walk to the calender with a view of the screen (Source: C. Kohlert [lecture], Klöckner Pentaplast)

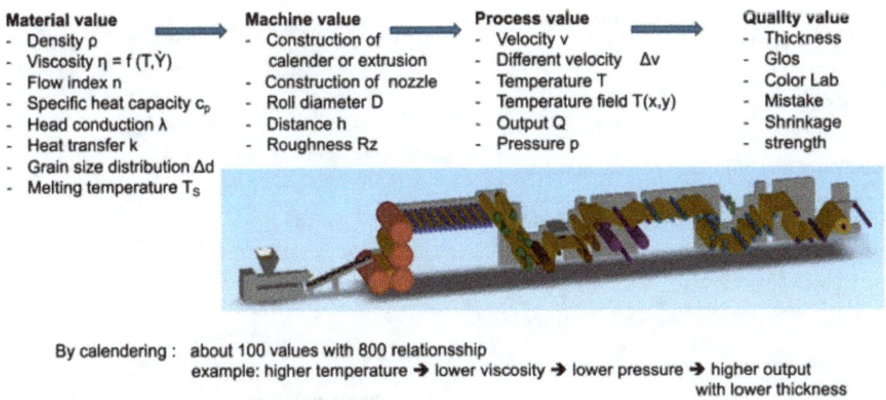

Figure 4.21 Start data entry for current actual values

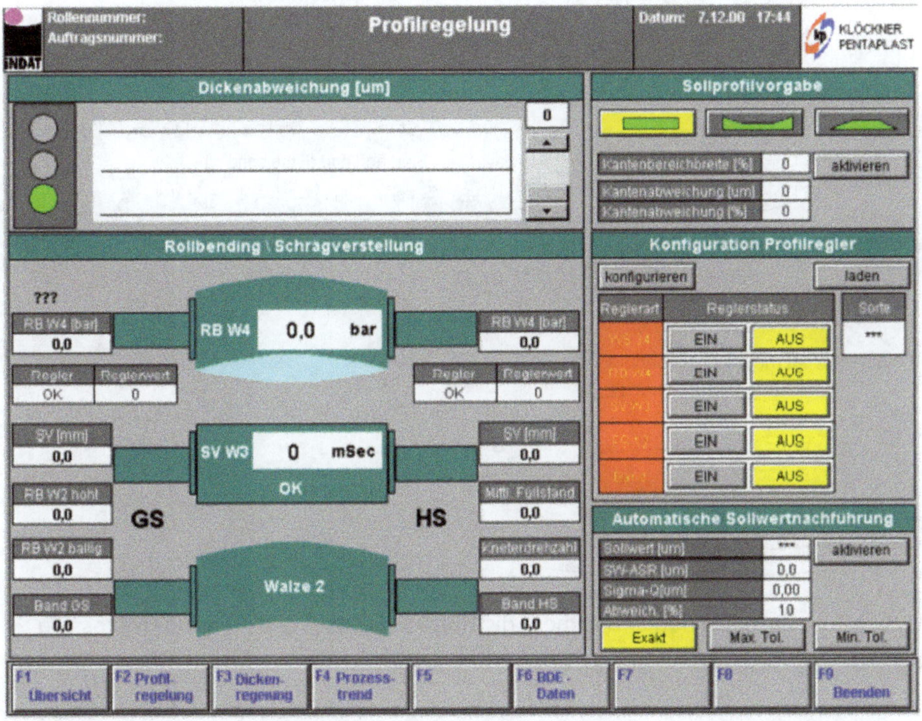

Figure 4.22 New entry of all process-relevant values (Source: C. Kohlert [presentation], Klöckner Pentaplast)

Basic equations for process optimization

Equation of the conservation of mass

$$\frac{\partial \varrho}{\partial t} + \frac{\partial}{\partial x}(\varrho v_x) + \frac{\partial}{\partial y}(\varrho v_y) + \frac{\partial}{\partial z}(\varrho v_z) = 0$$

The equation of motion in Cartesian coordinates

$$\frac{\partial P}{\partial x} = \frac{\partial p_{xx}}{\partial x} + \frac{\partial p_{yx}}{\partial y} + \frac{\partial p_{zx}}{\partial z} - \varrho\left(\frac{\partial v_x}{\partial t} + v_x\frac{\partial v_x}{\partial x} + v_y\frac{\partial v_x}{\partial y} + v_z\frac{\partial v_x}{\partial z}\right)$$

Heat balance

$$\varrho c_v \frac{dT}{dt} = -(\nabla \cdot q) - T\left(\frac{\partial P}{\partial T}\right)_\varrho (\nabla \cdot v) + (p : \nabla \cdot v)$$

c_v Specific heat at constant volume
q Heat flow vector
$q = -k \nabla T$
k Thermal conductivity of the liquid

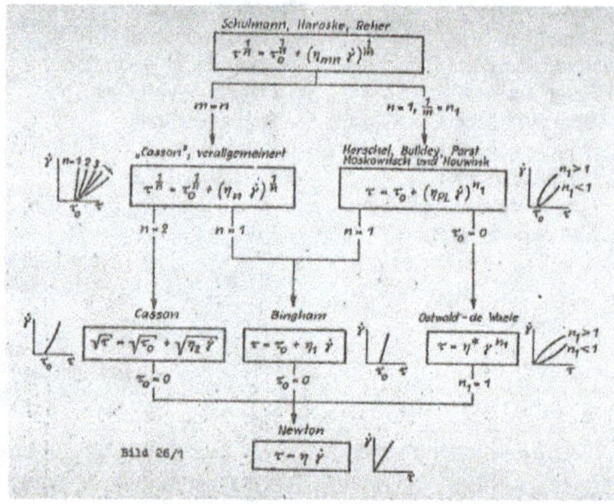

rheological state function

Figure 4.23 Calculation of the first roll gap over the total calender and succession for film thickness (Source: C. Kohlert [lecture], Klöckner Pentaplast)

❖ mathematical model of the bombage

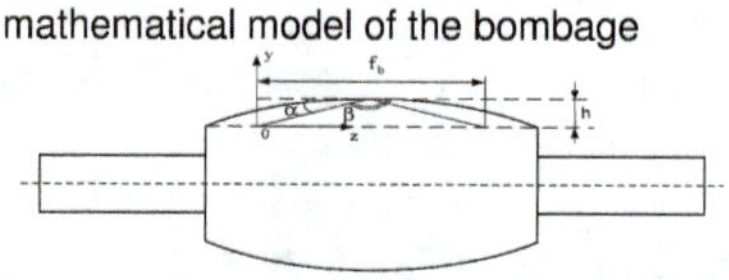

Thickness profile for film

$$S_b(z,h) = \frac{h}{1-\sin\alpha} \cdot \left\{ \sin\left[\alpha + \frac{\beta-\alpha}{w}\cdot\left(z + \frac{w-f_b}{2}\right)\right] - \sin\alpha \right\}$$

Start and end of the bending

$$\alpha = \frac{\pi}{180}\cdot 20 + \frac{7}{18}\cdot\pi\cdot\frac{(w-f_b)}{w}, \quad \beta = \frac{\pi}{180}\cdot 160 - \frac{7}{18}\cdot\pi\cdot\frac{(w-f_b)}{w}$$

❖ mathematical model of accrossing

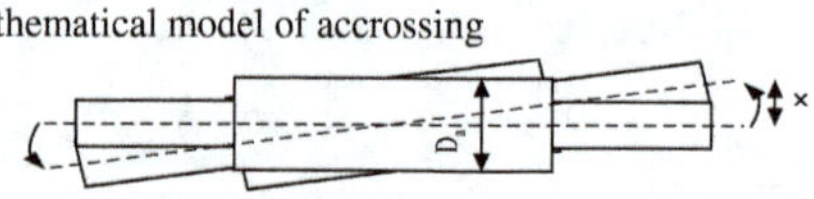

$$S_x(z,x) = \left(\sqrt{\left(x\cdot\frac{f_b}{w}\right)^2 + D_a^2} - D_a\right)\cdot\left(1 - \frac{2\cdot z}{f_b}\right)^2$$

❖ mathematical model of roll bending

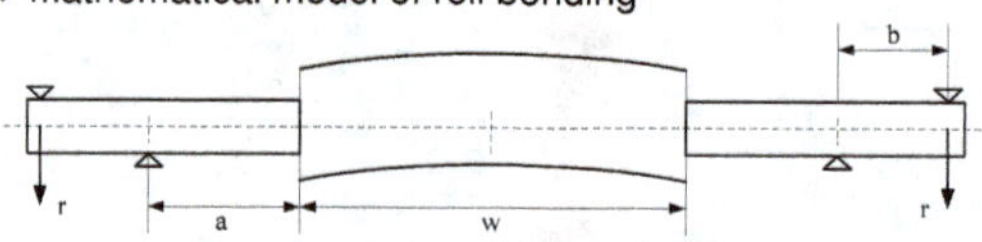

$$S_r(z,r) = \frac{r\cdot b}{2\cdot E\cdot I_e}\cdot\left[z_3^2 - w\cdot z_3 - a\cdot\left(w + \frac{I_e}{I_m}\cdot a\right)\right]$$

❖ mathematical model of roll deflection

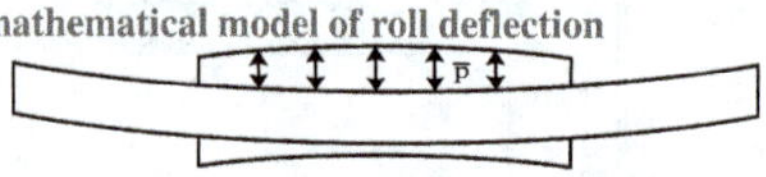

$$S_p(z) = d_2\cdot z^4 + d_1\cdot z + d_0, \quad d_k = f_k[S(z), p], \quad k = 0,1,2$$

Figure 4.24 Optimization of thickness distribution by roll bending and axis crossing (Source: C. Kohlert [lecture], Klöckner Pentaplast)

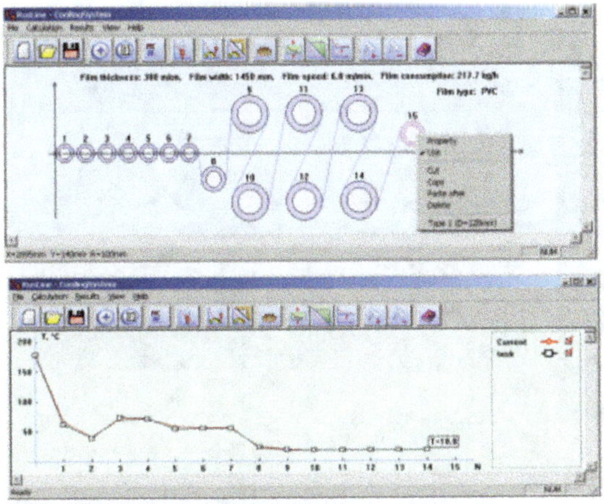

Temperature control is critical for transparency, the film shrinkage and the gloss of the film surface - crucial quality criteria.

The model allows the prediction of the optimum rolling temperatures.

Advantage:

- Reduce of change time
- Reduce complaint for false shrink, gloss or transparency

Figure 4.25 Calculation of the optimal film temperature curve in the post-calender units (pick-off and tempering zones) for high film transparency (Source: C. Kohlert [lecture], Klöckner Pentaplast)

Modell Muni-Rivlin

- based on two effects:

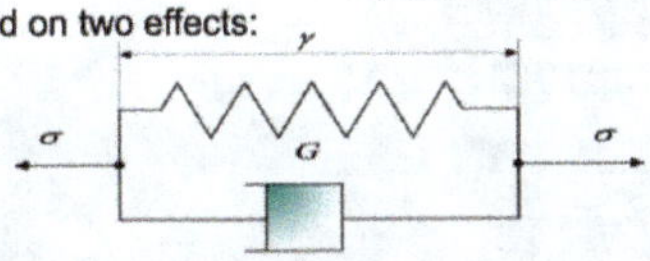

- Equation for the shrink between *i-1* and *i roll*:

$$C\cdot S_l^{6} + A\cdot S_l^{4} + \left(B - C - 2^{n+1}\left(\frac{U_{i-1}\cdot n}{L(1-n)}\right)^{n}\cdot\left[1-\frac{U_{i-1}}{Ui}\right]^{1-n}\cdot\frac{H_{i-1}W_{i-1}}{HW}\,\mu_0 e^{-b(T_{i+1}-T_i)}\right)\cdot S_l^{3} - A\cdot S_l^{2} - B = 0$$

- Total shrinkage is calculated from the sum of the orientations between the rollers of a calender line

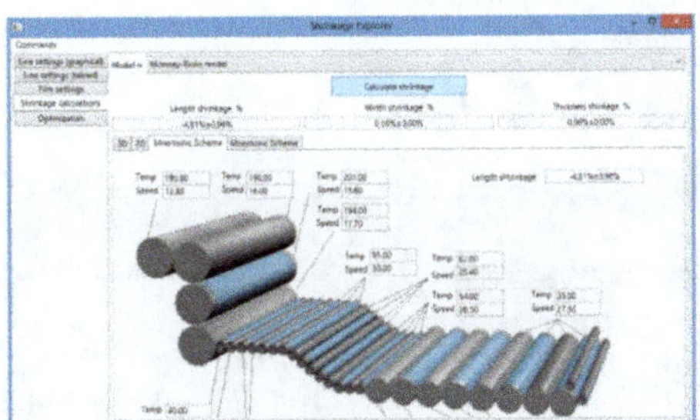

Advantage:

- Reducing changeover time
- Reduce of Complaint for false shrink

Figure 4.26 Calculation of the optimal longitudinal shrinkage values based on the film temperature curve and the roller speeds (Source: C. Kohlert [lecture], Klöckner Pentaplast)

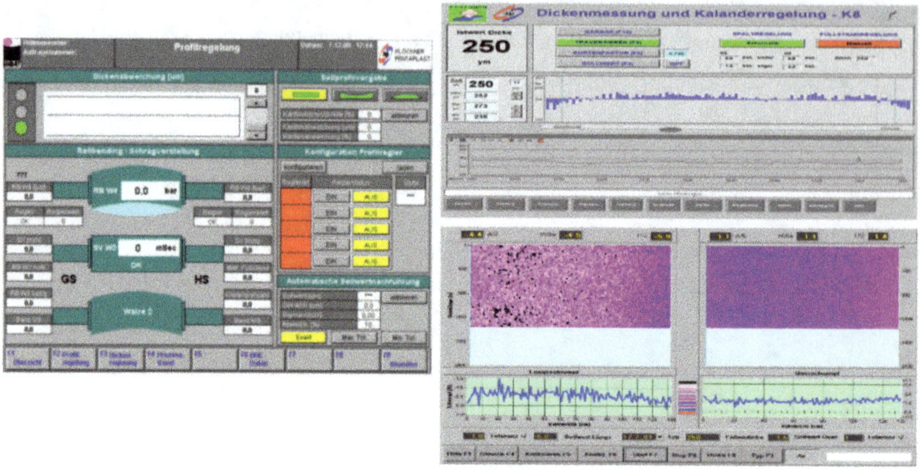

Figure 4.27 Output of final film thickness distribution and longitudinal shrinkage in the VR image (Source: C. Kohlert [lecture], Klöckner Pentaplast)

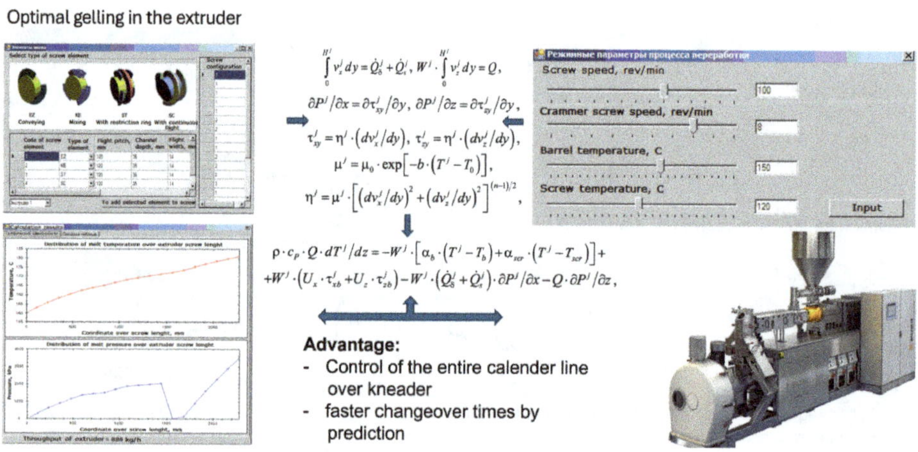

Figure 4.28 Calculation of the melting process in the bus kneader or planetary gear extruder (Source: C. Kohlert [lecture], Klöckner Pentaplast)

The advantages of this project are:

- Training of operators without using the entire calender line
- Optimization options for process engineers
- Minimizing changeover times
- Preparation of a calender automation for defined recipes and film thicknesses.

4.4 Modelling Examples

4.4.1 Software Package for Predicting the Quality of Polymer Films Based on Intelligent Analysis of Large Amounts of Production Data

Tamara Chistyakova and Andrey Polosin

Modern production of polymer films is characterized by a high degree of automation of technological processes. The comprehensive acquisition of production data provides information for film manufacturing. These are the details of the production order, the monitored parameters, and quality indicators of the film produced on any production line in the company. The production order generated by the operating data acquisition system contains the purpose, recipe, film thickness, roll length and width, as well as color, surface requirements, and the order number to produce the film.

The primary information from process parameter sensors is recorded by the SCADA system. The data set of controlled technological parameters includes:

- Revolutions of screw in the extruder feed hopper, revolutions of the extruder screw, and temperature of the tempered zones of the extruder barrel (in the phase of plastificate preparation).
- Circumferential speeds and temperatures of the calender rolls, and the roll bending load exerted on the last calender roll (at the phase of film calendering)
- Circumferential speeds and temperatures of the pick-off and cooling rolls (at the phase of film pick-off and film down cooling).

The main controlled indicators of film quality are:

- Thickness profile along the width of the web (information source: thickness gauge)
- Specific number of surface defects of various types; for example, black dots, inclusions of unmelted polymer, air bubbles, and holes (information source: camera/surface video surveillance system).

Production data is stored in an industrial (factory) database located on one of the servers of the plant's computer network, and obeys the "three V's" rule [1, 2]:

- Large data **v**olume
- High accumulation speed (**v**elocity)
- Diversity of sources and data types (**v**ariety).

The information arrays include:

- Numerical data (values of technological process parameters and film quality indicators, presented in the form of service codes, and not in real numbers, which makes them unsuitable for visualization and processing by mathematical methods)
- Text data (parameters of production tasks).

Production data is characterized by a large number (up to 80%) of zero, empty, and double values (due to malfunctions of technical automation devices and recording errors), and a lack of structure by production phases and types of parameters (input, control, and output parameters). In addition, the film quality indicators that are most important to customers are usually controlled automatically during production. The lack of monitoring of other indicators of film quality (e.g. thickness variation and degree of shrinkage) as well as plastic quality indicators (uniformity and thermal condition) means that it is not possible to obtain complete and reliable information about production and so prevent film defects.

Problem-oriented digital solutions are currently being developed and implemented to provide information technology support for the industrial production of multi-range high-tech products. This is intended to solve complex problems, such as the prediction of customer properties of products and energy resource-saving production control, based on the processing of large quantities. At the same time, the data obtained from different sources and the analysis of the generated data arrays are integrated using various artificial intelligence methods [3–9].

Reducing the incompleteness of information about quality indicators of semi-finished products and products is achieved using mathematical models of physical and chemical processes occurring during the production phases, in information processing and control systems [10, 11]. Therefore, it is relevant to develop a software package that enables the loading of arrays of unstructured parameter data to produce polymer films collected from various data sources.

Determination of tasks for processing production data: Architecture of the software package

Based on the information on characterization of the production of polymer films as a control object in Section 4.2.2, the following production data tasks are processed.

Loading and primary data processing:

For a given production line in the plant i_{PL} and data analysis time interval $t \in [t_0; t_K]$, using the connection parameters $C = \{L_c, P_c, A_{IDB}, N_{IDB}\}$, execute the following:

- Read, import, and store unstructured production data from an industrial database in the form of an array $D_{PROD}{}^0$
- Transform and filter production data with transformation into the $D_{PROD}{}^*$ arrays

- Structure data according to production phases and data types, and store it as structured data in the form of an array D_{PROD}
- Record the D_{PROD} array in the database of controlled and calculated production parameters for subsequent analysis and processing.

Statistical data analysis:

For a given production line i_{PL}, time interval $t \in [t_0; t_K]$, selected indicator of the quality of the polymer material $Q_{s,i}$ and control measures $U_{q,z}$ ($q = 1, \ldots, s$, $z = 1, \ldots, n_q$):

- Test the hypothesis about the normality of the distribution of the quality indicator $Q_{s,i}$ (using the goodness-of-fit criterion C_N)
- Assess the closeness of the relationship between the quality indicator $Q_{s,i}$ and the control measures $U_{q,z}$ (using the correlation coefficient)
- Generate a set of control measures whose correlations with the quality indicator are statistically significant $U^* = \{U_{q,v}^*, q = 1, \ldots, s, v = 1, \ldots, n_q^*\}$.

Predicting the quality of a polymer material:

For a given production line i_{PL}, time interval $t \in [t_0; t_K]$, selected quality indicator $Q_{s,i}$, and control measures influencing it U^*:

- Choose a method for processing the data read from the database of production parameters $D_S = \{U^*, Q_{s,i}\}$
- Create (train) a prediction model $Q_{s,i}^{\text{pred}} = f_{\text{PM}}(U^*, A_M)$ according to data D_{S1} in the time interval $t \in [t_0; t_1]$
- Assess the accuracy and adequacy of the prediction model based on data D_{S2} in the time interval $t \in (t_1; t_2]$
- Calculating the predicted values of the quality indicator $Q_{s,i}^{\text{pred}}$ for the period $t \in (t_2; t_K]$ according to data D_{S3}.

where:

t_0, t_K = start and end date and time within which the production data was read from the database

L_c, P_c = login and password for connecting to the database

A_{IDB} = address of the plant computer network server on which the database named NIDB is located

$Q_{s,i}$ = i-th indicator of the quality of the polymer material in the s-th phase

$U_{q,z}$ = z-th control measure in the q-th production phase (all control measures in the s-th phase and previous phases are analyzed)

n_q = number of control measures in the q-th phase

n_q^* = the number of control measures at the q-th phase affecting the quality indicator $Q_{s,i}$

$D_S = \{D_{S1}, D_{S2}, D_{S3}\}$ = array of values of selected production parameters in the time interval $t \in [t_0; t_K]$

A_M = coefficient vector of the prediction model, the values of which are determined during its training

D_{S1}, D_{S2}, D_{S3} = training, evaluation, and testing subsample of data.

The architecture of the software package for solving the problems posed when processing a lot of production data is shown in Figure 4.29. The symbols of the parameters used in Figure 4.29 are explained in Section 4.2.2 and Section 4.2.3.

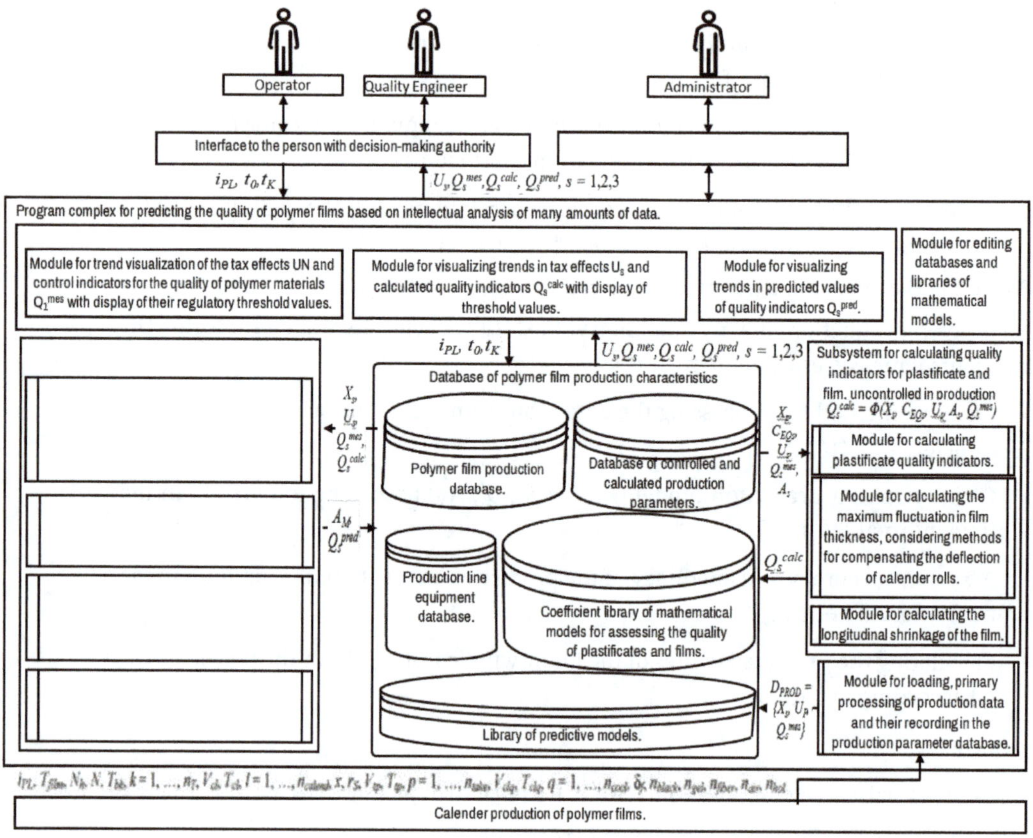

Figure 4.29 Architecture of a software package for the intelligent analysis of large amounts of data to produce polymer films

The software package includes a module for loading and primary data processing, a database of production characteristics, a subsystem for calculating uncontrolled quality indicators using mathematical models, a subsystem for quality forecasting using regression analysis and machine learning methods, a data visualization subsystem, and a user interface [12–15]. The production management personnel (operator, quality engineer) and the administrator interact with the software package. The management of production personnel determines the number of the production line and the time interval of the analyzed data for the formation and processing of the appropriate range of values of production parameters and the visualization of trends in control measures and quality indicators, which are both measured and calculated. The administrator sets parameters for connecting to a database to read production data from it, and edits databases of film types, line equipment, regulation ranges of production parameters, and a library of coefficients of mathematical models for assessing the quality of plastificate and film.

The setup of the film production properties software package and information interaction between the modules of the complex is carried out using a database.

For this purpose, the database contains:

- Data of recipes (including polymer type) and quality requirements for polymer films of various types
- Parameters of the properties of polymers in different physical states (vitreous, highly elastic, viscous flow state), necessary for calculating the quality indicators of plastificate and film
- Data on configurations and geometric parameters of units (extruders, calenders, integrated suction, and cooling devices) of different characteristics included in production lines
- Regulatory thresholds of technological process parameters, depending on the type of film
- Customizable coefficients of mathematical models to assess the quality of extrudate and film, depending on the type of film-forming polymers.

Algorithms for processing large amounts of production data to predict the quality of polymer materials:

To prepare production data for processing using machine learning methods, it is necessary to carry out transformation and filtering [16]. The algorithm for loading and primary processing of production data includes the phase of loading unstructured data, the phase of transformation, filtering, and structuring of data, and the phase of recording converted, filtered, and structured data into the production parameter database (Figure 4.30).

Loading unstructured data includes:

- Connecting to a database for parameter C
- Reading reference information about production parameters on the line with number i_{PL}
- Reading the values of unstructured production parameters in the time interval $t \in [t_0; t_K]$.

Production parameter reference information represents a comparison of the service names of the parameters used in the database and the actual names of the parameters used by production management personnel. The search for comparisons in the help file is automated. As a result, a file with comparisons between service and actual parameter names is generated.

The values of the production parameters are read out by executing an SQL query to the database. The data is read into a temporary text file in CSV format and saved as a two-dimensional array $D_{PROD}{}^0 = \{t_j, (P_{i,j})_{64}, i = 1, \dots, n, j = 1, \dots, m\}$.

where:

t_j = date and time of measurement ($t_j \in [t_0; t_K]$)

$(P_{i,j})_{64}$ = the value of the i-th production parameter measured at time t_j and represented as a service code (64-bit integer)

n = number of parameters

m = the number of time points that fall within a certain interval.

This technology allows you to double the data reading speed. This is particularly important because the data volume includes millions of values.

The production parameters are renamed to their actual names. To do this, you use the rename function, which works with a file comparison service and actual parameter names. It parses the service name of each parameter in the $D_{PROD}{}^0$ array and replaces it with the corresponding actual value from the mapping file.

Converting the values of each parameter consists of two steps:

1. The first 32 bits are separated from the 64-bit parameter value $(P_{i,j})_{64}$ using the conjunction operation. The second operand is the 32-bit number FFFFFFFF. Since the dimension of the first operand is larger (64 > 32), the most significant bits are padded with zeros. The resulting value $(P_{i,j})_{32}$ is a 32-bit integer: $(P_{i,j})_{32} = (P_{i,j})_{64} \wedge \text{FFFFFFFF}$, $j = 1, \dots, m$
2. The resulting value is converted into a real 32-bit number $P_{i,j}$ by reinterpreting the memory area occupied by the array $\{(P_{i,j})_{32}, j = 1, \dots, m\}$ [17].

Conversion operations are applied to all parameter values at once (on the array column $D_{PROD}{}^0$). By using batch data conversion, one can significantly increase (six-fold) the processing speed compared to a full search, where each parameter value is converted separately.

Data filtering involves finding and removing null, empty, and duplicate values for each parameter according to the following rules:

- If the parameter P_i has a zero or empty value at time t_j, it is removed from the array and the number of parameter values is reduced by one ($m_i = m - 1$ for the first deletion, $m_i = m_i - 1$ for each subsequent deletion)
- If $t_j = t_j + 1$ and $P_{i,j} = P_{i,j} + 1$, then the value $P_{i,j} + 1$ is removed, and the number of parameter values m_i is reduced by one.

The order of operations is based on the principle of reducing the amount of data processed (the largest number of values are removed according to the first filter rule). Batch processing operations are used when filtering data.

After filtering, if the parameter P_i does not have a single value ($m_i = 0$), it is removed, and the number of parameters n is reduced by one. This allows you to reduce the amount of RAM used and increase performance.

As a result, an array is formed containing real nonzero parameter values that are not duplicated:

$$D_{\text{PROD}}^* = \{t_j, \quad j = 1, \ldots, m, \quad P_{i,k}, \quad i = 1, \ldots, n^*, \quad k = 1, \ldots, m_i\} \tag{4.59}$$

where:

n^* = number of parameters with values ($n^* \leq n$)

m_i = the number of parameter values remaining P_i after filtering.

Structuring the data of the D_{PROD}^* array involves associating parameters with production phases and parameter types according to the actual names of the parameters (Figure 4.30). For example, if a parameter is included in the set of typical control measures at the s-th phase of film production using the calendering method, then it is included in the vector of control measures U_s. If the parameter characterizes the quality of the polymer material formed at the s-th phase of calender production, then it is included in the vector of controlled initial parameters—quality indicators Q_s^{mes} [17].

The result is an array $D_{\text{PROD}} = \{X_s, U_s, Q_s^{\text{mes}}, s = 1, 2, 3\}$, which is recorded in the database of controlled and calculated production parameters. The order of recording is strictly regulated: first, the values of the input parameters X_s are recorded, then the values of the control actions U_s are saved, and at the end the values of the quality indicators Q_s^{mes} are recorded. This order is determined by the relationships between entities in the parameter database, and is intended to ensure that recording errors do not occur. Recording is done using SQL queries to record arrays of parameter values. This allows the writing speed to be significantly (five-fold) increased compared to writing each individual value contained in the D_{PROD} array.

Incomplete information due to the lack of monitoring of several quality indicators of the polymer material at the production phases does not allow correct prediction of the customer properties of the film and so make informed quality management decisions. Therefore, the structured data D_{PROD} is supplemented with the values of quality indicators at each level Q_s^{calc} ($s = 1, 2, 3$), which are not controlled in production:

$$Q_1^{\text{calc}} = \{I_{\text{d}}, \gamma_{\text{av}}\}, \quad Q_2^{\text{calc}} = \{D_\delta\}, \quad Q_3^{\text{calc}} = \{S_{\text{fl}}\} \tag{4.60}$$

where:

I_{d} = index of thermal destruction [%]

γ_{av} = average degree of plastificate mixing

D_δ = maximum variation of film thickness over the entire width [µm]

S_{fl} = degree of longitudinal shrinkage of the film [%].

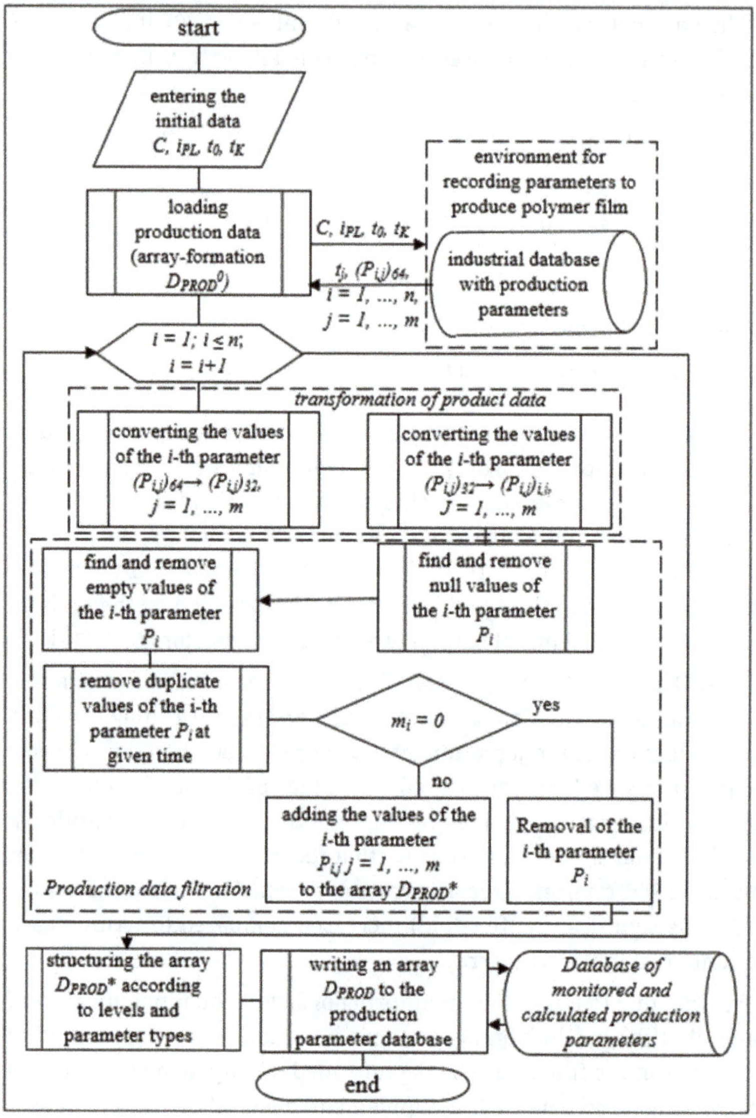

Figure 4.30 Algorithm for loading and primary processing of production data

Quality indicators are calculated using mathematical models depending on the control measures U_s measured on the production line, the parameters of the polymer properties P_{Ms} and the properties of the line units M_{Eqs} and C_{EQs} (Figure 4.31). These mathematical models are described in Section 4.2.3.

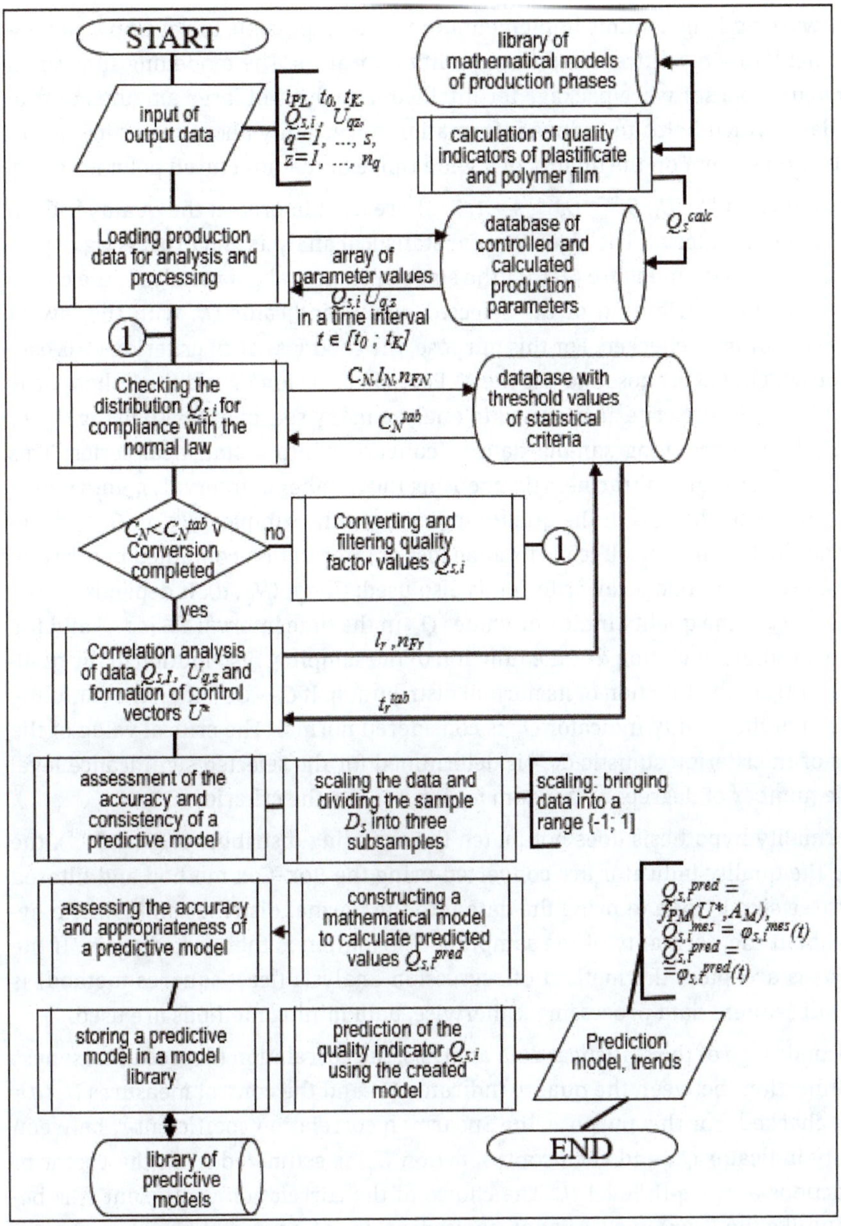

Figure 4.31 Algorithm for modelling, statistical analysis, and processing of production data to predict film quality indicators

Their software and algorithmic implementation made it possible to develop a subsystem for calculating plastificate and film quality indicators. The modelling subsystem is integrated into a software package for intelligent analysis of large amounts of production data, which helps to reduce information uncertainty about the efficiency of the production system on the quality of a wide range of manufactured polymer films.

Complete data sets $\{X_s, U_s, Q_s^{mes}, Q_s^{calc}, s = 1, 2, 3\}$ are used to predict the quality indicators of a polymer material. The algorithm for statistical analysis and processing of production data is shown in Figure 4.31. At the stage of statistical data analysis, the correspondence of the distribution of the selected quality indicator $Q_{s,i}$ with the law of normal distribution is checked. For this purpose, the goodness-of-fit criterion C_N is estimated, for which it is proposed to use the X^2 Pearson criterion C_P and the Kolmogorov criterion C_K [18]. These criteria are used to analyze many sets of data. Applying the C_P criterion involves grouping sample data to convert it into a statistical series. This grouping has a certain arbitrariness (in choosing the number of intervals n_k into which the entire range of changes in the quality indicator in the sample $Q_{s,i}^{min}$ to $Q_{s,i}^{max}$ is divided). And the grouping itself leads to some loss of information contained in the sample. Therefore, the Kolmogorov criterion is also used: $C_K = f_\lambda\ (N_{Qi}, d)$. It depends on the sample size N_{Qi} of the quality indicator values $Q_{s,i}$ in the time interval $t \in [t_0; t_K]$ and the maximum absolute deviation d of the function of the sampling distribution of the quality indicator from the function of its normal distribution. If $C_N < C_N^{tab}$ then the sampling distribution of the quality indicator $Q_{s,i}$ is considered normal. The critical value of the goodness-of-fit criterion statistic C_N^{tab} is determined for the selected significance level I_N and the number of degrees of freedom $n_{FN} = n_K - 3$ (for the criterion C_P).

If the normality hypothesis does not match the sampling distribution ($C_N \geq C_N^{tab}$), the values of the quality indicator are converted using the Box–Cox method and filtered by the "three sigma" rule to bring the data into the normal distribution [19]. The hypothesis about the normality of the sampling distribution is then tested again. If the hypothesis is accepted, the method of regression analysis (least-squares method) is used for subsequent data processing. Otherwise, data mining methods are used.

In the second stage of the statistical data analysis, the correlation (degree of closeness of the connection) between the quality indicator $Q_{s,i}$ and the control measures U_q (q = 1, ..., s) is checked. For this purpose, the Spearman correlation coefficient r_S between the quality indicator $Q_{s,i}$ and each control action $U_{q,z}$ is estimated from the vector of control actions at the q-th level U_q. The choice of the correlation coefficient r_S is because it (unlike the Pearson correlation coefficient) is not limited by the requirement of normality of the sampling distribution.

To test the hypothesis about the existence of a correlation, the significance t_r of the difference between the coefficient r_S and zero is assessed: $t_r = f_r\ (N_{Qi}, r_S)$ [20]. If $t_r \geq t_r^{tab}$, then the coefficient r_S is significantly (not randomly) different from zero and there is a connection between the quality indicator $Q_{s,i}$ and the control effect $U_{q,z}$. Therefore, the control action $U_{q,z}$ is included in the vector of control actions U^* to create a prediction model: $U_{q,v}^* = U_{q,z}$. The critical value of the Student t-test statistic t_r^{tab} is determined for the significance level l_r and the number of degrees of freedom $n_{Fr} = N_{Qi} - 2$.

In the data processing phase to predict quality, the sample $D_S = \{U^*, Q_{s,i}\}$ is divided into three subsamples:

- Training subsample D_{S1} (70% of the parameter values)
- Evaluation subsample D_{S2} (10% of the parameter values)
- Test subsample D_{S3} (20% of parameter values).

A mathematical model to predict the quality indicator $Q_{s,i}$ is built by processing the data from the training subsample. The choice of processing method depends primarily on the type of data distribution law, which is established at the stage of statistical data analysis.

If the indicator $Q_{s,i}$ is normally distributed, a parametric synthesis of a linear multivariate regression model is carried out [20, 21]:

$$Q_{s,i}^{\text{pred}} = a_0 + \sum_{q=1}^{q=s} \sum_{v=1}^{v=n_q} a_{q,v} U_{q,v}^*, \quad A_M = \left\{a_0,\ a_{q,v},\ q = 1, \ldots, s,\ v = 1, \ldots, n_q^*\right\} \tag{4.61}$$

where:

a_0, $a_{q,v}$ = regression model coefficients, the values of which are determined using the least-squares method.

If the distribution of the quality indicator $Q_{s,i}$ does not conform to the normal law, the following is used to create a prediction model:

- Recurrent neural network (for predicting the values of the quality indicator $Q_{s,i}$)
- Adaptive boosting of decision trees (to predict the presence/absence of film defects associated with the deviation of the quality indicator $Q_{s,i}$ beyond the regulatory constraints $Q_{s,i} \notin [Q_{s,i}^{min}; Q_{s,i}^{max}]$).

The choice of data mining method depends on the amount of data and the requirements for accuracy and cost-effectiveness of the forecast. A neural network provides high prediction accuracy but requires large amounts of data and long training time [22]. Its use is recommended when predicting the customer properties of pharmaceutical polymer films, which are subject to strict requirements. Decision trees are characterized by a high learning speed and interpretability of the results [23, 24]. However, small data changes can completely alter the tree, so using decision trees makes sense when the level of perturbation is low.

The neural network model is based on a five-layer recurrent network [25]. The network includes an input layer, a batch data normalization layer, a long short-term memory layer (LSTM layer) [26], a layer for implementing the exclusion of neurons from the network using the inverted dropout method, and an output layer. The activation function of neurons in the long short-term memory layer Ψ_h, $h = 1, \ldots, N_{\mathrm{LSTM}}$ is a hyperbolic tangent:

$$\psi_h = \frac{2}{1 + \exp\left[-2\left(\sum_{g=1}^{g=M_h} a_{g,h} w_{g,h} - \theta_h\right)\right]} - 1 \tag{4.62}$$

$$A_{\mathrm{M}} = \left\{a_{g,h},\ g = 1, \ldots, M_h,\ h = 1, \ldots, N_{\mathrm{LSTM}}\right\}$$

where:

N_{LSTM} = number of neurons in the long-term short-term memory layer

M_h = number of input signals of the h-th neuron of the long short-term memory layer

$a_{g,h}$ = weighting coefficient of the g-th input signal of the h-th neuron

$w_{g,h}$ = g-th input signal of the h-th neuron (input signals are normalized values of control actions contained in the vector U^*)

θ_h = threshold of the h-th neuron.

To calculate the weighting coefficients of the input signals $a_{g,h}$ in Equation 4.62, the back-propagation algorithm is used when training a neural network. To overcome "stuckness" in local minima of the error function of the neural network and increase the learning speed, corrections to the weighting coefficients are calculated at each training iteration using the Nesterov accelerated adaptive momentum algorithm [27, 28]. Computer experiments have shown that this algorithm provides a lower average relative error in calculating the quality indicators of a polymer film compared to the following algorithms:

- Root-mean-square propagation algorithm (the error is 20% less when calculating the specific number of black dots on the film)
- Adaptive momentum estimation algorithm (the error is 6% lower when calculating the number of black points) [29].

Accelerated learning is facilitated by using the batch normalization method. To regulate a neural network to prevent overtraining, a method of randomly excluding some neurons (by zeroing their output signals) is used in the network training phase. The inverted dropout method can be used to prevent neurons from adapting together to production data from the training subsample.

The accuracy and appropriateness of the created prediction model are determined based on the data from the evaluation subsample. The set of accuracy and adequacy indicators depends on the model type. To assess the quality of the regression model (Equation 4.61), the mean-square absolute error σ and the Fisher criterion C_{F} are used. Condition for model adequacy: $C_{\mathrm{F}} > C_{\mathrm{F}}^{\mathrm{tab}}$. The tabular value of the Fisher criterion $C_{\mathrm{F}}^{\mathrm{tab}}$ is determined for the significance level l_F, the number of degrees of freedom of the variance of the average model n_{FA} and the number of degrees of freedom of the residual variance of the constructed model n_{FR}. This method of assessing adequacy is because the processed production data is the result of a passive (industrial) experiment, characterized by the absence of parallel experiments.

The following indicators are used to evaluate the neural network model:

- Average absolute error Δ
- Average relative error δ
- Root-mean-square absolute error σ
- Area under the error curve (AUC) S_{ROC}.

The properties of the constructed model $\{i_{PL}, Q_{s,i}, U^*, A_M, \Delta, \delta, \sigma, C_F, S_{ROC}\}$ are stored in the library of prediction models. The created model is used to calculate the predicted values of the quality indicator $Q_{s,i}^{pred}$ according to the test subsample data in the time interval $t \in (t_2; t_K]$. To visualize the processing results, trends of the predicted values are displayed, and the quality indicators of the polymer film $\varphi_{s,i}^{pred}(t)$ become constructed. The proposed algorithms allow problems of primary processing to be solved, large amounts of data to produce polymer films to be statistically analyzed, and film quality indicators for different film grades and production line configurations to be forecast, considering the requirements for accuracy and cost-effectiveness of production data processing.

Test examples of the software package

The functional and load test of the software package was carried out based on data for one month of calendar production of pharmaceutical and food packaging films (based on rigid polyvinyl chloride) on calender line No. 2 of the Klöckner Pentaplast Rus LLC plant (St. Petersburg). The production line consists of a co-kneader with three thermal zones, an L-shaped calender with four rolls, a pick-off with 10 rolls, and a cooling unit with 15 rolls.

The volume of the initial unstructured data array amounted to 42.1 million values of production parameters (parameters of production tasks, technological process parameters, and film quality indicators).

Figure 4.32 shows some raw manufacturing data from an industrial database. The values of the production parameter (in this case the revolutions of the co-kneader's feed hopper screw) are stored in an industrial database in the form of service codes at different points in time. These codes are not suitable for statistical analysis and processing for quality prediction purposes. Manual conversion of these parameter values into real numbers is not possible because reinterpretation of data in memory areas is not possible without the use of computer technology.

	time timestamp without time zone	item_56af7804f700000000000001 bigint
1	2020-10-20 17:15:23	825728783872
2	2020-10-20 17:15:54	825728763904
3	2020-10-20 17:16:24	825728640767
4	2020-10-20 17:16:54	825728700672
5	2020-10-20 17:17:24	825728667392
6	2020-10-20 17:17:54	825728750592
7	2020-10-20 17:18:25	825728817152
8	2020-10-20 17:18:54	825728830464
9	2020-10-20 17:19:25	825728737279
10	2020-10-20 17:19:55	825728587520

Figure 4.32 Example of raw production data

The result of converting the given data using the primary data processing module included in the software package is shown in Figure 4.33. Each row of the data array contains the date and time of measurement of the parameter value in the production line (text data type) and the parameter value (real data type).

	DateTime	round(t1.Value, 1)
1	2020-10-20 17:15:23	12.3
2	2020-10-20 17:15:54	12.3
3	2020-10-20 17:16:24	12.2
4	2020-10-20 17:16:54	12.3
5	2020-10-20 17:17:24	12.2
6	2020-10-20 17:17:54	12.3
7	2020-10-20 17:18:25	12.4
8	2020-10-20 17:18:54	12.4
9	2020-10-20 17:19:25	12.3
10	2020-10-20 17:19:55	12.1

Figure 4.33 Example of converted production data stored in the productionparameters database included in the software package

The stress test of the primary data processing module was carried out on five production data samples of different sizes containing the values of 99 productionparameterss. The results of the load test are shown in Figure 4.34. Filtering reduces the amount of production data by an average of 80%. Figure 4.35 and Figure 4.36 show diagrams of the dependence of the conversion time and recording time of production data on the volume of the filtered data sample.

Scope of test data	4950000	9900000	19800000	29700000	39600000
Scope of production parameters used (after data filtration)	1263218	2386877	4016132	5052044	6137698
Degree of reduction in parameters after filtration, %	74.5	75.9	79.7	83.0	84.5
Time for loading the data, s	1	2	2	3	4
Time of data conversion, s	2	4	9	14	20
Time of data recording, s	30	52	84	99	115

Figure 4.34 Results of testing the primary data processing module

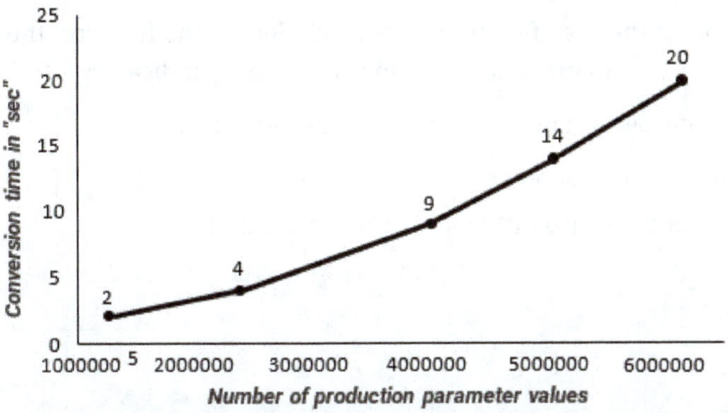

Figure 4.35 Dependence of the data conversion time on the sample size

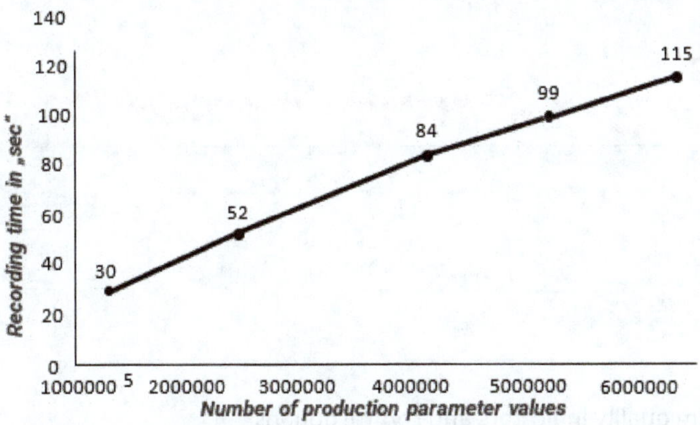

Figure 4.36 Dependence of the time of writing data to the database on the sample size

With an increase in the volume of production data by 386% (from 1.26 million values to 6.14 million values), the total time for its conversion and recording increases by 322% (from 32 s to 135 s) with a growth rate close to constant. This indicates an almost linear dependence of time on the amount of processed data, which makes it possible to predict the time required for primary processing of production data of a given amount.

The use of modern technologies (in particular, batch conversion and filtering, and SQL queries for recording arrays of production parameter values) leads to a significant (5–6-fold) increase in the speed of primary processing of production data.

Arrays of control measures for process and film quality indicators are read from the database of production parameters and displayed on the user interface in the form of trends. The operator can select the film quality display and control measures whose trends are displayed. The software package helps the operator to analyze the cause-and-effect relationships between technological process parameters and film quality indicators [30].

Figure 4.37 shows the trend of the specific number of black dots in the film and the trends of two control options that significantly influence this quality indicator:

- The number of revolutions per minute (rpm) of the co-kneader's barrel
- The rpm of the extruder's hopper screw.

Of course, the housing temperature also influences burns in the film.

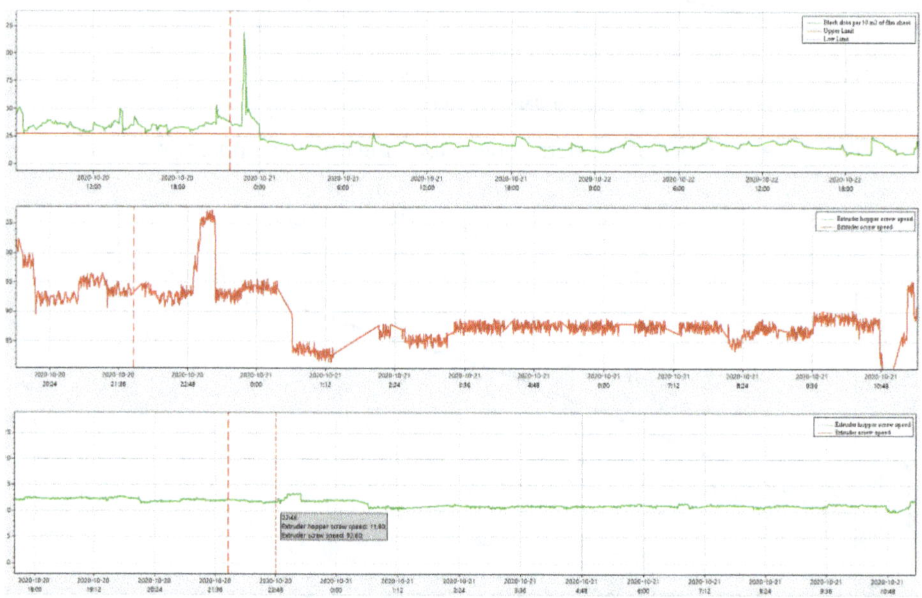

Figure 4.37 Trends in film quality indicators and control options

The calculation of plastificate and film quality indicators using mathematical models that are not measured in production also allows them to be visualized in the form of trends. Figure 4.38 shows the trend of the thermal decomposition index of plastic 1, calculated from the mathematical model of extrusion, and the maximum permissible value of the decomposition index 2, determined by the film type.

By analyzing this additional information, the operator detects overheating and thermal damage to the plastic in a timely manner (if the damage index exceeds the maximum permissible value) and changes the control measures on the co-kneader (which reduces the number of revolutions of the feed hopper screw as well as the barrel temperature). This can prevent corresponding film defects (black dots and yellow–brown burned stripes).

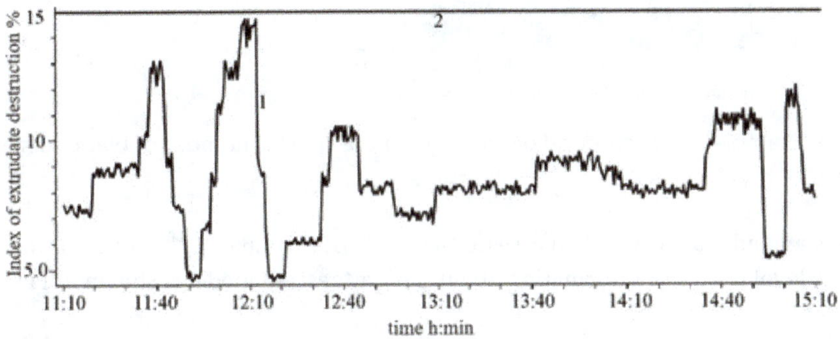

Figure 4.38 Trend of the extrudate quality indicator, calculated using a mathematical model

By analyzing these trends, the operator determines the technological process parameter, the change in which led to an increase in the number of such surface defects in the film. Based on this information, the operator makes a process control option that makes it possible to reduce the number of errors within the control range.

The result of testing the hypothesis about the normal distribution of the film quality indicator (using the example of the number of black dots per 10 m^2 of film) is shown in Figure 4.39. The original distribution of the number of errors deviates from the normal distribution. After transforming (using the Box–Cox method) and filtering (using the "three sigma" rule) the values of the number of black dots, the hypothesis of a normal distribution law is accepted. Testing with the Kolmogorov criterion at the significance level l_N = 0.3 results in $C_K = 0.74 < C_K^{tab} = 0.97$.

The agreement of the sample distribution (probability density histogram of distribution 1 in Figure 4.39) with the theoretical (normal) distribution (probability density diagram 2 in Figure 4.39) is confirmed graphically.

As a result of the correlation analysis of the production data, a vector of control options $U^* = \{N_h, N, T_{bk}, k = 1, 2, 3, V_{cl}, T_{cl}, l = 1, ..., 4\}$ was generated, which affect the specific number black dots in the film.

To process normally distributed production data, linear regression analysis was used. An empirical mathematical model was created:

$$n_{black}^{pred} = f_n \left(U^*, a_0, a_{q,v}, \ q = 1, \ v = 1, \ldots, 5, \ q = 2, v = 1, \ldots, 8\right) \qquad (4.63)$$

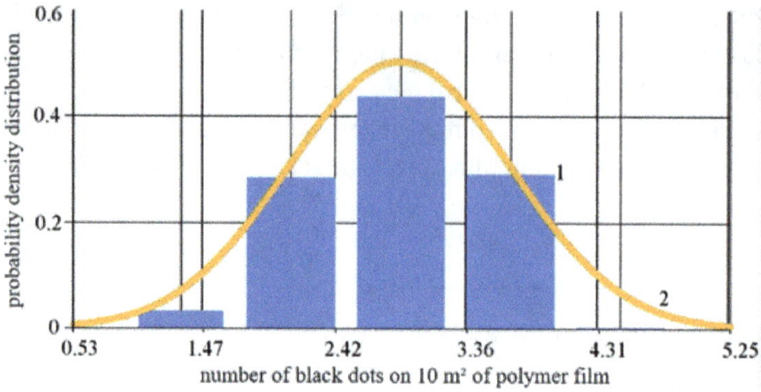

Figure 4.39 Example and theoretical distribution of the specific number of black dots in the film

Its accuracy and adequacy was then assessed, as $\sigma = 0.61$, $C_F = 1.69 > C_F^{tab} = 1.64$ (for the significance level $l_F = 0.05$). Information about the synthesized model is shown in Figure 4.40.

The model is adequate. Fischer's criterion 1.69 > 1.64

RMSE = 0.60933

Black dots per 10 m2 of film sheet $= -2.7627 \cdot 10^{0} + 0 \cdot 10^{0} \cdot X_1 + 8.3857 \cdot 10^{-2} \cdot X_2 + 2.1477 \cdot 10^{-2} \cdot X_3 + 1.2807 \cdot 10^{-1} \cdot X_4 - 1.4723 \cdot 10^{-2} \cdot X_5 - 3.9901 \cdot 10^{-1} \cdot X_6 + 5.6445 \cdot 10^{-1} \cdot X_7 + 5.0797 \cdot 10^{-1} \cdot X_8 - 6.6384 \cdot 10^{-1} \cdot X_9 - 1.3445 \cdot 10^{-1} \cdot X_{10} - 6.5548 \cdot 10^{-2} \cdot X_{11} + 0 \cdot 10^{0} \cdot X_{12} + 2.5207 \cdot 10^{-1} \cdot X_{13}$

X_1 - "Extruder barrel temperature in the third thermal zone ";
X_2 - "Extruder barrel temperature in the first thermal zone ";
X_3 - "Extruder barrel temperature in the second thermal zone";
X_4 - "Extruder hopper screw speed";
X_5 - "Extruder screw speed";
X_6 - "Temperature of the outer surface of the first calender roll";
X_7 - "Peripheral speed of the first calender roll";
X_8 - "Temperature of the outer surface of the second calender roll";
X_9 - "Peripheral speed of the second calender roll";
X_{10} - "Temperature of the outer surface of the third calender roll";
X_{11} - "Peripheral speed of the third calender roll";
X_{12} - "Temperature of the outer surface of the fourth calender roll";
X_{13} - "Peripheral speed of the fourth calender roll".

Figure 4.40 Type of equation, coefficient values, and assessment of accuracy and suitability of the regression model for predicting the number of errors

Based on the constructed model, a prediction was made about the specific number of black dots on the film. The trends of the measured (1) and predicted (2) values of the number of black dots are shown in Figure 4.41. For visualization in the form of a trend, a back-transformation of the predicted values of the number of black dots was carried out with the actual values of the film quality indicator.

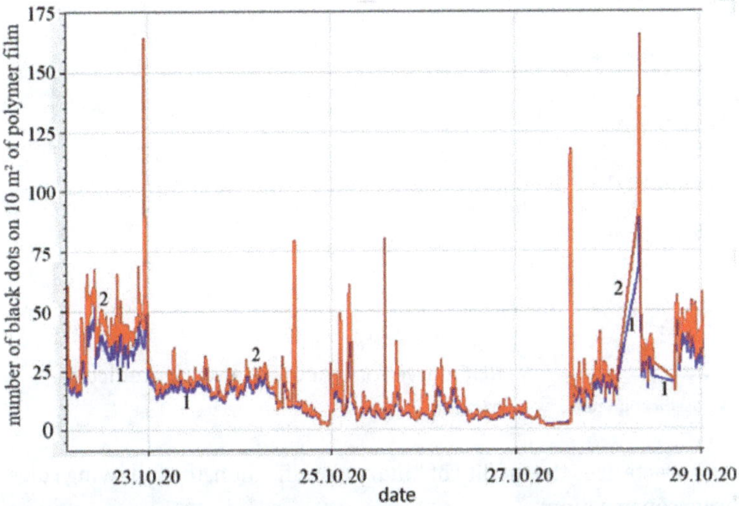

Figure 4.41 Trends in measured and predicted values (by linear regression model) of the specific number of black dots in the film

If the transformation and filtering of the quality indicator values does not lead to the normality of their distribution, a five-layer recurrent neural network is built to predict quality. Figure 4.42 shows the trends of the measured (1) and predicted (2) values of the specific number of black dots in the film. The prediction values were calculated using a neural network model with 128 neurons in the LSTM layer.

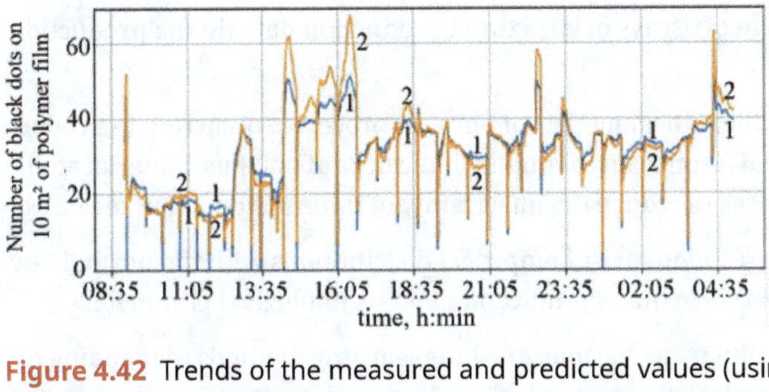

Figure 4.42 Trends of the measured and predicted values (using a neural network model with an LSTM layer) of the specific number of black dots in the film

The neural network model is characterized by the following indicators:

- Average absolute error $\Delta = 0.026$
- Average relative error $\delta = 2.8\%$
- Root-mean-square absolute error $\sigma = 0.033$
- Area under the error curve (AUC) $S_{ROC} = 0.998$ (Figure 4.43).

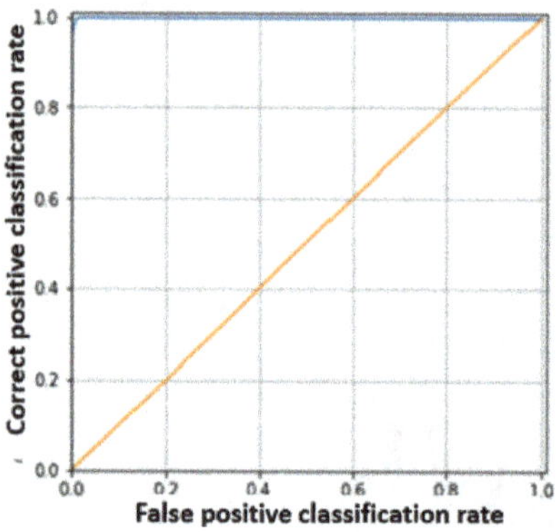

Figure 4.43
ROC curve for a neural network prediction model

Since the ROC curve characterizes the quality of binary classification, the following rules were introduced for its construction:

- If the specific number of such defects in the film does not exceed the legal threshold, there is no defect
- If the specific number of defects of a certain type in a film exceeds the legally prescribed threshold, a defect of the corresponding type exists.

Therefore, a software package was developed for downloading, primary processing, and statistical analysis of large amounts of production data from various sources and using these data to predict the quality indicators of a wide range of calendered polymer films. The software package enables:

- Generation and storage of arrays of structured production data (in the production parameter database)
- Calculation (using mathematical models of physical processes occurring in production units) of additional (unmeasured) quality indicators of polymer material at different production phases to reduce the uncertainty of information about the object
- Assessment of the correspondence of empirical distributions with the normal law and the correlation between quality indicators and technological parameters
- Prediction of film quality using various machine learning methods (depending on the type of distribution law, the amount of production data, the requirements for accuracy, and economics of forecasting).

The software package is tailored to the types of film produced and the configuration of the calender lines. The performance of the software package is confirmed by the results of its tests, which are based on data from the industrial production of polymer films for multi-product packaging.

4.4.2 Software Package for Processing Thickness Production Data to Detect Defects in Polymer Film Rolls

Tamara Chistyakova and Andrey Polosin

Customer complaints about poor-quality film rolls demands strict requirements for the quality of polymer films and the detection of defects in a timely manner. Defects can occur during film production and when winding the films into reels, as well as during delivery of the reels to the customer, processing there, and use of the films.

The biggest problem in this regard are defects caused by local variations (inconsistency) of thickness across the width of the film (flatness) and the kind of the rate of thickness change across the width. This is due to the wide range of films produced in thickness (100–1000 μm) and the significant width of the film, reaching 2.0–3.0 m. When producing the film, the thickness variation is reduced to an acceptable range (15–20 μm), which is usually 5–12% of the nominal thickness, according to the regulations. For this purpose, various methods are used in production to compensate for the deflection of calender rolls that occurs under the influence of spacing forces (axis crossing, roll bending) [31]. When winding up, thickness deviations add up with each new layer of film in the jumbo reel, the length of which is usually 400–20,000 m or significantly longer. This usually leads to deformation or damage to the film when it is wound up. Typical types of errors caused by differences in thickness (Figure 4.44) are:

- Telescopic effect (lateral projections of the layers of the central or inner part of the reel during its transportation or handling)
- Piston ring (local increase in thickness by 2–3 μm)
- Soft/hard edges combined with hard/soft areas
- Dark stripes in thickened areas of the film.

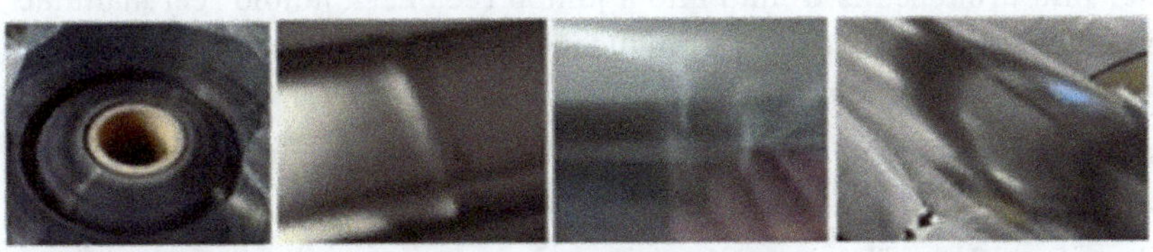

Telescoping effect | Piston ring | Soft edge combined with a hard area | Dark stripe on a thickened area

Figure 4.44 Typical types of errors in polymer film reels

When cutting the reels to the customer's required size, typical telescoping can occur, making the reel unusable for the customer. Piston rings cause cross-folds and charging. Soft/hard edges contribute to wrinkling and coating problems [32]. Poor flatness, which leads to telescoping of the reels, especially at the end of the reel, is caused by differences in thickness and an increase in the thickness gradient from one edge to the other. Therefore, the timely detection and elimination of these types of defects under continuous production conditions helps to increase the yield of high-quality films and save resources and energy.

In modern film production, high-precision radioisotope thickness gauges are usually used for continuous thickness control. They are installed after the cooling device and allow the formation of arrays of measured thickness values along the width and length of the film [33]. Such an array contains 20,000–30,000 (or more) thickness values: 100 or more thickness profiles across the width, depending on the length of the film in the jumbo reel, each profile containing approximately 200 thickness values. A productivity of around 1000 kg/h results in the production of 600–800 jumbo reels of film in one production month. The measured thickness is a random value due to random measurement errors. Therefore, to produce quantitative estimates of the thickness variation and rate of change of film thickness in each reel that characterize different types of defects, it is necessary to perform statistical processing on a series of thickness values.

Therefore, for the large-scale and multi-range production of polymer films, it is important to develop a software package adapted to the characteristics of the reels and the parameters of the processing methods, allowing you to identify the thickness variations, i.e. the thickness gradient.

The integration of the complex into the film quality management system software helps to increase the efficiency of the production system. This is achieved by reducing the incompleteness of information in management through the assessment and analysis of additional quality indicators that are not controlled in production but consider customer requirements.

Presentation of the problem of identifying defective film reels

The analysis of the types and parameters of polymer film reels, as well as the main types of reel defects, have made it possible to offer an informative description of the procedure for processing thickness data and identifying reel defects (Figure 4.45).

The polymer film produced is wound into a jumbo reel. Each jumbo reel manufactured during the period $t \in [t_0; t_K]$ has a unique identifier i_R in the factory (process input parameter X) and parameters R_M, including:

$\delta_{i,j}$ = the measured value of the film thickness in the i-th position along the reel length and in the j-th section along the reel width [µm]

n = number of width profiles per reel

m = number of thickness values in each profile

δ_0 = nominal (required by regulation) film thickness [µm]

δ_0^{min}, δ_0^{max} = maximum permissible thickness values according to the regulations (thickness tolerance limits) [µm]

W = film width [mm].

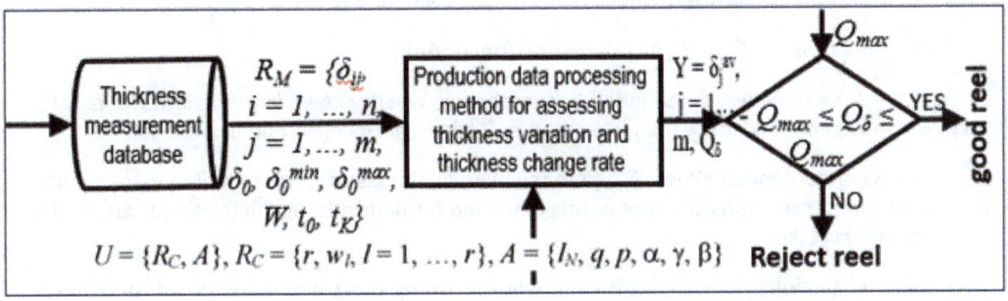

Figure 4.45 Information description of the procedure for processing thickness data and identifying defective film reels

The parameters R_M are generated (depending on the identifier i_R) from the database of thickness measurement results. The variable parameters of procedure U are:

- Parameters of consumption reels R_C in which jumbo reels are cut for customers
- Customizable parameters A of data processing methods.

It is necessary to take the R_C parameters into account because customer reels should not have a telescoping effect. The parameters of customer roles are the number of customer roles r in the jumbo reel and the width of every l-th customer reel w_l.

The parameters of mathematical methods A include:

l_N = significance level (probability of rejecting the correct hypothesis when testing the hypothesis about the normality of the thickness distribution)

q = number of thickness values for assessing the maximum thickness variation in film sections along the width of the jumbo reel

p, α, β = the number of thickness values used to estimate the average thickness in sections of film along the width of the jumbo reel in various ways

γ = proportion of the width of the edge section in the jumbo reel.

The output parameters of procedure Y are:

- $\delta_{j,av}$ (j = 1,..., m) = distribution of the average film thickness over the width of the jumbo reel
- Q_δ = indicators of thickness variation and rate of change of film thickness, which are not measured in production:

$$Q_\delta = \left\{ \begin{matrix} D_{\delta j}, (d\delta/dw)_j, \Delta_j^{gb}, \Delta_j^{se}, \Delta_j^{he}, \Delta_j^{db}, \; j = 1,\ldots,m, \; \Delta_{lj}^{ts}, \Omega_l, \\ j = s_{l-1}+1,\ldots,s_l, \; l = 1,\ldots,r \end{matrix} \right\} \quad (4.64)$$

where:

$\delta_j^{av} = f_\delta(\delta_{i,j}, i = 1,\ldots, n)$—average thickness in the j-th section, estimated by averaging the j-th thickness profile along the length of the jumbo reel [μm]

$D_{\delta j}$ = the maximum difference in thickness in the j-th section of the film [%]

$(d\delta/dw)_j$ = thickness gradient in the j-th section [μm or mm]

$\Delta_j^{gb}, \Delta_j^{se}, \Delta_j^{he}, \Delta_j^{db}$ = thickness deviations δ_j^{av} from the average thickness (across the width of the section/jumbo reel) to identify the piston ring, soft edges, hard edges, and dark streak [%]

$\Delta_{lj}^{ts}, \Omega_l$ = deviation of the thickness δ_j^{av} from the average thickness (corresponding to the width of the l-th customer reel) and the range of this deviation for determining the telescopic effect of the l-th customer reel [%]

s_l = number of sections along the width of the l-th customer reel (number of thickness values in the profile for a customer reel).

Based on the information description, the task of processing data on thickness and identifying defective reels is set, which is as follows. For given parameters of the jumbo reel R_M and customer reels R_C:

- Construct the distribution of the average thickness over the width of the jumbo reel $\delta_j^{av}, j = 1,\ldots, m$
- Test the hypothesis about its normality using criteria for checking deviations from the normal law C_N
- By processing using statistical methods, generate a set of film quality indicators Q_δ, compare them with the limit values $Q_{max} = \{\Delta^{gb}_{max}, \Delta^{se}_{max}, \Delta^{he}_{max}, \Delta^{db}_{max}, \Delta^{ts}_{max}, (d\delta/dw)_{max}\}$ and determine the presence of defective film reels according to the following rules:
  - If $\Delta_j^{gb} > \Delta^{gb}_{max}, j = 1,\ldots,m$, then there is a piston ring on the jumbo reel
  - If $\Delta_j^{se} < -\Delta^{se}_{max} \wedge \Delta_k^{se} > \Delta^{se}_{max}, j, k = 1,\ldots,g, j < k, j, k = m - g,\ldots,m, j > k$, then the jumbo reel has soft edges combined with hard interior areas
  - If $\Delta_j^{he} > \Delta^{he}_{max} \wedge \Delta_k^{he} < -\Delta^{he}_{max}, j, k = 1,\ldots,g, j < k, j, k = m - g,\ldots,m, j > k$, then the jumbo reel has hard edges combined with soft interior areas
  - If $\Delta_j^{db} < -\Delta^{db}_{max}, j = 1,\ldots, g \wedge \Delta_k^{db} < -\Delta^{db}_{max}, k = m - g,\ldots,m$, then there are dark stripes on the section of the jumbo reel with increased thickness (max. $\Delta_j^{db}, j = 1,\ldots,m$)
  - If $(|\Delta_{lb}^{ts}| > \Delta^{ts}_{max} \wedge |(d\delta/dw)_b| > (d\delta/dw)_{max} \wedge \Omega_l > 2\Delta^{ts}_{max}), b = s_{l-1}+ 1, s_l, l = 1,\ldots,r$, then the l-th customer reel has a telescopic effect.

Here is g the number of reels, which can be slit from the usable width of the jumbo reel: $g = f_g(m, \gamma)$.

Methods for analyzing and processing production data on film thickness

The mathematical software for solving the problem includes criteria for testing the normality hypothesis and methods for assessing film quality indicators. When testing the hypothesis that the data sample belongs to the normal law, the following criteria are applied:

- Criteria specifically developed for this purpose
- Goodness-of-fit criteria of type χ^2 (usually the Pearson criterion)
- Non-parametric goodness-of-fit tests (Kolmogorov criterion with Bolshev correction (which reduces the dependence on sample size), Cramer–Mises–Smirnov ω^2 criterion, and Anderson–Darling Ω^2 criterion).

The hypothesis about the normal distribution of film thickness is complex, since only the determination range of the parameters (mathematical expectation m_δ and variance σ_δ^2) of the normal distribution ($m_\delta > 0$, $\sigma_\delta^2 > 0$) is known. These parameters are estimated based on the sample analyzed, for example using the maximum likelihood method. This is particularly important when using non-parametric tests, since the classical results for these criteria obtained when testing simple hypotheses when the parameters of the normal law are known cannot be used. The distributions and tabular (critical) values of the statistics of non-parametric tests differ. Applying χ^2-type criteria involves grouping a sample of data to convert it into a statistical series. This grouping shows a certain arbitrariness (in the choice of the number of grouping intervals, which affects the significance of the criteria) and itself leads to a certain loss of information (up to 5%) in the sample [34].

A comparison of the performance of the criteria (the probability of rejecting a false hypothesis) shows that when testing complex hypotheses for small samples, criteria of the type χ^2 and non-parametric tests are inferior to special criteria. Among the latter, the Geary criterion G [35] and the David–Hartley–Pearson criterion D [36] have a high discriminatory power:

$$G = \sum_{j=1}^{j=m} \left| \delta_j{}^{\text{av}} | - |\bar{\delta} \right| / \left(m\sqrt{s_\delta^{2*}} \right) \tag{4.65}$$

$$s_\delta^{2*} = \sum_{j=1}^{j=m} \left(\delta_j{}^{\text{av}} - \bar{\delta} \right)^2 / m, \bar{\delta} = \sum_{j=1}^{j=m} \delta_j{}^{\text{av}} / m$$

$$D = \left(\delta^{\max} - \delta^{\min} \right) / \sqrt{s_\delta^2}, s_\delta^2 = \sum_{j=1}^{j=m} \left(\delta_j{}^{\text{av}} - \bar{\delta} \right)^2 / (m-1) \tag{4.66}$$

where:

$\overline{\delta}$ = average film thickness in the jumbo reel [μm]

s_{δ}^{2*} = based estimate of thickness variation [μm^2]

$\delta^{\min} = \min_{j=1,\ldots,m} \delta_j{}^{\text{av}}$ = the smallest thickness in the thickness distribution [μm]

$\delta^{\max} = \max_{j=1,\ldots,m} \delta_j{}^{\text{av}}$ = the largest thickness in the thickness distribution [μm]

s_{δ}^{2} = unbiased estimate of thickness variation [μm^2].

Given the sample size m = 200 (number of thickness values in the profile) and the significance level l_N = 0.05 (usually accepted when analyzing technical systems), the significance of the criteria has the following values:

- The power of the Geary criterion is 0.527
- The power of the David–Hartley–Pearson criterion is 0.390
- The power of the Pearson criterion is 0.250
- The power of the Kolmogorov criterion is 0.235 [34].

As an alternative hypothesis, the hypothesis is that the sample belongs to a logistic distribution that comes closest to the normal law. Therefore, the library of criteria for testing the hypothesis about the normality of the thickness distribution includes the Geary criterion (as in the main criterion), the David–Hartley–Pearson criterion, and the Pearson criterion.

The last two criteria are used to control the correctness of the result of testing the normality hypothesis using the Geary criterion. This is necessary due to the asymmetry of the distribution of the statistics of the Geary criterion and the high approximation error of this distribution by the normal law.

The Geary test (Equation 4.65) and David–Hartley–Pearson (Equation 4.66) tests are two-tailed. The condition for accepting the hypothesis about the normal distribution has the formula:

$$C_{\text{N}\ (l_{\text{N}}/2;m)}^{\text{tab}} \leq C_{\text{N}} \leq C_{\text{N}\ (1-l_{\text{N}}/2;m)}^{\text{tab}} \tag{4.67}$$

The critical value $C_{\text{N}}^{\text{tab}}$ of the test statistic (Equation 4.65 or Equation 4.66) is determined from the table of quantiles of the statistical distribution at a given significance level l_N and sample size m.

Pearson's χ^2 criterion is right-tailed, so the hypothesis of a normal distribution law is accepted if:

$$C_{\text{N}} \leq C_{\text{N}\ (l_{\text{N}};\theta)}^{\text{tab}} \tag{4.68}$$

The critical value C_N^{tab} of the statistics of this criterion is determined for a given significance level l_N and the number of degrees of freedom $\theta = N - 3$. The number of degrees of freedom depends on the number of data grouping intervals N of the same length. To select the number of intervals, a two-step procedure is used [37]:

(a) The initial value is estimated using the generally accepted empirical formula:

$$N_0 = 3.3 \ln m + 1 \tag{4.69}$$

(b) The final value is selected from the condition:

$$mP_h\left(m_\delta, \sigma_\delta^2\right) \geq 10, \quad h = 1, \ldots, N, \quad N \leq N_0 \tag{4.70}$$

The theoretical probability that thickness values fall in the h-th interval $P_h\left(m_\delta, \sigma_\delta^2\right)$ is estimated using the normal law. If the condition shown in Equation 4.70 is violated for any of the intervals, that interval and the adjacent interval are combined into one. In this case, the number of intervals N and the number of degrees of freedom θ decrease accordingly.

To assess the quality indicators of film reels, the analysis of which makes it possible to identify typical types of defects, the following methods were proposed [37–39].

(1) Method for assessing the maximum difference in thickness in film sections along the width of the jumbo reel (adjustable method parameter, q):

$$D_{\delta j} = \left(\delta_j^{max} - \delta_j^{min}\right) / \delta_j^{min} \times 100 \quad j = 1, \ldots, m \tag{4.71}$$

$$\delta_j^{extr}\Big|_{extr=\{min;max\}} = \begin{cases} extr_{k=j,\ldots,j+q-1}\delta_k^{av}, & j = 1, \ldots, m-q+1 \\ extr_{k=j,\ldots,m}\delta_k^{av}, & j = m-q+2, \ldots m \end{cases} \tag{4.72}$$

where:

δ_j^{extr} = maximum thickness in the film section starting in the j-th section and covering q sections (q analyzed values of average thickness δ^{av}) [μm].

When the number of sections making up a section of film becomes less than q (the right edge of the jumbo reel), to calculate the maximum thickness using Equation 4.71, the values of the average thickness δ^{av} in all remaining sections up to the other edge of the jumbo reel must be calculated. The right end of the reel is analyzed.

(2) Method for assessing the thickness deviation from the average thickness of film sections to identify the piston ring on the jumbo reel (adjustable method parameter, p):

$$\Delta_j^{gb} = \left(\delta_j^{av} - \delta_j^{gb}\right) / \delta_j^{gb} \times 100 \quad j = 1, \ldots, m \tag{4.73}$$

$$\delta_j^{gb} = \begin{cases} \sum\limits_{k=1}^{k=j+p} \delta_k^{av}/(j+p), & j = 1,\ldots,m-p-1 \\ \bar{\delta}, & j = m-p,\ldots,m \end{cases} \tag{4.74}$$

where:

δ_j^{gb} = average thickness of the film section starting at the left end of the jumbo reel and comprising sections $j+p$ ($j+p$ thickness values δ^{av} for calculating the value δ_j^{gb} as an average over an expanding window) [µm].

If the number of sections j becomes equal to or greater than $m-p$, the thickness δ_j^{gb} calculated according to Equation 4.74 is the average thickness of the film in the jumbo reel.

(3) Method for assessing the deviation of the thickness from the average thickness of film sections to identify soft edges and hard inner sections in the jumbo reel (adjustable method parameter, α):

$$\Delta_j^{se} = \left(\delta_j^{av} - \delta_j^{se}\right)/\delta_j^{se} \times 100, \quad j = 1,\ldots,m \tag{4.75}$$

$$\delta_j^{se} = \begin{cases} \sum\limits_{k=j}^{k=j+\alpha-1} \delta_k{}^{av}/\alpha, & j = 1,\ldots,m-\alpha+1 \\ \sum\limits_{k=j}^{k=m} \delta_k{}^{av}/(m-j+1), & j = m-\alpha+2,\ldots,m \end{cases} \tag{4.76}$$

where:

δ_j^{se} = the average thickness of the film section starting in the j-th section and comprising α sections (α thickness values δ^{av} for calculating the value δ_j^{se} as a sliding average) [µm].

If the number of sections composing the section becomes less than α, the thickness values δ^{av} in all remaining sections up to the right end of the jumbo reel are used for calculation according to Equation 4.76.

(4) Method for assessing the thickness deviation from the average thickness of film sections to identify hard edges and soft interior areas on the jumbo reel (adjustable method parameter, β):

$$\Delta_j^{he} = \left(\delta_j^{av} - \delta_j^{he}\right)/\delta_j^{he} \times 100, \quad j = 1,\ldots,m \tag{4.77}$$

$$\delta_j^{he} = \begin{cases} \sum\limits_{k=1}^{k=j+\beta} \delta_k{}^{av}/(j+\beta), & j = 1,\ldots,m-\beta-1 \\ \bar{\delta}, & j = m-\beta,\ldots,m \end{cases} \tag{4.78}$$

where:

δ_j^{he} = the average thickness of the section starting from the left end and comprising the sections $j+\beta$ ($j+\beta$ thickness values δ^{av} constituting the expanding window) [µm].

If the number of sections j becomes equal to or greater than $m-\beta$, the thickness δ_j^{he} calculated according to Equation 4.78 is equal to the average thickness of the film in the reel $\overline{\delta}$.

(5) Thickness deviation evaluation method for identifying dark streaks:

$$\Delta_j^{db} = \left(\delta_j^{av} - \overline{\delta}\right)/\overline{\delta} \times 100, \quad j = 1, \ldots, m \tag{4.79}$$

(6) Method for assessing the deviation of the thickness from the average film thickness in a customer reel, the deviation range, and the thickness gradient for detecting the telescoping of a customer reel:

$$\begin{aligned} &\Delta_{l,j}^{ts} = \left(\delta_j^{av} - \overline{\delta}_l\right)/\overline{\delta}_l \times 100, \quad j = s_{l-1}+1, \ldots, s_l, \\ &s_l = f_s\left(s_{l-1}, w_l, \Delta w_{mes}\right), \quad l = 1, \ldots, r \end{aligned} \tag{4.80}$$

$$\overline{\delta}_l = \sum_{j=s_{l-1}+1}^{j=s_l} \delta_j{}^{av}/(s_l - s_{l-1}), \quad \Omega_l = \left(\Delta_l^{ts}\right)^{max} - \left(\Delta_l^{ts}\right)^{min}, \quad l = 1, \ldots, r \tag{4.81}$$

$$(d\delta/dw)_j = \begin{cases} \left(-\delta_{j+2}{}^{av} + 4\delta_{j+1}{}^{av} - 3\delta_j{}^{av}\right)/(2\Delta w_{mes}), & j = 1 \\ \left(\delta_{j+1}{}^{av} - \delta_{j-1}{}^{av}\right)/(2\Delta w_{mes}), & j = 2, \ldots, m-1 \\ \left(3\delta_j{}^{av} - 4\delta_{j-1}{}^{av} + \delta_{j-2}{}^{av}\right)/(2\Delta w_{mes}), & j = m \end{cases} \tag{4.82}$$

where:

$\overline{\delta}_l$ = average film thickness in the l-th customer reel [μm]

$\Delta W_{mes} = W/(m-1)$ = thickness measurement step along the width [mm]

$\left(\Delta_l^{ts}\right)^{min}, \left(\Delta_l^{ts}\right)^{max}$ = the smallest and largest deviation in thickness δ^{av} from the thickness $\overline{\delta}_l$ [%].

To estimate the film thickness gradient, the expression system Equation 4.82 includes three-point finite differences that provide a second-order approximation error. In the end sections of the reel, the pitch is calculated using one-sided differences (right and left differences) and in inner sections using the central (symmetrical) difference.

Architecture and operational examples of the software package

The architecture of the software package for processing thickness data and identifying defective film reels is shown in Figure 4.46.

The complex interacts with two categories of users:

- Management of production personnel (production manager, quality engineer)
- Administrator.

The administrator works with a database of polymer film thickness measurement results, with an interface that loads arrays of measured film thickness values into jumbo reels (from CSV files generated by the thickness gauge), and enters parameters of the jumbo reel.

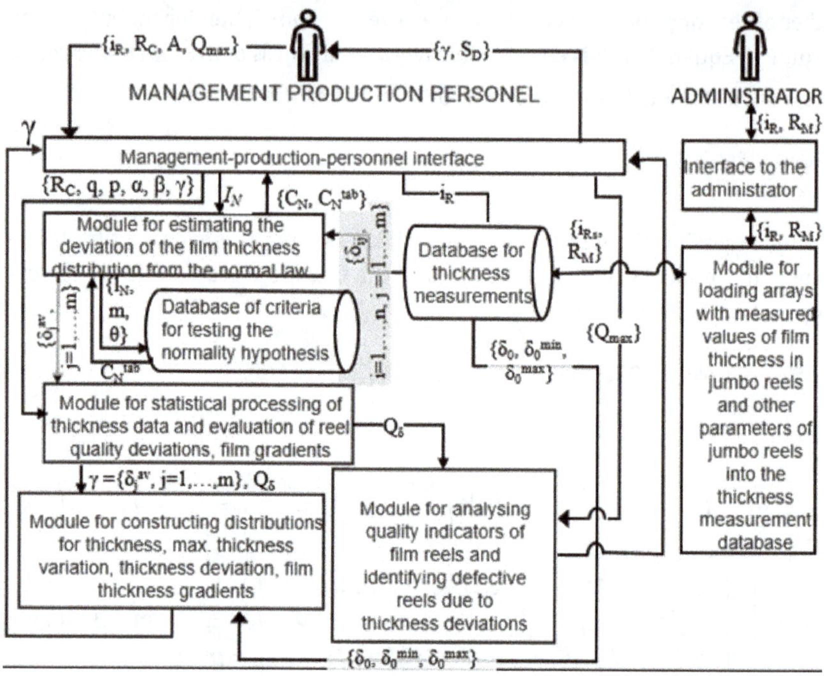

Figure 4.46 Architecture of a software package for processing thickness data and identifying defects in polymer film rolls

The database of thickness measurement results contains for each jumbo reel its unique identifier i_R, a set of measured thickness values, the film width W, date and time of the start and end of film production t_0 and t_K, and regulatory requirements for thickness δ_0, δ_0^{min}, and δ_0^{max}.

The process is then as follows:

1. The management production personnel select the jumbo reel identifier i_R and enter the customer role parameters R_C. Using the i_R identifier, a series of thickness values and regulatory thickness requirements are read from the measurement results database

2. While processing and analyzing a series of thickness values across the entire surface of the jumbo reel, an average thickness profile is created across the width of the roller, and the hypothesis about the normality of the thickness distribution is tested

3. The critical values of the statistics of the criteria for testing the normality hypothesis are read from the criteria database depending on the significance level and the sample size (number of data grouping intervals)

4. If the data conforms to the normal law, statistical processing is carried out, estimating the indicators of thickness variation and the rate of change of film thickness Q_δ using Equation 4.71–Equation 4.82.

5. The software package is tailored to the parameters q, p, α, β, and γ of the processing methods and the limit values Q_{max} of the thickness fluctuations and the rate of change of the film thickness.
6. The analysis of indicators Q_δ, which consists in comparing their values for the evaluated film reel with the limit values Q_{max}, allows identification of a defect if $Q_\delta \not\subset [-Q_{max}; Q_{max}]$, about which a message S_D is generated
7. The results are visualized in the form of thickness profiles, indicators of thickness variation, and the rate of change of thickness across the width of the rolls with display of their limits.

The software package was tested using data from jumbo reels made from polymer films for pharmaceutical, food, and technical packaging (δ_0 = 32–750 µm, W = 1310–2030 mm). The reels are manufactured at the factory of Klöckner Pentaplast Europe GmbH & Co. KG (Montabaur) and have customer complaints for telescoping, piston rings, soft/hard edges, and dark stripes.

The results of the analysis and processing of data on film thickness, using the example of identifying the telescope effect, are shown in Figure 4.47–Figure 4.49.

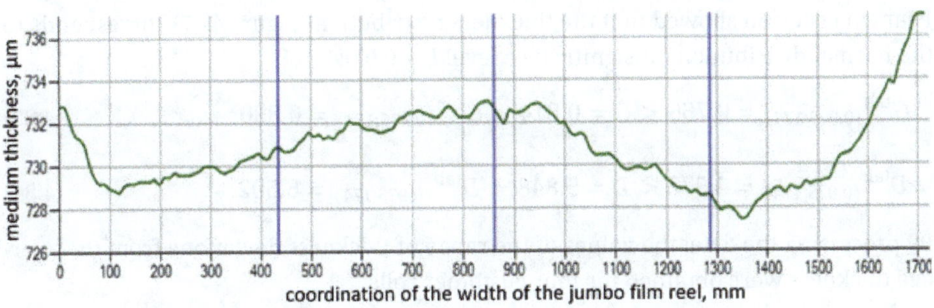

Figure 4.47 Film thickness profile across the width of the parent roll

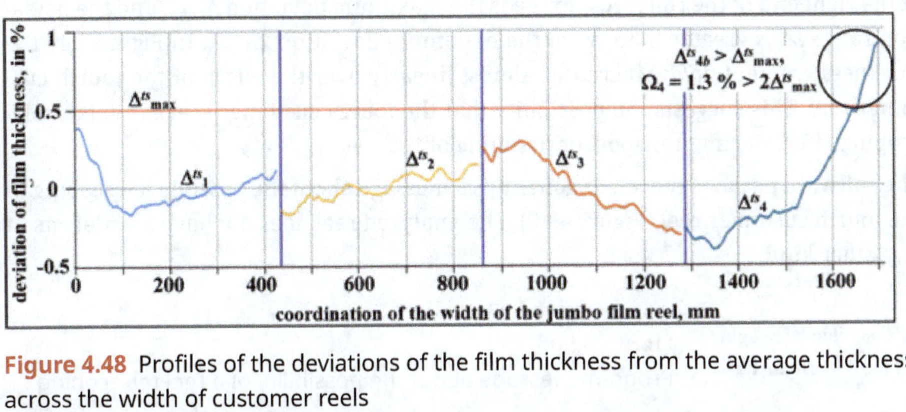

Figure 4.48 Profiles of the deviations of the film thickness from the average thickness across the width of customer reels

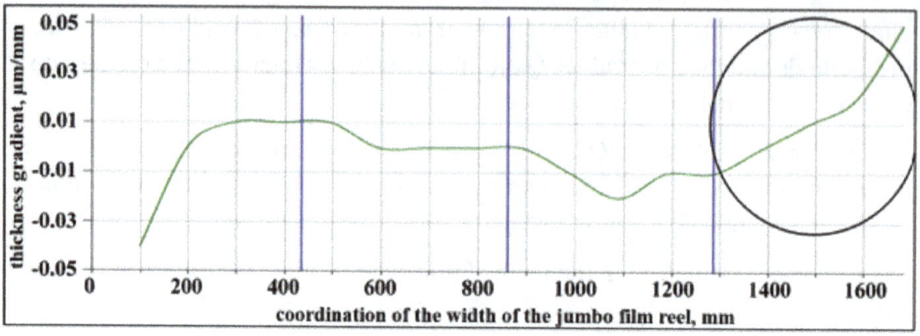

Figure 4.49 Smoothed film thickness gradient profile across the width of the jumbo reel

The jumbo reel is characterized by the following parameters:

- n = 113, m = 174, r = 4 (the jumbo reel is cut into four customer rolls, the boundaries of which are shown by purple vertical lines in Figure 4.47)
- W = 1712 mm, δ_0 = 750 µm, δ_0^{min} = 712 µm, δ_0^{max} = 788 µm.

The fourth customer reel (1286–1712 mm) exhibits a telescope effect.

The statistical analysis of the data using the Geary criterion and the David–Hartley–Pearson criterion showed that the thickness distribution (Figure 4.47) corresponds to the normal distribution (at significance level l_N = 0.05):

$$G^{tab}{}_{(0.025;174)} = 0.766 < G = 0.814 < G^{tab}{}_{(0.975;174)} = 0.830 \tag{4.83}$$

$$D^{tab}{}_{(0.025;174)} = 4.572 < D = 5.848 < D^{tab}{}_{(0.975;174)} = 6.502 \tag{4.84}$$

By processing the data, the values of the range of thickness deviations from the average thickness were obtained for four customer rolls:

$$\Omega_1 = 0.55\%,\ \ \Omega_2 = 0.31\%,\ \ \Omega_3 = 0.60\%,\ \ \Omega_4 = 1.3\%$$

From these data and Figure 4.48 for the fourth customer reel, the thickness deviation at the right end of the roller $\Delta^{ts}{}_{4b}$ exceeds the maximum deviation $\Delta^{ts}{}_{max}$, and the deviation range Ω_4 is greater than twice the maximum deviation $2\Delta^{ts}{}_{max}$. In Figure 4.49 the thickness gradient $d\delta/dw$ increases almost linearly over the width of the fourth customer reel. This suggests the possibility that the fourth customer reel exhibits a telescoping effect during transport or handling of the reel.

The software package therefore generates a message about the possible telescoping of the fourth customer reel (Figure 4.50). The analyzed reel does not have any defects of any other kind.

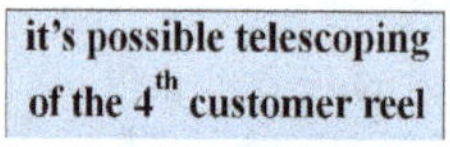

Figure 4.50 Program message about the possibility of a reel telescoping

The results of processing data on film thickness using the example of identification of another type of failure—a piston ring—is shown in Figure 4.51 and Figure 4.52.

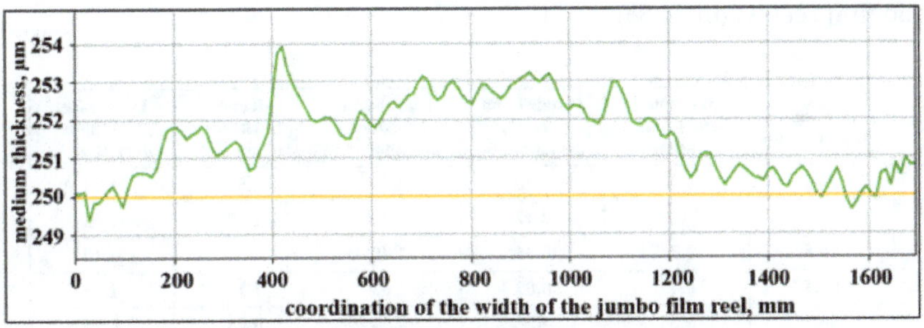

Figure 4.51 Film thickness profile across the width of the jumbo reel

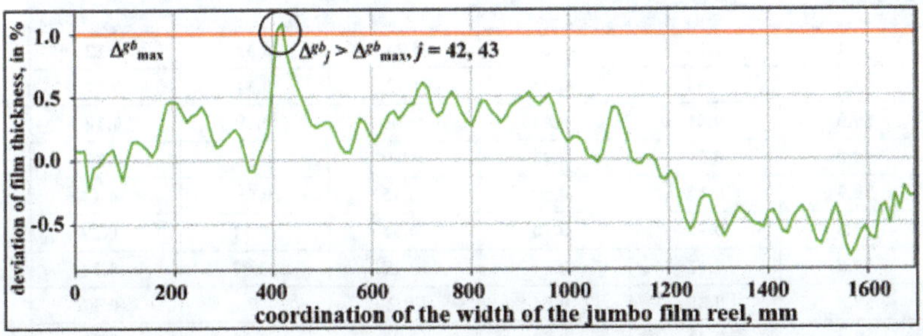

Figure 4.52 Profile of the deviation of the film thickness from the average thickness of the film sections along the width of the jumbo reel

The jumbo film reel, which has a piston ring, is characterized by the followingparameterss: number of width profiles n = 150, number of thickness values m = 174, W = 1695 mm, δ_0 = 250 µm, δ_0^{min} = 237 µm, δ_0^{max} = 263 µm. From Figure 4.51 it is clear that the film thickness meets the legal restrictions.

Figure 4.52 shows that there is a section of the film in which the thickness deviation $\Delta^{gb}{}_j$, j = 42, 43 exceeds the maximum deviation $\Delta^{gb}{}_{max}$. This indicates the presence of a piston ring in this area. The software package displays a message with the coordinates of the beginning and end of this film section (Figure 4.53). These coordinates are calculated by multiplying the numbers of the starting and ending sections of the section by the step of measuring the film thickness ΔW_{mes}.

Gauge band is in the area
412 – 421 mm of the jumbo reel

Figure 4.53 Program message about the piston ring on a roll with the coordinates of the beginning and end of the defective film section

The corresponding interface of the software package displays the calculated values of the average thickness, the maximum thickness fluctuation, the thickness deviations from the average thickness, and the thickness gradient in each section along the width of the jumbo film reel (Figure 4.54).

Coordinate along the film width, mm	Average thickness, µm	Maximum thickness variation, %	Gauge band thickness deviation, %	Dark band thickness deviation, %	Soft edges thickness deviation, %	Hard edges thickness deviation, %
0	60.9	1.0	0.40	1.54	1.54	1.54
9	60.9	1.0	0.42	1.54	1.57	1.54
19	60.8	0.91	0.36	1.45	1.51	1.45
28	60.6	0.49	0.02	1.03	1.12	1.03
38	60.3	0.54	-0.39	0.54	0.65	0.54
47	60.4	0.79	-0.14	0.72	0.84	0.72
57	60.6	0.88	0.21	1.0	1.14	1.01
66	60.4	0.72	0.08	0.82	0.97	0.82
75	60.2	0.43	-0.21	0.46	0.63	0.47
85	60.1	0.36	-0.39	0.21	0.39	0.22
94	60.0	0.44	-0.43	0.12	0.31	0.14
104	60.0	0.41	-0.41	0.10	0.29	0.12
113	60.0	0.35	-0.43	0.03	0.23	0.05
123	59.9	0.17	-0.57	-0.15	0.05	-0.13
132	59.8	0.12	-0.70	-0.31	-0.11	-0.29
141	59.8	0.12	-0.64	-0.27	-0.07	-0.24
151	59.8	0.30	-0.56	-0.2	0	-0.17
160	59.8	0.43	-0.59	-0.25	-0.05	-0.21
170	59.8	0.43	-0.64	-0.32	-0.13	-0.28
179	59.8	0.34	-0.52	-0.23	-0.04	-0.19

Figure 4.54 Interface for tabular visualization of the results of processing production data by thickness

Therefore, a software package was developed that implements methods for statistical processing of production data on the thickness of polymer films to identify typical types of defects in film reels associated with thickness deviations from the average thickness. The software package enables:

- Loading of arrays with film thickness values into jumbo reels
- Statistical processing and evaluation of the indicators of thickness fluctuation and the rate of change of film thickness
- Visualization of the thickness distribution, thickness variation indicators, and thickness gradient across the width of the film
- Identification of typical types of reel defects (telescoping, piston ring, hard/soft edges, and dark streaks) and generation of notifications to production management personnel about reel defects.

The software package is tailored to the characteristics of jumbo and customer reels, as well as the parameters of statistical methods for processing production data on thickness and threshold values for indicators of different film thicknesses. The performance of the software package is confirmed by the results of testing using data from reels of industrially produced polymer packaging films. The defects present in the analyzed reels were correctly identified by the program. In conclusion, the use of the software package leads to a reduction in the processing time of production data.

4.4.3 Software Package for Optimal Planning of Polymer Film Production

Tamara Chistyakova, Andrey Polosin

The most important task when planning flexible production of multi-product polymer films is the formation of an optimal sequence for placing and processing orders for film production on production lines.

The difficulty of solving the planning problem lies in the need to consider the various technological features of the production process, the changing range of products, as well as different line configurations and types of equipment. In addition, calender-based production planning is characterized by a high computational cost (belongs to the class of NP tasks of large dimensions), since the number of possible planning options is determined by the volume of the package of orders and the number of production lines for their implementation [40]. So, if one determines the order of timely execution of only 10 orders on only one production line, the number of plan options is 3,628,800. With 58 orders, this value increases to 2.35×10^{78}.

In real production plants, up to 15 production lines can often be operated at the same time, and the average number of orders for a planning period of one month is between 10 and 100. Therefore, when solving the problem of planning optimization, consideration of many possible planning options is required. This entails the need to use modern optimization methods when solving production planning problems that can find the optimal solution in a large search space within an acceptable computing time [41–43].

In this context, it is relevant to develop a software package that is adapted to various characteristics of the production process, has ergonomic user interfaces that allow variability in the use of functionality for different planning situations, and provides an effective solution to planning problems of different nature.

Presentation of the problem of optimal planning of film production

The information description of the procedure for planning industrial film production includes the following data.

Vector of input parameters, *X*, consisting of the following parameters:

Order characteristics

$$\tilde{O} = \left\{\overline{O}_i : \overline{O} = \left(N^{\mathrm{O}}, K^{\mathrm{O}}, P^{\mathrm{O}}, Q^{\mathrm{m}}, \tilde{R}, \tilde{G}, \mathrm{Cs}\right), i = 1, \ldots, N\right\} \tag{4.85}$$

Production line properties

$$\tilde{E} = \left\{\overline{E}_j : \overline{E} = \left(N^{\mathrm{E}}, K^{\mathrm{E}}, V^{\mathrm{P}}, \overline{\mathrm{To}}^{\mathrm{E}}, \overline{C}_{\mathrm{E}}\right), j = 1, \ldots, \mathrm{Me}\right\}$$

Planning parameters

$$\widetilde{Pd} = \left(\tau^{\mathrm{O}}, Y^{\mathrm{cr}}\right)$$

A vector of variable parameters that describes different options for distributing jobs along production lines and the order in which they are executed (different options for production plan, *Q*):

$$\begin{aligned} Q = \{Q_i : Q = (j, k, \tau_{\mathrm{O}_i}, \tau_j), \\ i = 1, \ldots, N, \quad j = 1, \ldots, \mathrm{Me}, \\ k = 1, \ldots, L_j, L_j \in N\} \end{aligned} \tag{4.86}$$

A vector of output parameters describing the planning result, i.e., the optimal distribution of orders along the production lines and in time, and an extreme of the objective function Y^{cr} (optimal plan, Q^{opt}):number 2

$$\begin{aligned} Q^{\mathrm{opt}} = \{Q_i : Q = (j, k, \tau_{\mathrm{O}_i}, \tau_j), \\ j = 1, \ldots, \mathrm{Me}, \quad k = 1, \ldots, L_j, L_j \in N, \\ i = 1, \ldots, N\} \end{aligned} \tag{4.87}$$

where:

$N^{\mathrm{O}}, K^{\mathrm{O}}$ = number and code of the order

P^{O} = film type according to the order

Q^{m} = order quantity of film (e.g. mass)

$\tilde{R}$ = film recipe

$\tilde{G} = \{W^{\mathrm{m}}, \mathrm{Th}^{\mathrm{m}}\}$ = geometric parameters (width W^{m}, thickness Th^{m}).

Cs = customer

N = number of orders

$N^{\mathrm{E}}, K^{\mathrm{E}}$ = designation and code of the production line

V^{P} = productivity

$\overline{\mathrm{To}}^{\mathrm{E}} = \left\{\overline{\mathrm{To}}^{\mathrm{E}}{}_r, r = 1, \ldots, R\right\}$ = technological operations carried out on the line

$\overline{C}_{\mathrm{E}}$ = line equipment settings

Me = number of production lines

$\tau^{O} = [\tau^{b}, \tau^{e}]$ = planning period, starting with planning start τ^{b} and end of planning τ^{e}

Y^{cr} = target function

$k = 1,\ldots,L_j$ = sequence of processing the i-th order on the j-th line

L_j = number of orders for the j-th line

$\tau_{Oi} \in [\tau^{b}, \tau^{e}]$ = start of the i-th order (including the date and processing time in production)

τ_i = processing time of the i-th order

The main target function for optimal scheduling is the total time to execute the orders and reconFigure the equipment to produce the required film types for each j-th line:

$$F_j(Q) = \sum_{k=1}^{k=L_j-1} \omega^{Ch}{}_{O_{j,k},O_{j,k+1}} \tau^{Ch}(O_{j,k}, O_{j,k+1}) + \sum_{k=1}^{k=L_j} \omega^{O}{}_{k} \tau^{O}{}_{j,k} \tag{4.88}$$

where:

$\omega^{Ch}{}_{O_{j,k},O_{j,k+1}}$ = a weighting coefficient that determines the priority of the reconfiguration of the line from k-th order execution to execution of the $(k + 1)$-th order

$\tau^{Ch}(O_{j,k}, O_{j,k+1})$ = time to reconFigure the line from executing the k-th order to execute the $(k + 1)$-th order [min]

$\omega^{O}{}_{k}$ = a weighting coefficient that determines the priority of fulfilling the k-th order

$\tau^{O}{}_{j,k}$ = execution time of the k-th order [min]

The time for reconfiguration of equipment $\tau^{Ch}(O_{j,k}, O_{j,k+1})$ is calculated based on formalized expert information on the parameters of reconfiguration of equipment from one type of film to another type of film and technological regulations for the operation of the equipment.

The film production time for the k-th order is determined based on the ordered film quantity $Q^{m}{}_{k}$ and the productivity of the j-th line $V^{P}{}_{j}$ when executing the k-th order (productivity depends on the characteristics of the order).

If there is information about production cost indicators when solving a planning problem, then the objective function is the cost of implementing the production plan on the j-th line:

$$F^{P}{}_{j}(Q) = \sum_{k=1}^{k=L_j-1} \omega^{P}{}_{O_{j,k},O_{j,k+1}} \tau^{Ch}(O_{j,k}, O_{j,k+1}) + \sum_{k=1}^{k=L_j} \omega^{P}{}_{k} \tau^{O}{}_{j,k} \tag{4.89}$$

where:

$\omega^{P}{}_{O_{j,k},O_{j,k+1}}$ = cost of reconfiguration of equipment from k-th order fulfilment to $(k + 1)$-th order fulfilment [c.u./h]

$\omega^{P}{}_{k}$ = operating costs of the equipment when the plan is fulfilled [c.u./h]

Based on the information description of the planning process, the optimal planning problem is posed, which is as follows. For given input parameters X, it is necessary to find optimal values of the following variable parameters in the search space for feasible solutions and within the planning period $[\tau^b, \tau^e]$, which provide an objective function selected as an extremum:

$$Q^{opt} = \{Q_i : Q = (j, k, \tau_{O_i}, \tau_j), \quad j = 1, \ldots, \text{Me}, \ k = 1, \ldots, L_j, L_j \in N, i = 1, \ldots, N\} \tag{4.90}$$

When solving a problem for one production line, the objective function (Equation 4.89 or Equation 4.90) is minimized. When solving a problem for several production lines, the total time for order fulfilment and reconfiguration of equipment on the production line for which it is maximum is minimized [44]:

$$F_C\left(Q^{opt}\right) = \min_{Q \in D \subset R^{|Q|}} \max_{j=1,\ldots,\text{Me}} \left(F_j\left(Q\right)\right) \tag{4.91}$$

where:

D = set of permissible values of variable parameters Q

$R^{|Q|}$ = solution search space

The solution to the optimization problem is a vector Q^{opt} in which the value of the objective function for each production line $F_j\left(Q^{opt}\right)$ and the total production cycle time $F_C\left(Q^{opt}\right)$ is minimal. This ensures that the production systems are evenly loaded.

When solving the optimal planning problem, the following restrictions are considered:

- Each order must be executed within the specified scheduling interval $\tau^b \leq \tau_i \leq \tau^e$
- Each order must be assigned to the type of production line capable of producing the desired type of film
- Each production line can only produce one order at a time
- In the optimal production plan, the production of all orders from a given order package is planned [44].

Architecture of the software package

The developed software package (Figure 4.55) offers an effective solution to the problem of production planning. It includes a subsystem for solving the optimal planning problem, information support, and a subsystem for visualizing planning results [45–49].

Interaction with users in the software package is supported by ergonomic interfaces. The production manager interface allows selection of orders to produce polymer films and production lines for placing orders, set a planning period, and select an objective function to optimize planning. The database administrator interface allows you to maintain up-to-date databases with information about the product range, characteristics of production lines, and orders. Through a special interface, a production process specialist (process engineer) fills out forms on various production indicators,parameterss for reconfiguring production line equipment and information about the current state of the equipment. The expert information obtained forms the basis for the creation of a rule base for the reconfiguration of production lines [50, 51].

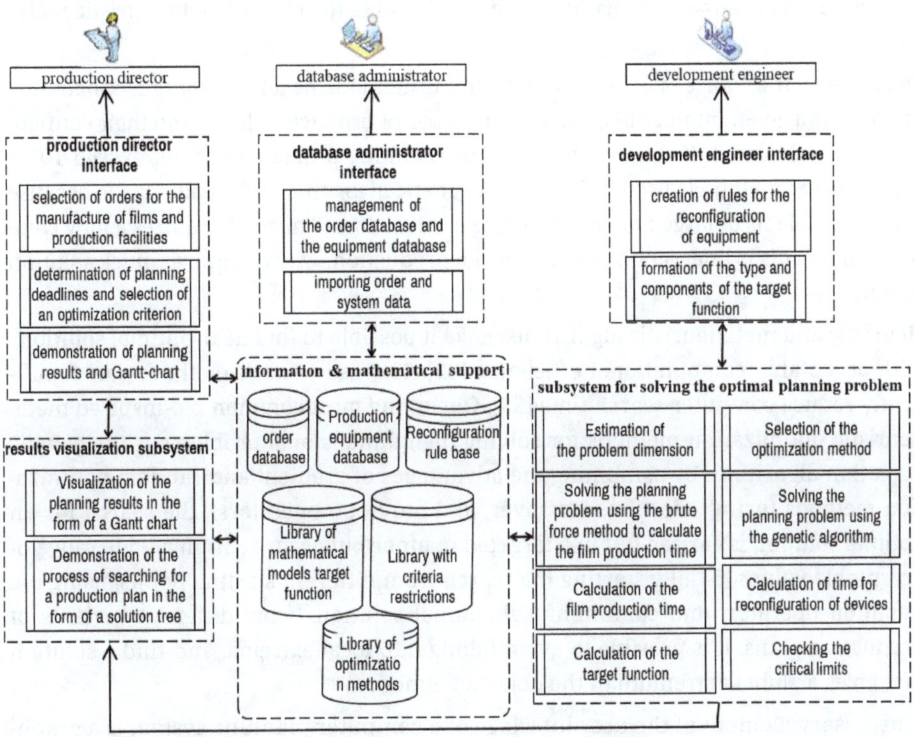

Figure 4.55 Architecture of a software package for optimal planning of the production of polymer films

The information support of the software package includes:

- Supplemented databases with orders and equipment
- A basis of rules for reconfiguring production line equipment to fulfil a wide range of orders
- Libraries with objective functions and criteria constraints.

When solving the problem of optimal planning, as the number of orders increases, the number of production lines increases, and the number of their possible transformations to produce films of various types increases, the number of options for the production schedule grows exponentially.

Various optimization methods are used to solve planning problems of different dimensions:

- "Brute force" methods, which belong to the class of exact methods that make it possible to reliably find the global extremum of the objective function
- Genetic evolutionary algorithms, which belongs to the class of metaheuristic methods.

To decide on the choice of an optimal scheduling method, the total number of scheduling options for a given number of orders, the number of production lines, and their configurations is calculated. If the dimension of the planning problem does not exceed 10^6, a "brute force" method is used. Otherwise, a genetic algorithm is applied. This is because solving problems of large dimensions using a "brute force" method requires a long computer time. This is unacceptable in industrial production, which requires quick solution finding.

Heuristic and metaheuristic algorithms make it possible to find near-optimal solutions with acceptable computational efficiency [52]. The use of such methods can significantly reduce computer search time [53]. One of the most common bio-inspired metaheuristic optimization methods for solving high-dimensional problems is the genetic evolution algorithm. By combining the advantages of gradient and random optimization methods (using selection, crossover, and mutation operators), the search for an optimal solution is carried out in a targeted manner but occurs within a wide range of acceptable values [54]. By starting the search from different starting points, using the mutation operator, and combining suboptimal solutions from different variants of production plans, it is possible to avoid falling into local extrema, and find a solution that gives a global extremum of the objective function.

A necessary element of the user interface of a computer planning system is a graphical visualization of the generated optimal production plan [55]. Therefore, at the end of the process of finding a solution using the chosen optimization method, the production manager is offered a schedule for the execution of orders on production lines in the form of a Gantt chart [56]. This shows the order processing order on each production line with the date and time of the start and end of film production (Figure 4.56). In addition, the production manager can see the value of the objective function and the solution tree of the optimization problem and demonstrate an improvement in the value of the objective function with each iteration of the planning algorithm.

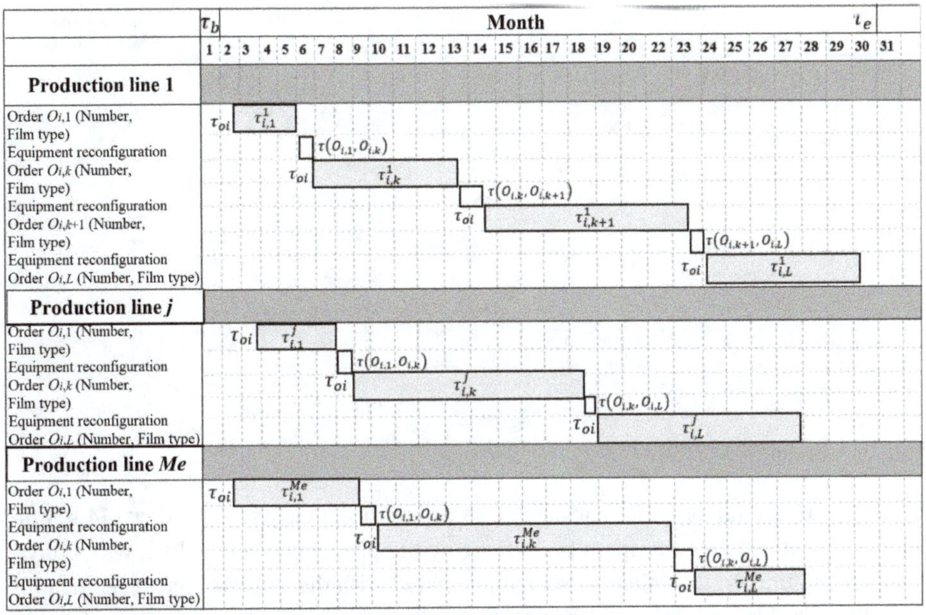

Figure 4.56 Form for visualizing planning results in the form of a "Gantt chart"

where:

$\tau^{j}_{i,k}$ is the execution time of the i-th order on the j-th line

k = order execution sequence number

Test example of the software package

To assess the effectiveness of the use of the software package in solving optimal planning problems, the production plan prepared by specialists at a plant to produce polymer films (Germany) was compared with the plan created using the software package.

Input data for production planning:

- 58 orders to produce various types of packaging films
- One extrusion line to produce multi-layer tubular films
- More than 200 different combinations of film parameters
- More than 100 rules for reconfiguring equipment to produce a wide range of films
- A planning period of 1.5 months.

The objective function for optimization is the total time for executing orders and reconfiguring the production line. Based on the number of orders, a genetic algorithm is used to solve the production planning problem. The optimal sequence of order execution in the production line generated by the program is shown in Figure 4.57.

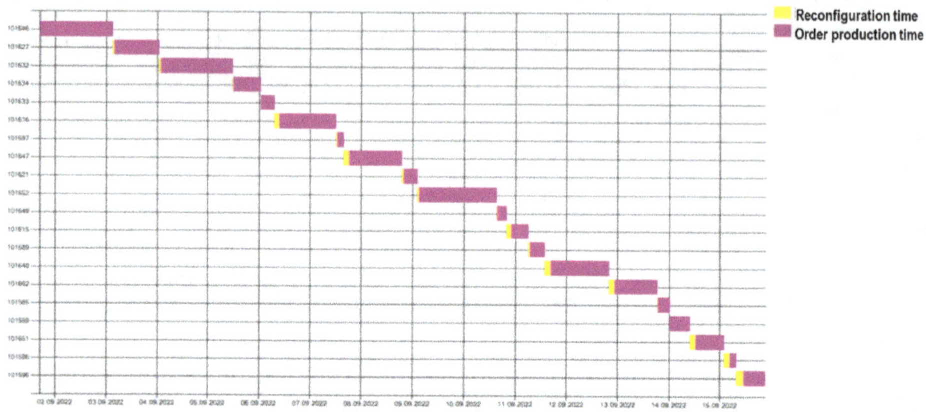

Figure 4.57 Optimal production plan in the form of a "Gantt chart"

According to the factory, one package of 58 orders previously took 44 days, 14 h, and 33 min, including the time spent reconfiguring the production line, which was 4 days, 11 h, and 45 min. In the calendar schedule created with the software package, the jobs took 42 days, 1 h and 33 min to execute, including the reconfiguration time of 1 day, 22 h and 45 min. Therefore, by using the software package, the overall time for processing batches of orders is reduced by 5–7% (depending on the range of films, the number of orders and the number of production lines).

The reduction in production time is achieved by finding a sequence of jobs that is close to optimal, which significantly reduces the time for reconfiguration of equipment between jobs. Therefore, the previously described software package was developed to solve the problem of optimizing the planning of industrial production of multi-product polymer films [57]. This allows creation of almost optimal production plans and the ability to carry out their graphic visualization in the form of a "Gantt chart". To optimally solve planning problems of different dimensions, the software package implements the "brute force" method and a genetic algorithm. The software package is tailored to the spectrum of polymer films produced, the properties of the production lines, and planning optimization criteria. Tests of the software package based on data from the industrial production of multi-product polymer films confirmed its performance.

In conclusion, the use of a software package increases the efficiency of film production planning, reduces labor intensity and time spent on planning, increases productivity and utilization of production lines, and reduces production costs.

4.4.4 Anti-counterfeiting

Christian Kohlert

Triangle method

The counterfeiting of many products for high-quality products in the technology, pharmaceutical, and food industries also requires film manufacturers to take measures to thwart or ensure the traceability of their products. The literature reports product losses of 5–7%, resulting in revenue losses of nearly \$1 trillion. In addition, life-threatening pharmaceutical products and business risks in the electronics and music industries, as well as losses of R&D investments due to counterfeiting, are also a factor. This not only impacts cash flows but also reduces jobs and necessary investments.

A method developed by KP [76] uses pigments in the micrometre range introduced into the film, which are almost invisible due to their low concentration, and which only through UV light or can be captured photographically in the dark (Figure 4.58).

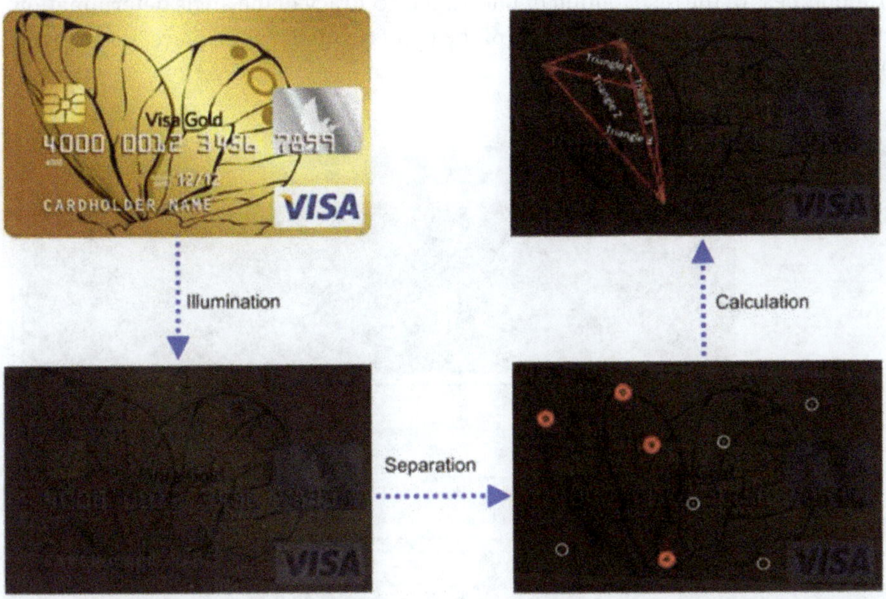

Figure 4.58 Points in an overlay film on a credit card with angle detection

Using a mass balance, you can define the number of points on a given area and determine triangular relationships from the most luminous points. Saving the angles α and β allows for later checking and thus checking for originality. For packaging films with a thickness range of 100–500 μm, only concentrations of 0.1% are required, so that based on the amount of film, the pigment costs are in the range of €0.01/kg of film. This app can be installed on any smartphone, and given sufficient memory, each phone can store millions of such data sets (Figure 4.59).

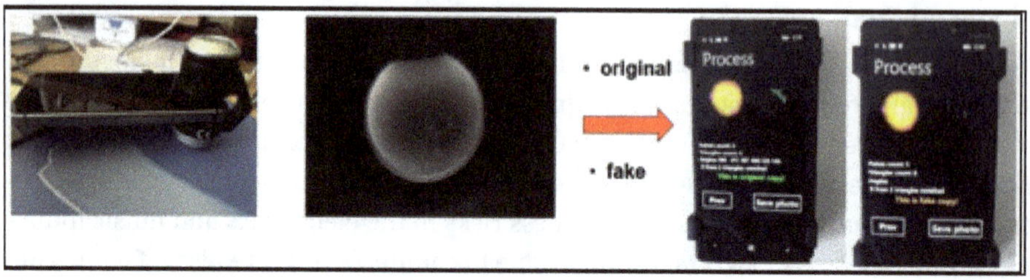

Figure 4.59 Recording the points with a smartphone and magnifying glass, and checking and evaluating the original or fake

A second app can then record the image again as a control app and check by evaluating all existing triangles whether the saved variant is included (Figure 4.60). Due to the randomness of the distribution of points, the accuracy of the angle determination, and the additional storage of the shape of the points (round, oval, square, star-shaped, etc.), forgery is impossible.

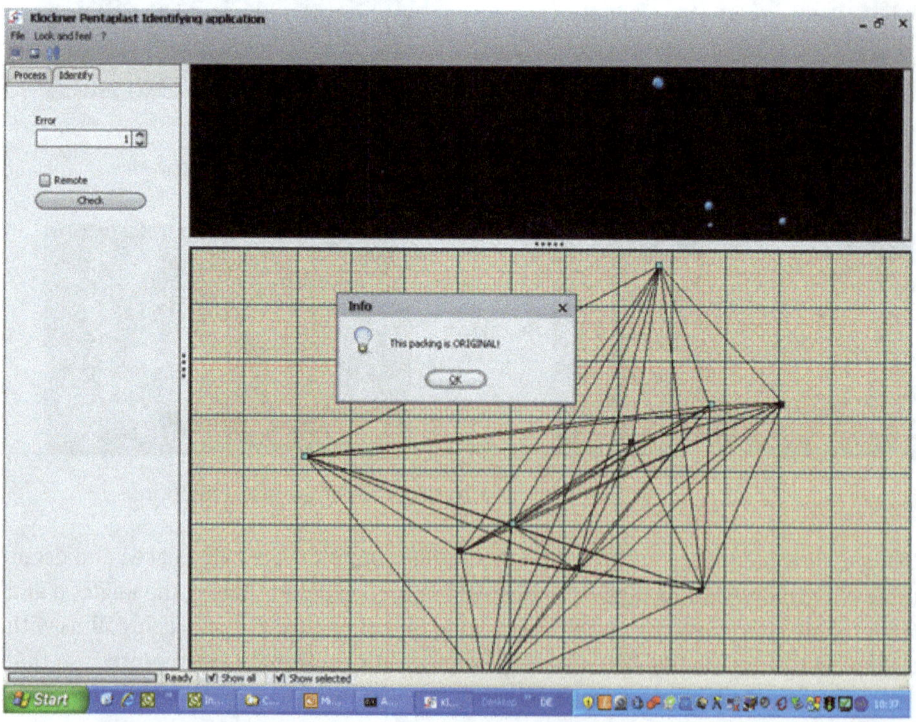

Figure 4.60 Control app with a statement about originality

The only disadvantage of this variant is that the location of the check must correlate with the location of the initial recording. This method can of course also be used for paintwork. Figure 4.61 shows the entire anti-counterfeiting process using the triangle method on a smartphone.

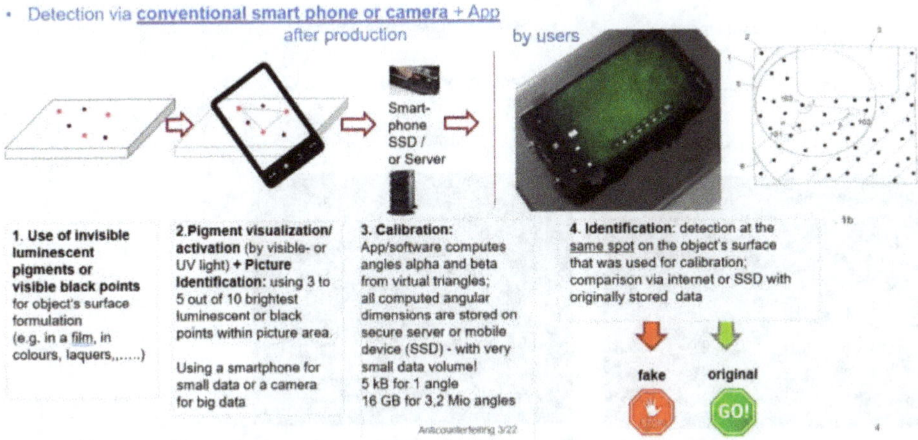

Figure 4.61 Anti-counterfeiting procedure using a triangle variant

Light decay method

To ensure anti-counterfeit protection for an entire batch, such UV or photoluminescent color pigments are also used in a slightly higher concentration of around 1%. By supplying energy via a UV lamp or in the dark via the flash on the cell phone, luminosity is produced, which disappears after a certain time (Figure 4.62). This time course is recorded and saved as a light decay curve, and repeated during an authenticity check for comparison with the saved curve.

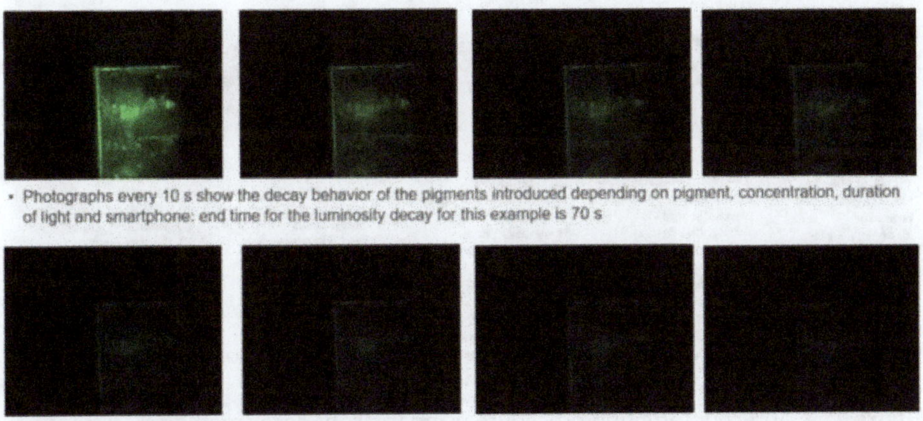

Figure 4.62 Light decay from luminous films

Due to the choice of pigment mixture, the concentration, the flash intensity and flash time, the choice of smartphone and the choice of evaluation time, a forgery requires enormous effort and is therefore almost impossible. An example of comparing two control recordings is shown in Figure 4.63.

Figure 4.63 Comparison of two samples based on the original (yellow line) using the light decay method

Figure 4.64 shows the entire process of anti-counterfeiting using the light decay method via smartphone. The advantage is that the recording point does not have to be identical to the control point. The disadvantage is that it is necessary to use the same type of smartphone for recording and checking, as well as the possibility of complicated and time-consuming forgery.

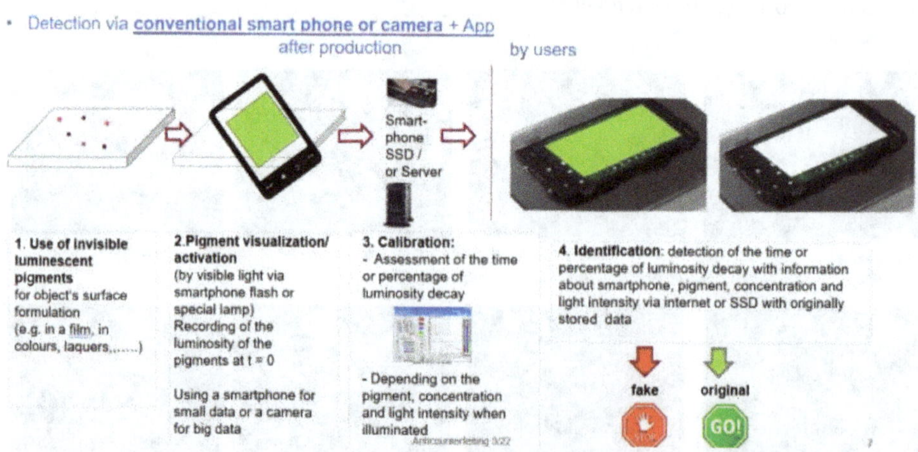

Figure 4.64 Anti-counterfeiting procedure using the light decay method

References for Section 4.4

[1] S. Sonawane, D. Patel, M. Kevadiya, R. Modi, J. Moradiya, A. Thomas, Big data by 3V's and its importance, *International Journal of Research in Engineering, Science and Management*, 2018, vol. 1, no. 12, pp. 11–12

[2] X. Wu, X. Zhu, G.-Q. Wu, W. Ding, Data mining with big data, *IEEE Transactions on Knowledge, and Data Engineering*, 2014, vol. 26, no. 1, pp. 97–107

[3] M. Kohlert, O. Hissmann, Applied Industry 4.0 in the polymer film industry, Proceedings of the 16th TAPPI European PLACE Conference, Basel, 2017, pp. 183–190

[4] M. Kohlert, T. B. Chistyakova, Advanced process data analysis and on-line evaluation for computer-aided monitoring in polymer film industry, Известия Санкт-Петербургского государственного технологического института (технического университета), 2015, no. 29, c. 83–88

[5] M. Kohlert, A. König, Large, high dimensional, heterogeneous multi-sensor data analysis approach for process yield optimization in polymer film industry, *Neural Computing and Applications*, 2015, vol. 26, no. 3, pp. 581–588

[6] M. Kohlert, A. König, Advanced polymeric film production data analysis and process optimization by clustering and classification methods, *Frontiers in Artificial Intelligence and Applications*, 2012, vol. 243, pp. 1953–1961

[7] T. Chistyakova, M. Teterin, A. Razygraev, C. Kohlert, Intellectual analysis system of big industrial data for quality management of polymer films, *Advances in Neural Networks*, ISNN 2016: 13th International Symposium on Neural Networks Proceedings, Cham: Springer, 2016, pp. 565–572, (Lecture Notes in Computer Science. Theoretical Computer Science and General Issues; vol. 9719)

[8] A. König, A. Gratz, Advanced methods for the analysis of semiconductor manufacturing process data, *Advanced Techniques in Knowledge Discovery and Data Mining*, Heidelberg: Springer, 2005, pp. 27–74

[9] M. Kohlert, O. Hissmann, Defects put to good use, *Kunststoffe International*, 2018, vol. 108, no. 6-7, pp. 34–37

[10] T. B. Chistyakova, A. N. Polosin, Computer modelling system for energy- and resource saving control of multi-assortment polymeric film production, *Cyber-Physical Systems: Design and Application for Industry 4.0*, Ed. by A. G. Kravets, A. A. Bolshakov, M. V. Shcherbakov, Cham: Springer Nature Switzerland, 2021, pp. 119–131, (Studies in Systems, Decision and Control; vol. 342)

[11] T. B. Chistyakova, A. M. Araztaganova, C. Kohlert, Computer system for thermal shrinkage polymer films obtaining processes control, *Herald of Technological University*, 2016, vol. 19, no. 17, pp. 101–104

[12] Т. Б. Чистякова, М. А. Тетерин, Программный комплекс для мониторинга и управления качеством полимерных пленок международной промышленной корпорации, Динамика сложных систем, XXI век, 2018, Т. 12, no. 3, c. 52–62

[13] T. B. Chistyakova, F. Kleinert, M. A. Teterin, Big data analysis in film production, *Cyber-Physical Systems: Advances in Design & Modelling*, Ed. by A. G. Kravets, A. A. Bolshakov, M. V. Shcherbakov, Cham: Springer Nature Switzerland, 2020, pp. 229–236, (Studies in Systems, Decision and Control; vol. 259)

[14] T. B. Chistyakova, A. N. Polosin, V. P. Meshalkin, A. S. Razygraev, Computer system for resource-saving control of multi-assortment production of polymeric materials based on processing big industrial data, *ChemChemTech*, 2021, vol. 64, no. 8, pp. 90–98

[15] T. B. Chistyakova, A. N. Polosin, M. A. Teterin, F. Kleinert, Computer system for processing industrial information for controlling the production of multi-assortment polymeric films, Proceedings of IV International Conference on Control in Technical Systems (CTS), [S. l.]: IEEE, 2021, pp. 184–187

[16] V. Gudivada, A. Apon, J. Ding, Data quality considerations for big data and machine learning: going beyond data cleaning and transformations, *International Journal on Advances in Software*, 2017, vol. 10, no. 1, pp. 1–20

[17] А. Н. Полосин, Г. К. Платоненко, М. А. Тетерин, Ф Кляйнерт, Программный комплекс для первичной обработки больших неструктурированных данных производства пленок, Математические методы в технологиях и технике, 2022, no. 12, ч. 2, c. 247–254

[18] G. Kauermann, H. Küchenhoff, C. Heumann, *Statistical Foundations, Reasoning and Inference for Science and Data Science*, Cham: Springer Nature Switzerland, 2021, 356 pp

[19] А. Дамрин, Т. Б. Чистякова, Архитектура программного комплекса статистической обработки больших промышленных данных для анализа ключевых показателей качества пленки, Математические методы в технологиях и технике, 2022, no. 12, ч. 2, с. 255–261

[20] А. О. Дамрин, Т. Б. Чистякова, Алгоритмы статистической обработки больших промышленных данных для прогнозирования ключевых показателей качества полимерных пленок, Математические методы в технологиях и технике, 2023, no. 9, с. 82–87

[21] L. Fahrmeir, T. Kneib, S. Lang, B. Marx, *Regression: Models, Methods, and Applications*, Heidelberg: Springer, 2013, 712 p

[22] S. Haykin, *Neural Networks and Learning Machines*, 3rd ed, New York: Pearson Prentice Hall, 2008, 936 p

[23] R. E. Schapire, The boosting approach to machine learning: An overview, *Nonlinear Estimation and Classification*, Ed. by D. D. Denison et al., New York: Springer Science + Business Media, 2003, pp. 149–171 (Lecture Notes in Statistics; vol. 171)

[24] T. Hastie, R. Tibshirani, J. Friedman, *The Elements of Statistical Learning: Data Mining, Inference, and Prediction*, New York: Springer Science + Business Media, 2009, 745 pp

[25] T. B. Chistyakova, M. A. Teterin, Data mining system for predicting quality of polymeric film, *Journal of Physics: Conference Series*, 2021, vol. 1902, 012045

[26] S. Hochreiter, J. Schmidhuber, Long short-term memory, *Neural Computation*, 1997, vol. 9, no. 8, pp. 1735–1780

[27] Ю. Е. Нестеров, Метод решения задачи выпуклого рограммирования со скоростью сходимости O(1/k2), Доклады Академии Наук СССР, 1983, Т. 269, no. 3, с. 543–547

[28] T. Dozat, Incorporating Nesterov momentum into Adam, Proceedings of the International Conference on Learning Representations, Puerto Rico, 2016, pp. 2013–2016

[29] М. А. Тетерин, Н. Д. Грищенков, Т. Б. Чистякова, Прогнозирование качества полимерных пленок с использованием рекуррентной нейронной сети, Математические методы в технологиях и технике, 2022, no. 2, с. 105–111

[30] Т. Б. Чистякова, М. А. Тетерин, А. С. Разыграев, Свидетельство о государственной регистрации программы для ЭВМ no. 2016662524 Российская Федерация. Программный комплекс для оценки значимости технологических параметров и трендового анализа производства полимерных материалов: no. 2016660083: заявл. 28.09.2016: опубл. 14.11.2016, К. Колерт

[31] Т. Б. Чистякова, С. И. Куликов, А. Н. Полосин, К. Колерт, ПО для управления толщиной каландрованных тонких материалов, Автоматизация в промышленности, 2007, no. 1, с. 8–11

[32] J. K. Good, D. R. Roisum, Winding: Machines, Mechanics and Measurements, Lancaster: Destech Pubns, 2008, 398 pp

[33] H. Gartner, A. Schnabel, C. Kohlert, Control systems for calenders—from mixer to take-off, *Kunststoffe Plast Europe*, 1997, vol. 87, no. 6, pp. 24–27

[34] Б. Ю. Лемешко, С. Б. Лемешко, С. Н. Постовалов, Е. В. Чимитова, Статистический анализ данных, моделирование и исследование вероятностных закономерностей, Новосибирск: Изд-во Новосиб. гос. техн. ун- та, 2011, 888 с

[35] R. C. Geary, Testing for normality, *Biometrika*, 1947, vol. 34, no. 3-4, pp. 209–242

[36] H. A. David, H. O. Hartley, E. S. Pearson, The distribution of the ratio, in a single normal sample, of range to standard deviation, *Biometrika*, 1954, vol. 41, no. 3-4, pp. 482–493

[37] А. Н. Полосин, Т. Б. Чистякова, Ю. И. Ивашинникова, Математические методы и программный комплекс для обработки производственных данных по толщине для дентификации брака рулонов полимерной пленки, Математические методы в технологиях и технике, 2023, no. 10, с. 91–99

[38] F. Kleinert, Analysis of inline thickness profile: Case study: presentation of private report, Klöckner Pentaplast Europe GmbH, Co. KG, Montabaur, 2022, 17 slides

[39] T. B. Chistyakova, A. N. Polosin, V. G. Ksenofontov, A. S. Razygraev, Mathematical models and data processing methods for improving the efficiency of high-tech production of polymeric films,

Proceedings of the International Conference on Information Processes and Systems Development and Quality Assurance (IPSQDA-2023), [S. l.]: IEEE, 2023 pp. 23–29

[40] О. С. Логунова, М. Б. Аркулис, Автоматизированная система оперативного календарного планирования многостадийного производства: математическая модель и программная реализация, Вестник Череповецкого государственного университета, 2021, no. 3, c. 18–37

[41] B. R. Sarker, A. Diponegoro, Optimal production plans and shipment schedules in a supply-chain system with multiple suppliers and multiple buyers, *European Journal of Operational Research*, 2009, vol. 194, no. 3, pp. 753–773

[42] M. Baldea, I. Harjunkoski, Integrated production scheduling and process control: a systematic review, *Computers & Chemical Engineering*, 2014, vol. 71, pp. 377–390

[43] M. Parente, G. Figueira, P. Amorim, A. Marques, Production scheduling in the context of Industry 4.0: review and trends, *International Journal of Production Research*, 2020, vol. 58, no. 17, pp. 5401–5431

[44] T. B. Chistyakova, O. E. Shashikhina, I. G. Kornienko, A. A. Plekhanov, Methods and technologies for improving the efficiency of multi assortment production optimal planning, Creativity in Intelligent Technologies and Data Science: Proceedings of the 5th International Conference (CIT&DS 2023), Ed. by A. G. Kravets, M. V. Shcherbakov, P. P. Groumpos, Cham: Springer Nature Switzerland, 2023, pp. 251–261, (Communications in Computer and Information Science; vol. 1909)

[45] О. Ю. Комягина, Т. Б. Чистякова, Программный комплекс для оптимального планирования производства многоассортиментных полимерных пленок, Вестник Саратовского государственного технического университета, 2012, Т. 1, no. 2, c. 379–384

[46] T. B. Chistyakova, A. S. Razygrayev, R. V. Makaruk, C. Kohlert, Decision support system for optimal production planning polymeric materials using genetic algorithms, Proceedings of the XIX IEEE International Conference on Soft Computing and Measurements (SCM'2016), [S. l.]: IEEE, 2016, pp. 257–259

[47] T. B. Chistyakova, O. E. Shashikhina, C. Kohlert, Computer system for optimal planning of multiassortment polymer films industrial production, Proceedings of the 2nd International Conference on Control Systems, Mathematical Modelling, Automation and Energy Efficiency (SUMMA), [S. l.]: IEEE, 2020, pp. 561–565

[48] О. Е. Шашихина, Т. Б. Чистякова, Компьютерная система для оптимального планирования полимерных производств, Известия Санкт-Петербургского государственного технологического института (технического университета), 2021, no. 59, c. 94–100

[49] T. B. Chistyakova, O. E. Shashikhina, I. G. Kornienko, A. A. Plekhanov, Optimal planning software package for use in the control system of flexible extrusion production of polymer materials, Proceedings of IV International Conference on Control in Technical Systems (CTS), [S. l.]: IEEE, 2021, pp. 188–191

[50] Т. Б. Чистякова, О. Е. Шашихина, Интеллектуальный программный комплекс моделирования процесса планирования многоассортиментных промышленных производств, Прикладная информатика, 2022, Т. 17, no. 5, c. 41–50

[51] T. B. Chistyakova, O. E. Shashikhina, I. V. Novozhilova, Intelligent information system for optimizing the production planning process, Proceedings of the International Conference "Quality Management, Transport and Information Security, Information Technologies" (IT&QM&IS-2022), [S. l.]: IEEE, 2022, pp. 267–270

[52] В. В. Курейчик, С. И. Родзин, Вычислительные модели эволюционных и роевых биоэвристик (обзор), Информационные технологии, 2021, Т. 27, no. 10, c. 507–520

[53] T. Tasic, B. Buchmeister, B. Acko, The development of advanced methods for scheduling production processes, *Journal of Mechanical Engineering*, 2007, vol. 53, no. 12, pp. 844–857

[54] А. П. Карпенко, Эволюционные операторы популяционных алгоритмов глобальной оптимизации. Опыт систематизации, Информационные и математические технологии в науке и управлении, 2018, no. 2, c. 22–30

[55] Е. А. Скорнякова, В. М. Васюков, В. Ш. Сулаберидзе, Особенности создания пользовательского интерфейса автоматизированной системы производственного планирования, Вестник

Казанского государственного технического университета им. А.Н. Туполева, 2019, Т. 75, no. 1, с. 52–56

[56] А. А. Плеханов, Т. Б. Чистякова, И. Г. Корниенко, Методы и технологии визуализации результатов оптимального планирования распределенного многоассортиментного производства полимерных материалов, Математические методы в технологиях и технике, 2023, no. 9, с. 76–81

[57] Т. Б. Чистякова, О. Е. Шашихина, К. А. Захаренко, А. А. Плеханов, К. Колерт, Свидетельство о государственной регистрации программы для ЭВМ no. 2021613630 Российская Федерация. Программныйкомплексдляоптимальногопланированияпроизводствамногоассортиментных полимерных пленок: no. 2021612447: заявл. 25.02.2021: опубл. 11.03.2021

[58] C. Kohlert, B. Schmidt, W. Egenolf, T. Zistjakova, Patent DE no. 102008032781. Verpackungsfolie für Produktauthentifizierung, Authentifizierungsverfahren und -system: no. 102008032781.6: Anmeldetag 11.07.2008: Veröff. 21.01.2010

[59] C. Kohlert, B. Schmidt, W. Egenolf, T. Zistjakova, Patent EP no. 2318286. Packaging Film for Product Authentication, Authentication Method, and System: no. 09776915.2: filing date 02.07.2009: publ. 25.01.2012

[60] C. Kohlert, B. Schmidt, W. Egenolf, T. Zistjakova, Patent PT no. 2318286. Packaging Film for Product Authentication, Authentication Method, and System: no. 09776915.2: filing date 02.07.2009: publ. 25.01.2012

[61] C. Kohlert, B. Schmidt, W. Egenolf, T. Zistjakova, Patent US no. 8578509. Packaging Film for Product Authentication, Authentication Method, and System: no. 12/997895: filing date 02.07.2009; publ. 05.11.2013

[62] C. Kohlert, B. Schmidt, W. Egenolf, T. Zistjakova, Patent CN no. 102089221 Packaging Film for Product Authentication, Authentication Method, and System: no. 2009801266792.4: filing date 02.07.2009: grant date 26.06.2013

[63] К. Колерт, Б. Шмидт, В. Эгенольф, Т. Чистякова, Патент no. 2490191 Российская Федерация. Упаковочная пленка для аутентификации продуктов, а также способ и система аутентификации: no. 2011105020/12: заявл. 02.07.2009: опубл. 20.08.2013

[64] C. Kohlert, M. Kohlert, T. Chistyakova, A. Ivanov, I. Sadykov, Counterfeit-proofing based on the principle of randomness, *Kunststoffe International*, 2010, vol. 100, no. 7, pp. 32–35

[65] M. Kohlert, C. Kohlert, A. König, Automated anticounterfeiting inspection methods for rigid films based on infrared and ultraviolet pigments and supervised image segmentation and classification, Lecture Notes in Computer Science (including subseries Lecture Notes in Artificial Intelligence and Lecture Notes in Bioinformatics), 2010, vol. 6276, pp. 321–330

[66] Т. Б. Чистякова, И. А. Садыков, К. Колерт, А. Б. Иванов, Методы кодирования и идентификации упаковок фармацевтической продукции для защиты от фальсификации, Информационные технологии, 2011, no. 7, с. 52–57

[67] Т. Б. Чистякова, И. А. Садыков, К. Колерт, Программный комплекс для защиты полимерных упаковок фармацевтической продукции от фальсификации, Известия Санкт- етербургского государственного технологического института (технического университета), 2011, no. 11, с. 64–68

[68] T. Chistyakova, R. Makaruk, I. Sadykov, C. Kohlert, Methods and technologies for protecting pharmaceutical products in polymer packaging from counterfeiting, *Smart Innovation, Systems and Technologies*, 2021, vol. 220, pp. 119–130

[69] Т. Б. Чистякова, Д. С. Тимошин, И. А. Кузнецов, А. С. Разыграев, Комплекс методов и средств синтеза мобильных приложений для защиты полимерных упаковок от подделки, Математические методы в технике и технологиях: сборник трудов международной научной конференции. В 12 томах. Том 8, Санкт-Петербург: Изд-во Политехн. ун-та, 2017, с. 38–44

[70] К. А. Захаренко, С. А. Кочанов, Т. Б. Чистякова, К. Колерт, Математические модели и алгоритмы для разработки мобильных программных комплексов защиты полимерной упаковки от подделки, Математические методы в технологиях и технике, 2022, no. 12, ч. 2, с. 231–234

[71] Т. Б. Чистякова, К. Колерт, Д. С. Тимошин, И. А. Кузнецов, А. С. Разыграев, Свидетельство о государственной регистрации программы для ЭВМ no. 2018618913 Российская Федерация.

Программный комплекс для проектирования мобильных приложений по защите полимерных паковок от подделок: no. 2018615434: заявл. 28.05.2018: опубл. 23.07.2018

[72] Т. Б. Чистякова, К. Колерт, И. А. Садыков, О. С. Албуткина, А. И. Масликов, Свидетельство о государственной регистрации программы для ЭВМ no. 2010614239 Российская Федерация. Программный комплекс для обеспечения защиты полимерных изделий от подделки: no. 2010612502: заявл. 12.05.2010: опубл. 30.06.2010

[73] Т. Б. Чистякова, И. А. Садыков, И. Ю. Волков, С. А. Тарасов, К. Колерт, Свидетельство о государственной регистрации программы для ЭВМ no. 2015610979 Российская Федерация. Программный комплекс идентификации полимерных упаковок с использованием мобильных устройств: no. 2014660431: заявл. 15.10.2014: опубл. 21.01.2015

[74] К. А. Захаренко, С. А. Кочанов, Т. Б. Чистякова, К. Колерт, Мобильное приложение для кодирования и идентификации упаковочных полимерных пленок, Математические методы в технологиях и технике, 2022, no. 2, с. 78–81

[75] С. А. Кочанов, Р. Э. Власов, К. Колерт, Учебно-исследовательский комплекс кодирования и идентификации полимерных упаковочных материалов для защиты продукции от фальсификации, Т. Б. Чистякова, К. А. Захаренко, Математические методы в технологиях и технике, 2023, no. 8, с. 41–47

[76] Patent DE 10 2008 032 781 A1; WO 2010/003585 A1 from January 14, 2010

5 Measurement and Control Algorithms

Christian Kohlert

5.1 Measurement Technology

The development of the production of polymer films has been closely accompanied by new measuring methods. These developed from hand measurements of thickness, to off-line measurements in the laboratory, to in-line measurements in production facilities. Today's trend is to collect all measurement data and evaluate large data sets ("big data" and "data mining") for automation and process optimization.

The state of the art in all manufacturing processes is in-line thickness measuring systems on a radiometric basis (krypton-85) for measuring film thickness and width, as well as surface defects via CCD cameras. More recent developments include fill level measuring systems using pulse transit time measurements, shrinkage measuring systems using polarized light, gloss and color measuring systems, and systems for measuring film width (see Figure 5.1 and Figure 5.2).

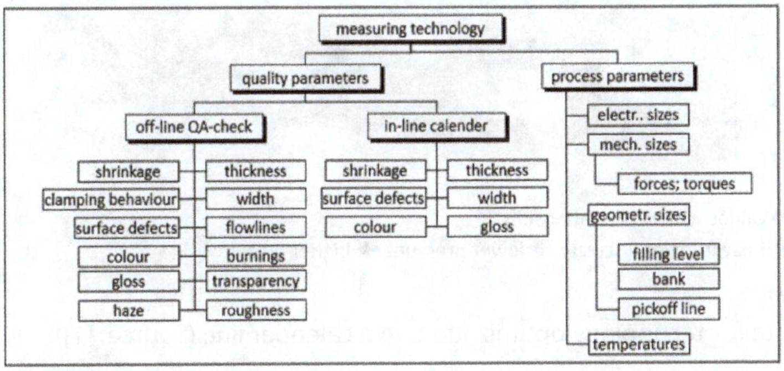

Figure 5.1 Measurement technology status when calendering polymer films (Source: [1])

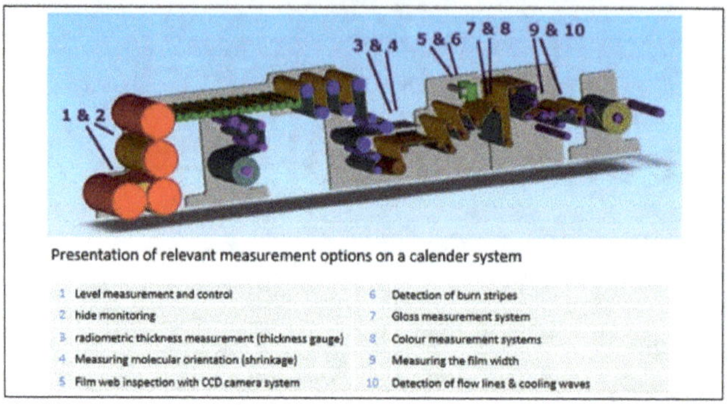

Figure 5.2 Modern measuring systems on calender lines (Source: [1])

Figure 5.3 shows the large number of factors influencing the throughput with the necessary quality. The data ranges from material, machine, and process data to quality data, some of which can be measured or determined mathematically.

For a calender line there are over 100 different data variables, which are linked to each other with around 800 connections. For example, opening a window in winter lowers the ambient temperature and thus lowers the film surface temperature, which in turn leads to an increase in viscosity and pressure on the rollers, which as a result increases the film thickness and shrinkage. Figure 5.4 shows a selection of the measurable variables on the aggregates of a calender line from the mixer to the aftertreatment. Figure 5.5 shows standard and special measurement technology which is part of the state of the art, and is also used for special calenders with a focus on special films.

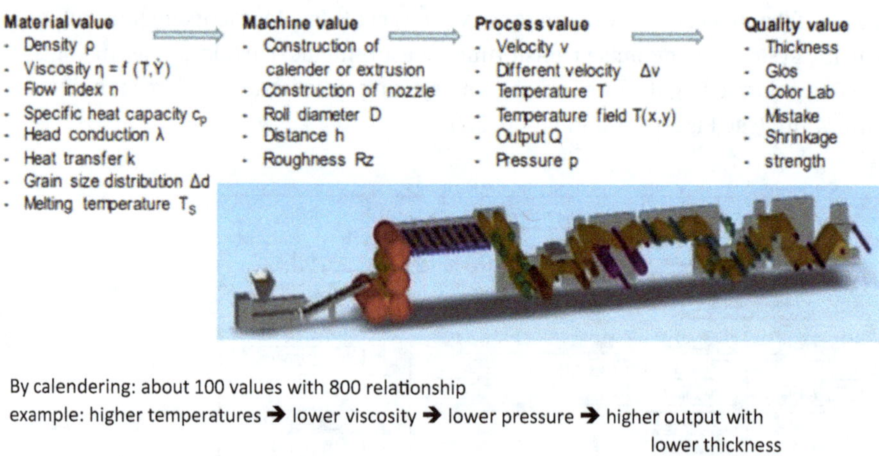

Figure 5.3 Usable data for technology optimization on a calender line (Source: [1])

part of the unit	measurable / current variables	part of the unit	measurable / current variables
mixer	• rotational speed • torque • initial weight • mixing temperature (start / end) • mixing time • internal pressure / suction vacuum	calender rollers	• temperature** • rotational speed / friction* • torque • power consumption • positioning steps gap adjustment • gap pressure DS (HS • Axis crossing / roll bending roll#2 & roll#4 (pressure) • level in the first roller gap
fluxing unit	• rotational speed intake hopper screw & kneader screw* • torque intake hopper screw & kneader screw* • fluxing power (kneader-power) • knife speed • temperature zone 1 to 4** • level in the intake hopper • vacuum suction	pickoff unit	• temperature** • rotational speed* • torque • power consumption • Film thickness • Film quality (molecule orientation, OCS)
		post calender equipment	• Same as pickoff unit

* Rotational speeds of linear speeds in m/min are current and can be documented with deviations etc. via PLC.
** For roll temperatures, flow and return temperatures are available on the units.

Figure 5.4 Measurable quantities on the individual aggregates of a calender line (Source: [1])

The data available on a calender line is usually stored in a decentralized manner on the individual measuring systems (Figure 5.6). Figure 5.7 shows an example of what such measurement data can be used for. This ranges from direct error evaluation to process analyses to possible further processing for process optimization.

Ultimately, all this data must be brought together in a server to enable data evaluation considering all influencing factors (Figure 5.8). Through this combination, quality data such as occurrence of "black dots" can be traced back to influencing factors, which can then be improved with knowledge of this connection (Figure 5.9). Such measurement systems with data collection and evaluations then enable production optimization with regard to the customer's quality requirements (Figure 5.10).

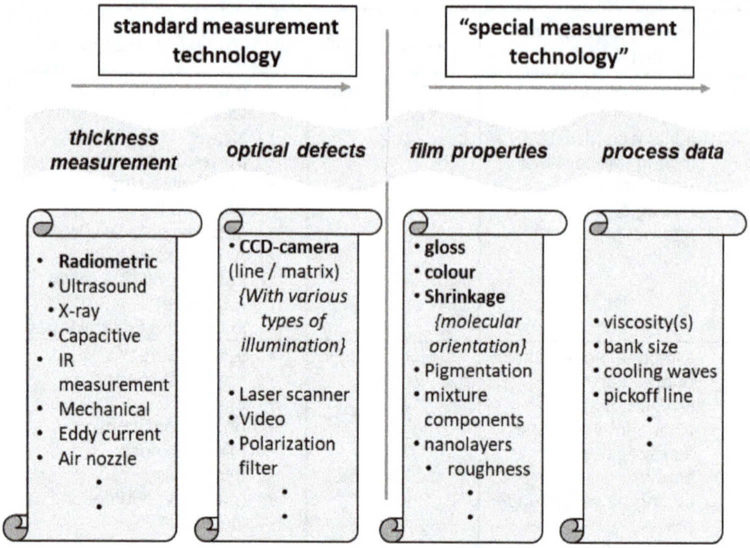

Figure 5.5 Standard and special measurement technologies on a calender line (Source: [1])

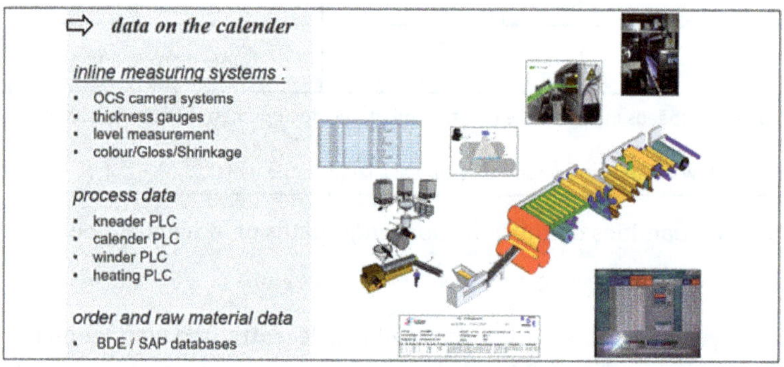

Figure 5.6 Available data on a calender line (Source: [1])

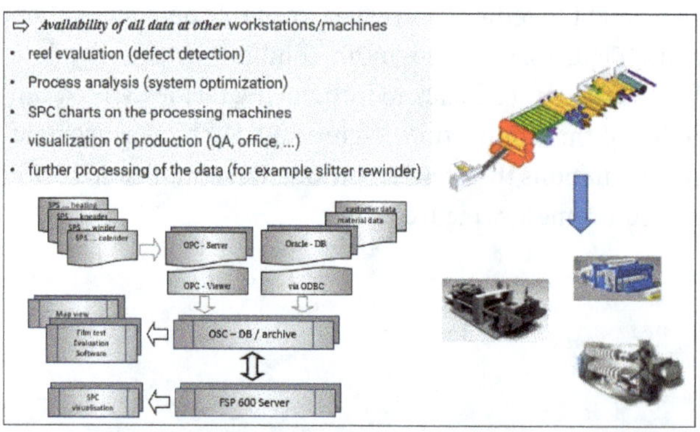

Figure 5.7 Use of the measurement data (Source: [1])

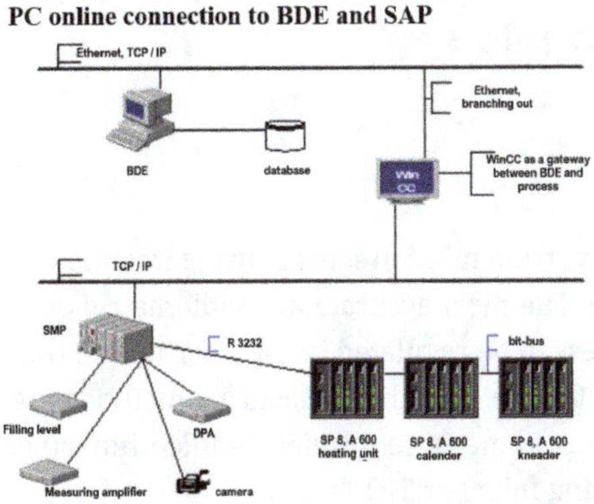

Figure 5.8 Data connections (Source: [1])

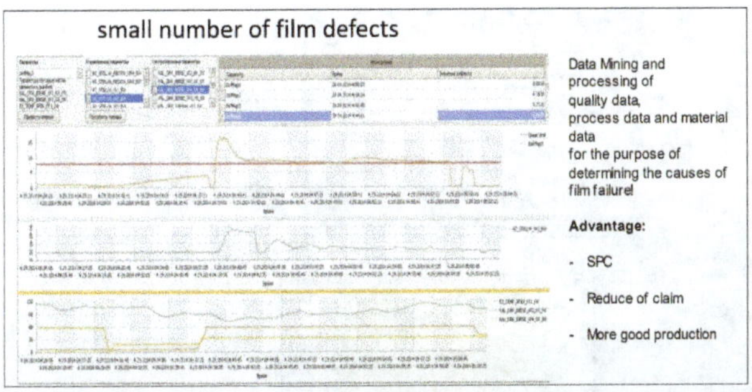

Figure 5.9 Data evaluation depending on influencing factors (Source: [1])

What are the benefits of inline measurement systems and data analysis?

production	⇒	Clear information about the current quality
	⇒	Direct influence on production through: *Detection - marking - blocking and reaction(s)*
	⇒	Optimization of production
	⇒	Further development of the product / process
quality management	⇒	Archiving and data management
	⇒	DIN (ISO) standard compliance / auditability
controlling	⇒	Cost savings (material, machine, 'personnel')
	⇒	Exact production benchmark with trends
... all ...	⇒	satisfied customers

Figure 5.10 Benefits of the measurement systems and analyses (Source: [1])

5.2 On-line Measurement Technology

5.2.1 Thickness Measurement

Thickness measuring systems include capacitive, triangulation sensors using laser beams or radiometric and X-ray devices (Figure 5.11). The most accurate are radiometric ones using krypton-85, which allow the film thickness to be regulated in the micrometre (micron) range, with a measurement accuracy of 0.1 µm. All other measuring methods are either inaccurate or require complex operations to measure unfilled vs filled films (i.e. they have measurement inaccuracies when using fillers or TiO_2).

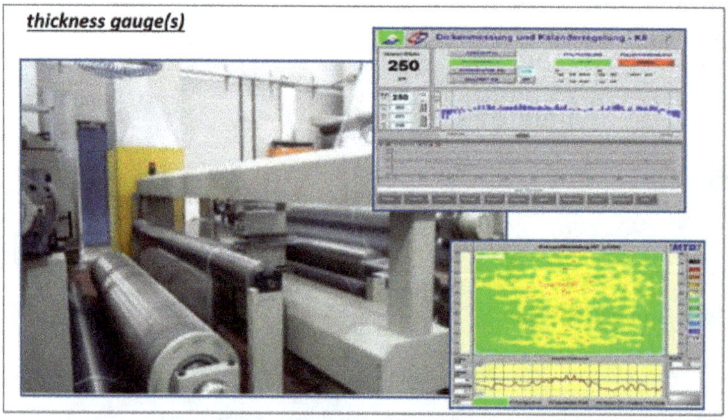

Figure 5.11 Thickness measuring system (Source: [1])

5.2.2 Measurement of Surface Defects

CCD cameras with recording rates of up to 100 kHz and usually 4000 pixels are now used to measure surface defects such as black dots, specks, holes, streaks, or wrinkles. This means that at a speed of 180 m/min (3000 mm/s) a length recording of 30 µm is covered, and with a measuring width per camera of around 1 m, a measuring width of 250 µm is covered. This would result in a measuring spot of 7500 µm² or 0.0075 mm². Since this figure should be doubled to ensure reliable detection, errors of 0.015 mm² or more could be reliably detected in this example. This can of course be optimized with the number of cameras and their recording speed (Figure 5.12).

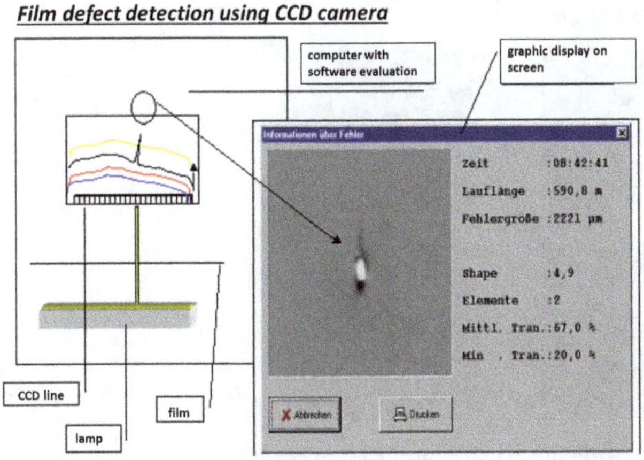

Figure 5.12 Film defect detection using a CCD camera (Source: [1])

These camera systems are available for transmitted light or reflected light on covered films. As a rule, the cameras are installed at a distance where they cover around 1 m of film width. For example, for a 2500 mm wide film three cameras are needed (Figure 5.13). By recalculating the grey values of the individual pixels of a CCD camera, the errors can be made clearly visible to the eye, thereby increasing the confidence of the operator staff (Figure 5.14). Various error classifications are shown in Figure 5.15. By using top and lower thresholds for the grey levels of the video signal, normal noise in a video signal can be hidden.

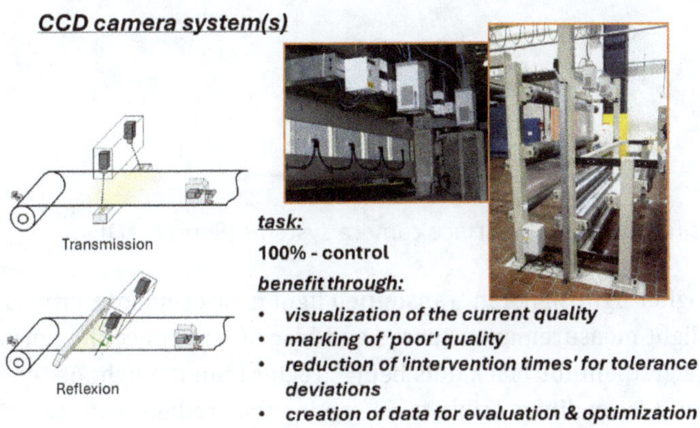

Figure 5.13 CCD camera systems (Source: [1])

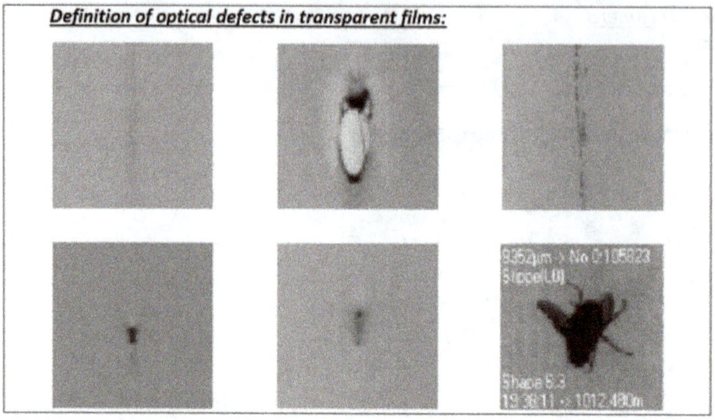

Figure 5.14 Optical errors detected with a CCD camera (Source: [1])

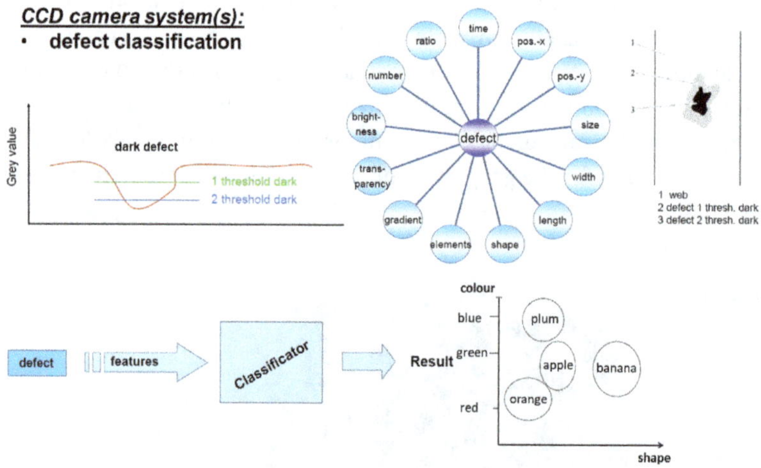

Figure 5.15 Error classification in CCD surface camera systems (Source: [1])

Holes always send a higher light signal in transmitted light measurements, or no signal at all in reflected light measurements. Specks and black dots reduce the normal light intensity, with the gradient for black dots being greater than the light decay for specks. This means that it is readily possible to determine the gradient with two different thresholds, and thus distinguish between black dots and specks. In addition, unacceptable and acceptable error sizes can be agreed with customers based on the error sizes and their form. Films free from the smallest surface defects cannot be produced inexpensively.

Many optimizations of surface error measurements occur by optimizing the lighting technology. This ranges from simple neon tubes for simple measurements to cold light lamps and LED lamps. The light color also has an influence on the measurement accuracy. The best way to detect incipient burns is with blue light (Figure 5.16). To validate such optical systems in accordance with ISO standards, i.e. to check the required measurement accuracy, samples with prefabricated validation points are often used (Figure 5.17).

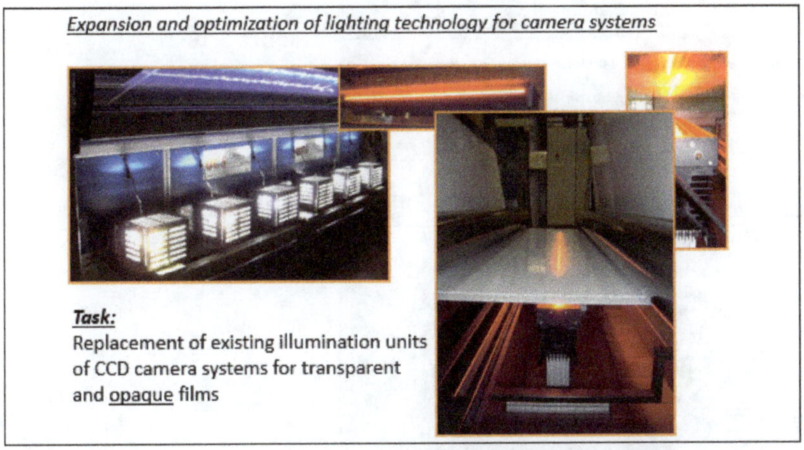

Figure 5.16 Illumination technology for CCD cameras (Source: [1])

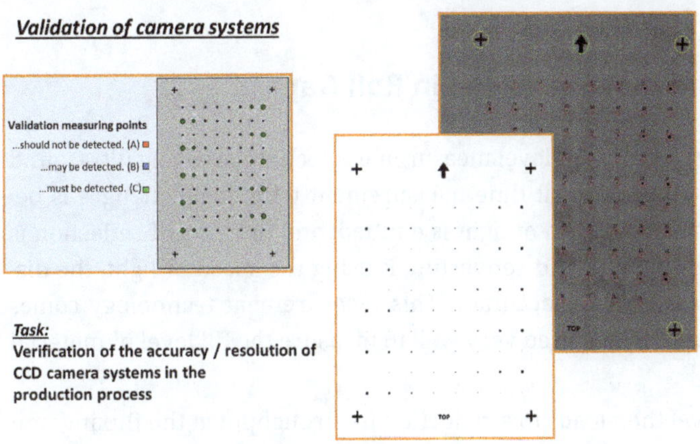

Figure 5.17 Validation of camera systems (Source: [1])

To make it easier for the operator to evaluate the large number of different surface defects, a summary quality indicator can be determined based on the significance of the evaluation of individual defects. This shows the average rating of the film quality (Figure 5.18). The representation can be used in color or with a film grade (Figure 5.19).

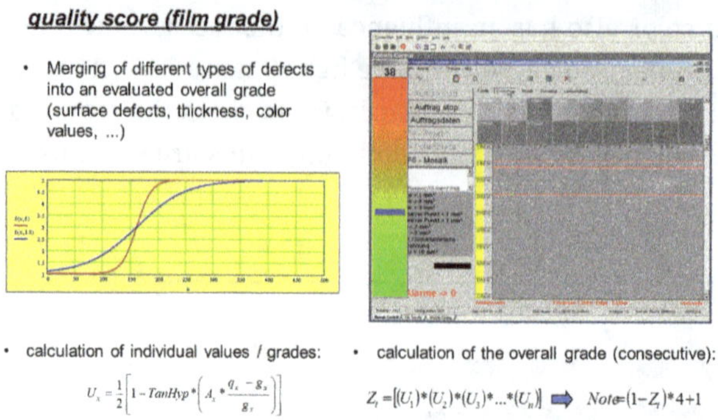

Figure 5.18 Quality indicator (Source: [1])

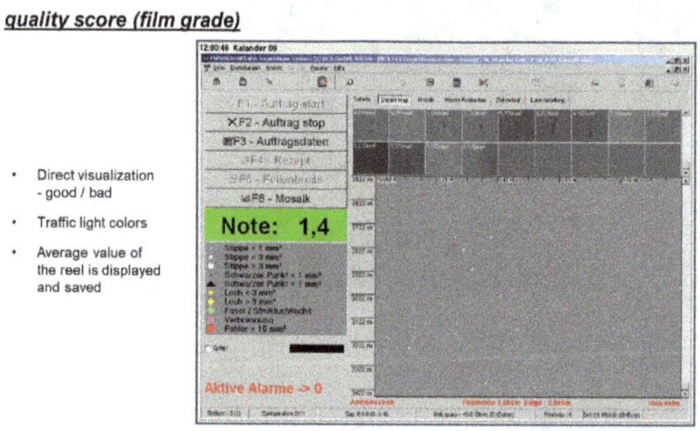

Figure 5.19 Film grade (Source: [1])

5.2.3 Measuring the Material Level in Roll Gaps

In addition to the conventional point level measurement techniques using ultrasound or triangulation sensors, pulse transit time measurement with angle changes is becoming increasingly popular. A beam of light is emitted, and the time of reflection is measured. After halving this time and converting it using the speed of light, the distance can be defined to millimetre accuracy. This measurement technology comes from safety technology and can be used very well to measure the fill level of material in the roll gap.

An increase in the fill level then leads to a reduction in throughput at the fluxing unit and vice versa. In the event of a drastic drop in the material level in the first roll nip, an alarm or an emergency stop of the entire calender line can be implemented to avoid damage to the calender rolls due to roll contact (Figure 5.20 and Figure 5.21).

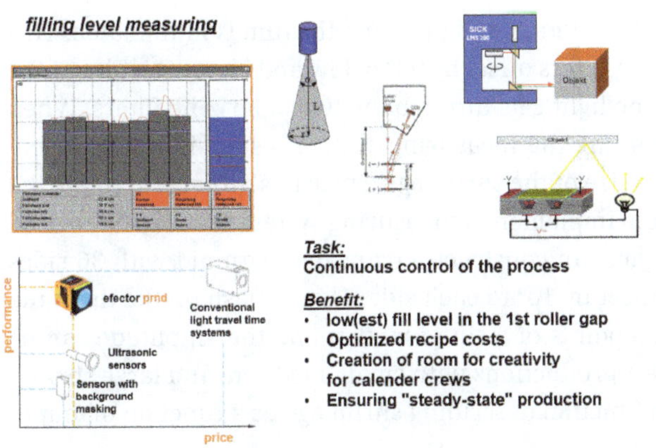

Figure 5.20 Level measurement technology (Source: [1])

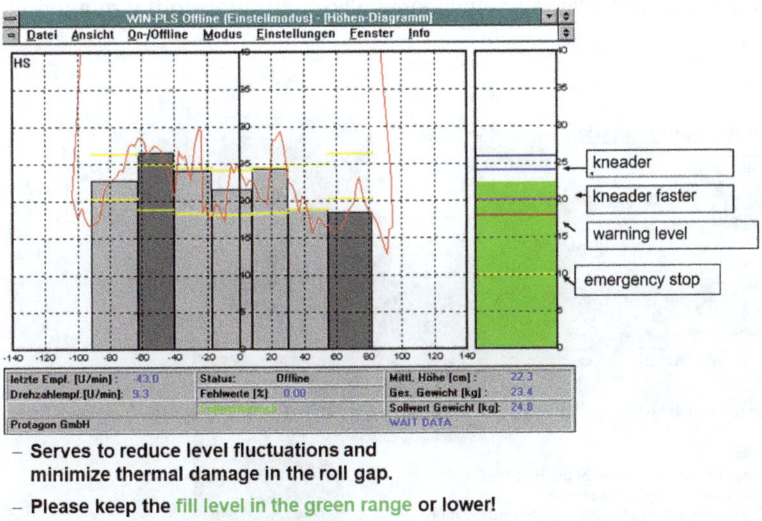

Figure 5.21 Level measurement and control (Source: [1])

5.2.4 Measuring Film Shrinkage

In addition to the desired film thickness, film shrinkage is an important quality factor, as it defines material optimization during subsequent deep-drawing. If it is too large, there may be material problems at the edge of the deep-drawn products; if it is too small, there will be too much edge waste.

The shrinkage correlates with the rotation of light when the film is illuminated. This can be measured using polarizing filters offset by 90° below and above the film. If the molecules are oriented evenly, no light gets through the two polarizing filters. When the molecules are oriented, rotating the measuring system results in a light maximum, which correlates with the size of the shrinkage. Since it is difficult to rotate the calender line, and it is difficult to implement a measuring system that can be rotated under production conditions, you can construct a truncated pyramid with 36 sides and attach a polarizing filter offset by 10° to each side. Then a flash on the tip of the truncated pyramid results in 36 points of a sine function and the amplitude can be determined mathematically. Test productions with small, medium, and large shrinkage then result in a calibration function describing shrinkage as a function of amplitude.

Unfortunately, this system is only suitable for amorphous polymers, since the generally unknown crystallinity of semi-crystalline polymers also contributes to the rotation of light (Figure 5.22).

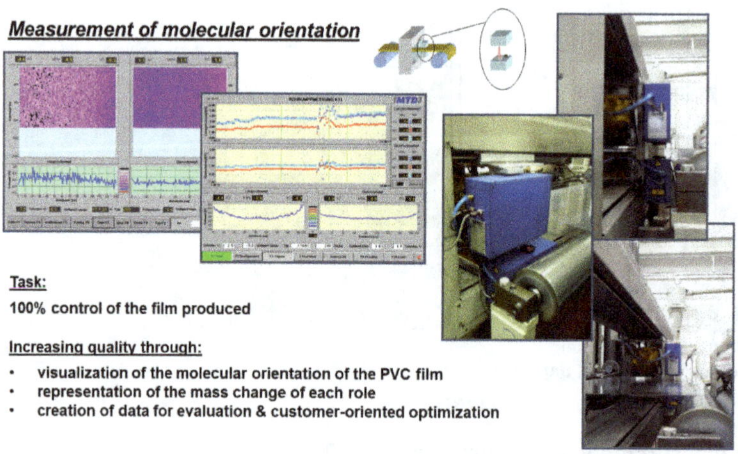

Figure 5.22 Measuring the molecule orientation (Source: [1])

5.2.5 Measurement of Gloss and Color

Modern calendering systems also have measuring devices from laboratory practice for on-line measurements of gloss and color. These are slightly modified to inform the operator directly of the customer requirements for gloss and color that must be met, so that they can intervene immediately in the event of deviations (Figure 5.23).

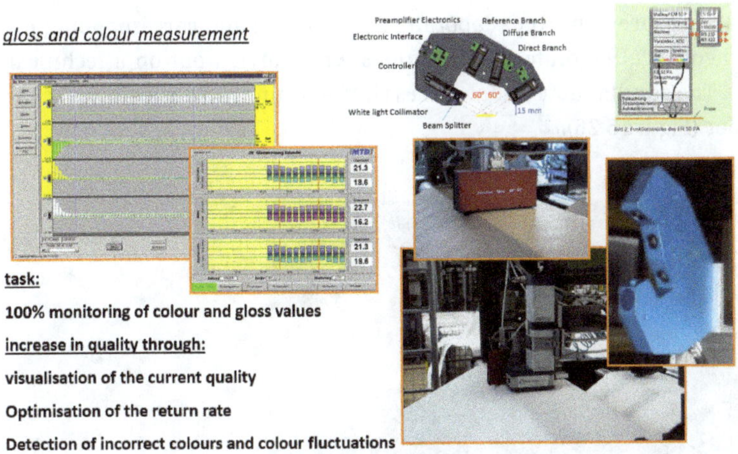

Figure 5.23 Measurement of gloss and color (Source: [1])

5.2.6 Measurement of Material Viscosities

The viscosity of the recipe to be processed is crucial for the calendered thickness and the shrinkage properties, i.e. for the most important quality parameters of a polymer film. Normally, the viscosity is determined using rheological laboratory measuring devices such as capillary viscometers or Weissenberg viscometers for certain rheological state functions depending on the shear rate and process temperature.

Since the degree of thermal damage and the exact values for deformation rate and temperature are of course not known during the processing process, it is not possible to specify an exact viscosity. Therefore, a process viscometer within the calendering process is very desirable. At Schnabel, the possibility of a process viscometer is being tested by measuring the torque on the cutting rollers that cut the material (Figure 5.24).

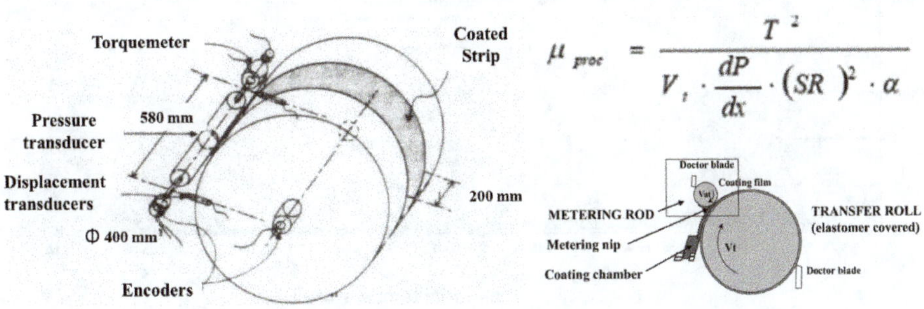

Figure 5.24 Schematics of a cutting roller for measuring the torque depending on different viscosities (Source: [2])

Using experimental measurements on a laboratory rolling mill and corresponding investigations in a rheological laboratory, calculations were carried out on a technical scaler (Figure 5.25). The results show initial success in determining the rheological conditions on a calender (Figure 5.26).

Figure 5.25 Cutting roller for determining process viscosity on a technical calender (Source: [2])

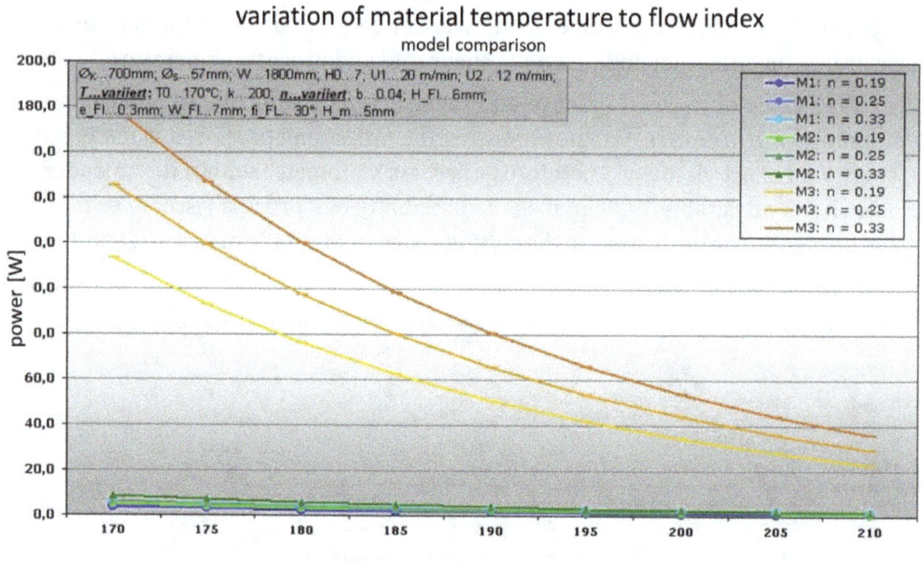

Figure 5.26 Influence of the material temperature for different materials (flow index) on the performance of a test cutting roller (Source: [2])

5.2.7 Measurement of Surface Temperatures

Using infrared cameras, it is possible to determine the surface temperatures of the film very precisely, and so draw conclusions about temperature problems on previous rollers (Figure 5.27). Since temperature has a direct effect on shrinkage and film thickness, uniformity of temperature is one of the most important basic requirements for achieving uniform film properties across the width and length of the film. Using such measurements, blockages in peripheral bores of the calender rolls can also be defined.

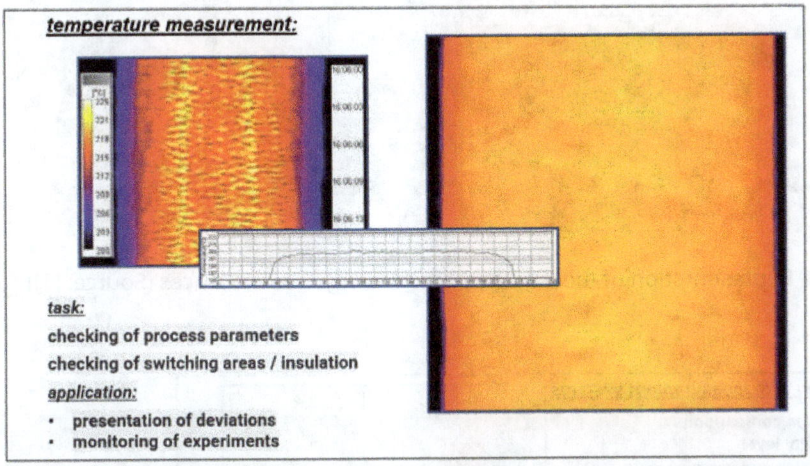

Figure 5.27 Temperature measurement of the film using an IR line or matrix camera (Source: [1])

5.2.8 Presentation of the Measurement Results

Since the measurement results are important for both the operator and for quality control, they are usually summarized on a monitor and displayed in various places (Figure 5.28).

The measurement results should also be made available to the operator and process engineer as a trend or statistical evaluation for long-term evaluation (Figure 5.29).

The aim should be to record, collect and archive all data on a system, display it and make trends and dependencies available in a timely manner. This is the only way errors can be identified and avoided at an early stage (Figure 5.30).

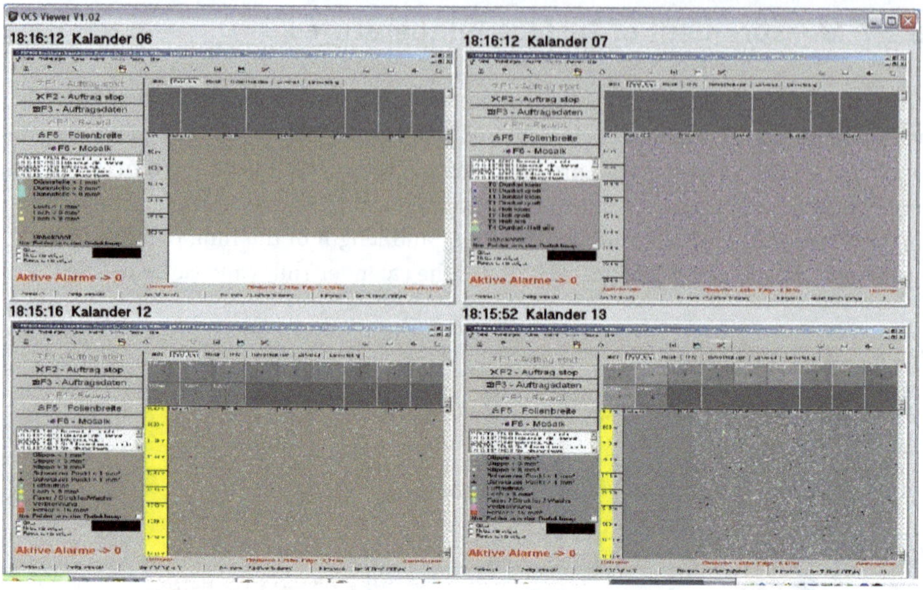

Figure 5.28 Representation of measurement technology in many places (Source: [1])

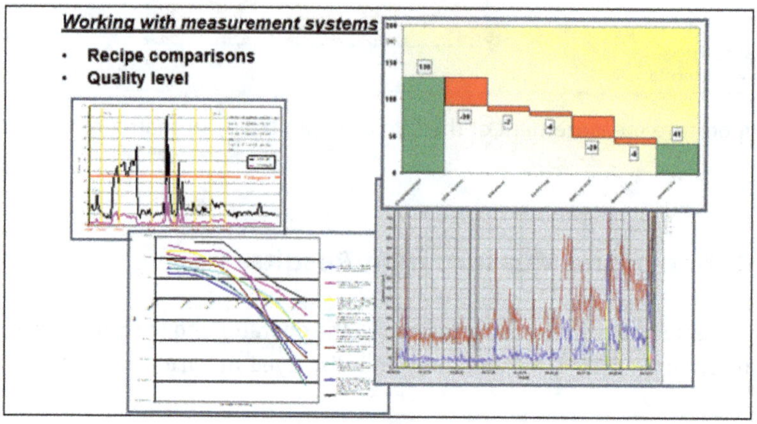

Figure 5.29 Working with the measurement results (Source: [1])

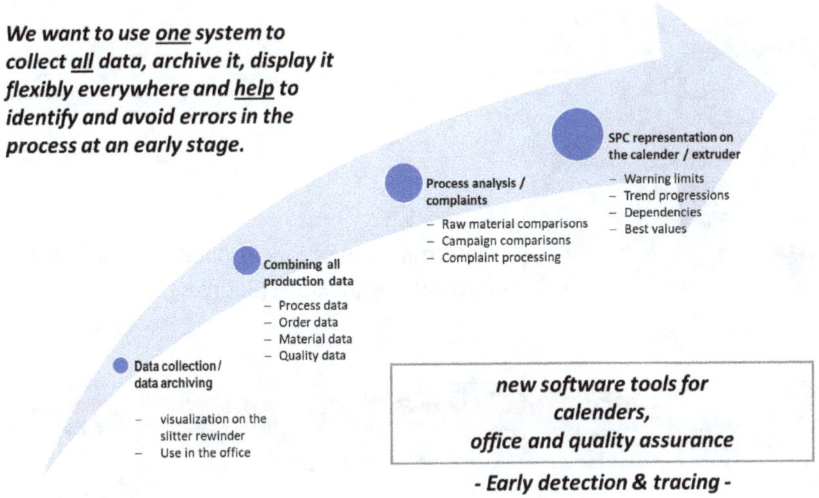

Figure 5.30 Data acquisition until presentation (Source: [1])

5.3 Off-line Measurement Technology

5.3.1 Determination of the Causes of Thickness Fluctuations

The measurement of longitudinal thickness fluctuations with simultaneous entry of the diameters and linear speeds of the individual rolls of the calender, as well as the successor, allow the causes of such fluctuations to be recalculated using a Fourier analysis. These can result from diameter fluctuations or bearing problems of individual rollers (Figure 5.31 and Figure 5.32). In Figure 5.33 a problem in the 5th reel of a 5-roller calender is shown.

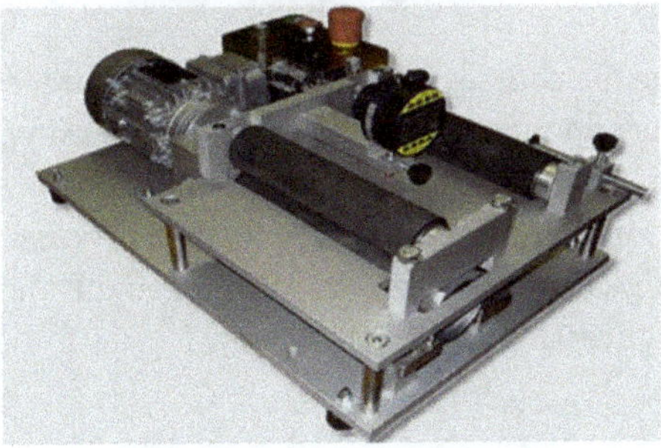

Figure 5.31 Determination of longitudinal film thickness fluctuations (fixed) (Source: [3])

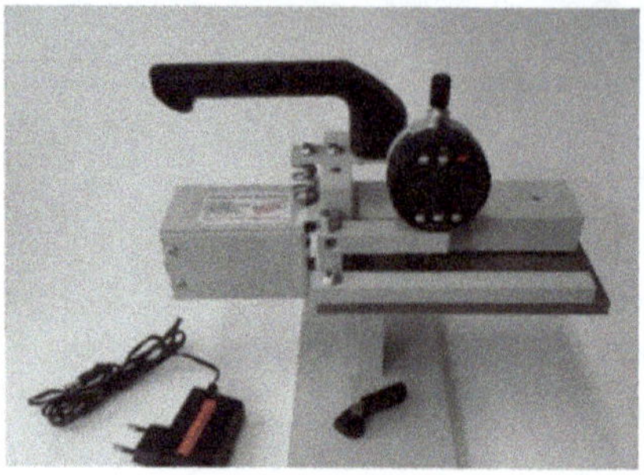

Figure 5.32
Determination of longitudinal film thickness fluctuations (flexible) (Source: [3])

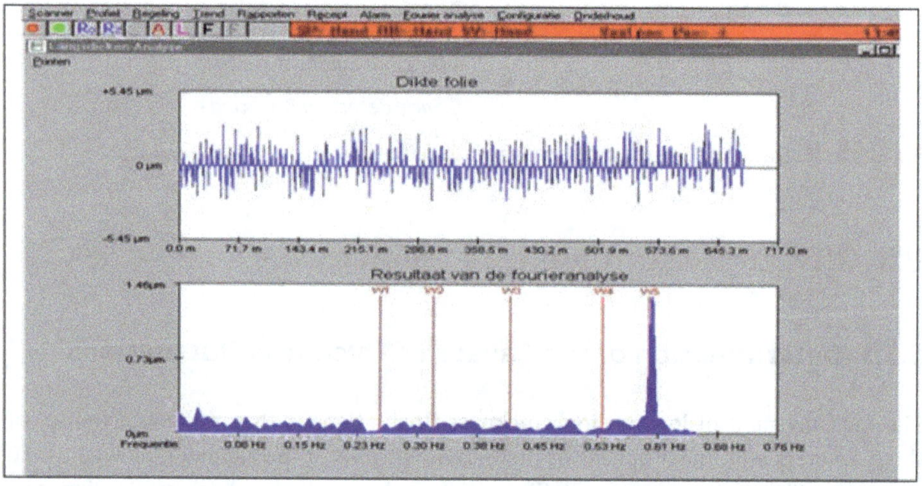

Figure 5.33 Longitudinal thickness profile with assignment of thickness fluctuations to the individual reels of a calender line

5.3.2 Measurement of Shrinkage and Shrinkage Force

Figure 5.34 shows a heating oven to determine the course of shrinkage of a film under the influence of heat using a camera recording and the effective shrinkage force using a force sensor, and Figure 5.35 shows creation of the sample patterns using a punch. Both the shrinking curve and the shrinking force are important for further processing steps, such as the constant width of the edge trim and deep-drawing speed during the deep-drawing process (Figure 5.36 and Figure 5.37).

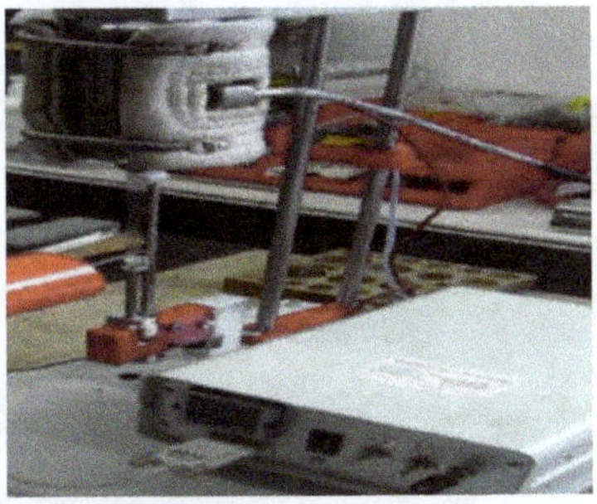

Figure 5.34
Shrinkage measuring oven (Source: [3])

Figure 5.35
Sample creation for shrinkage measuring furnace (Source: [3])

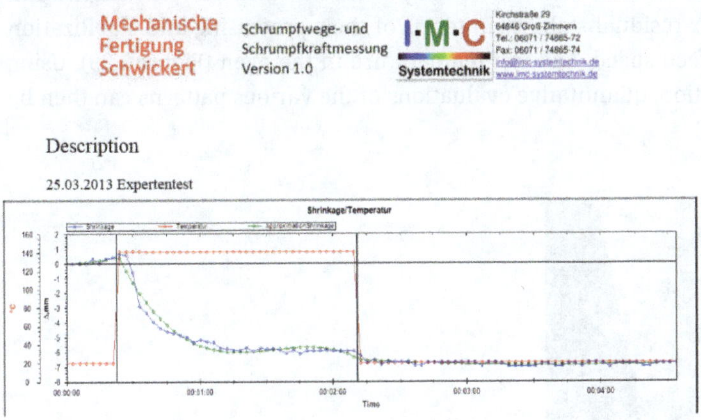

Figure 5.36 Shrinkage curve of a calendered film

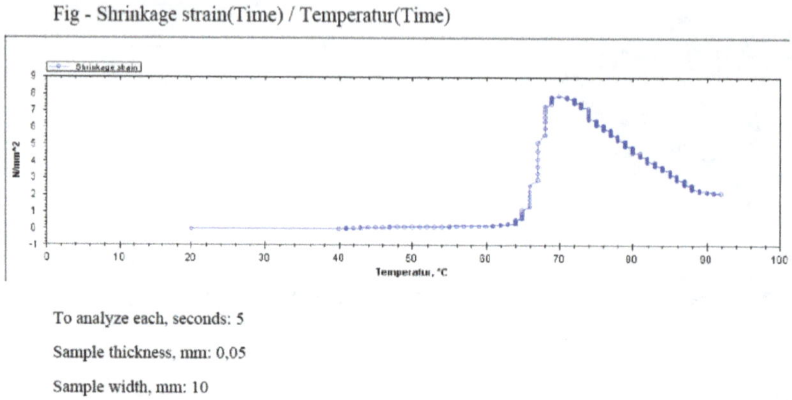

Figure 5.37 Shrinkage force curve of a calendered film

5.3.3 Residual Stability of PVC Films

The residual stability of PVC films is crucial for optimal further processing using heat, for example when deep-drawing. Too little residual stability then leads to combustion symptoms such as brown discoloration, whereas a residual stability that is too high is uneconomical.

Figure 5.38 shows an automatic residual stability oven, in which various samples can be evaluated for their residual stability in terms of their processing and stabilization using film patterns, feed speed, and test temperature in the oven (Figure 5.39). Using camera-based evaluation, quantitative evaluations of the various patterns can then be achieved (Figure 5.40).

Figure 5.38 Automatic residual stability oven (analogous to the Mathis oven) (Source: [3])

Figure 5.39 Residual stability representation of various samples after the automatic residual stability oven

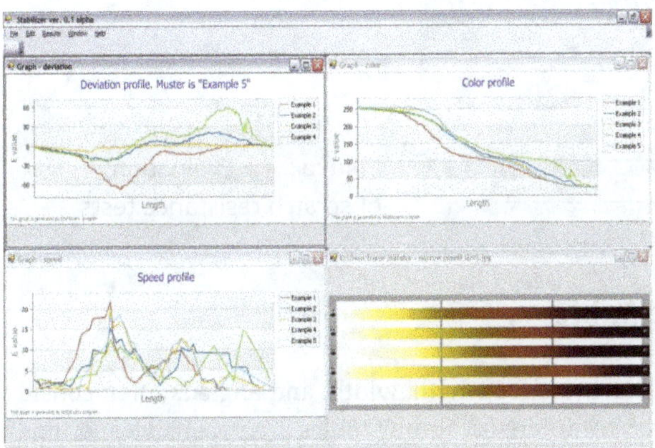

Figure 5.40 Stability pattern evaluation using a camera

5.3.4 Scratch Resistance Ratings

Many film surfaces are susceptible to scratches from use or transport, but different types of polymers or coatings can lead to improvements based on customer requests. To be able to evaluate this quantitatively, measuring systems have been developed (Figure 5.41). Using a camera evaluation, improvements can be shown as a comparison (Figure 5.42).

Figure 5.41
Scratch resistance measuring device (Source: [3])

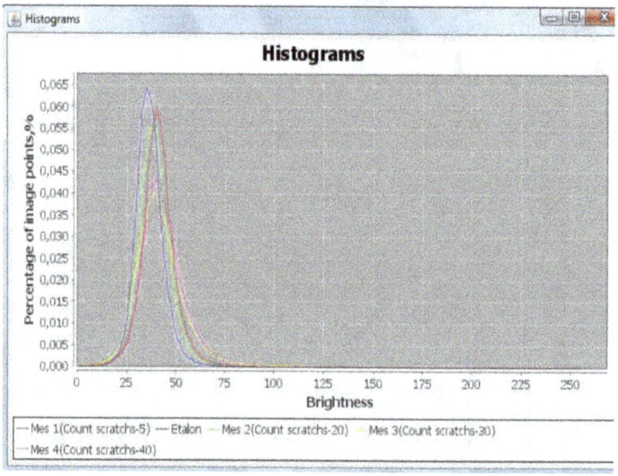

Figure 5.42 Evaluation of scratch resistance tests

5.3.5 Format Measuring Table

When it comes to formats, the exact size for length, width, and angle is often considered a quality criterion. However, traditional scanning tables are expensive and time-consuming to use. The format measuring table from MFS (Figure 5.43) allows the length, width and four angles of different formats to be determined quickly and inexpensively, thanks to the precisely assigned position of four chambers.

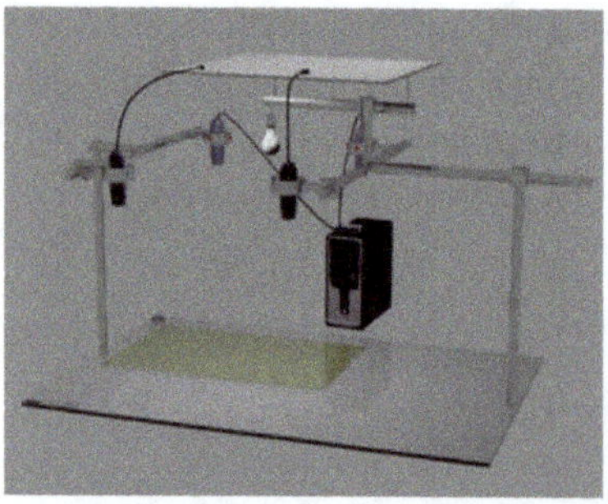

Figure 5.43
Format measuring table (Source: [3])

5.4 Control Algorithms

Based on measurement technology and computer technology, there have been many developments in the field of control technology in the last few decades (Figure 5.44).

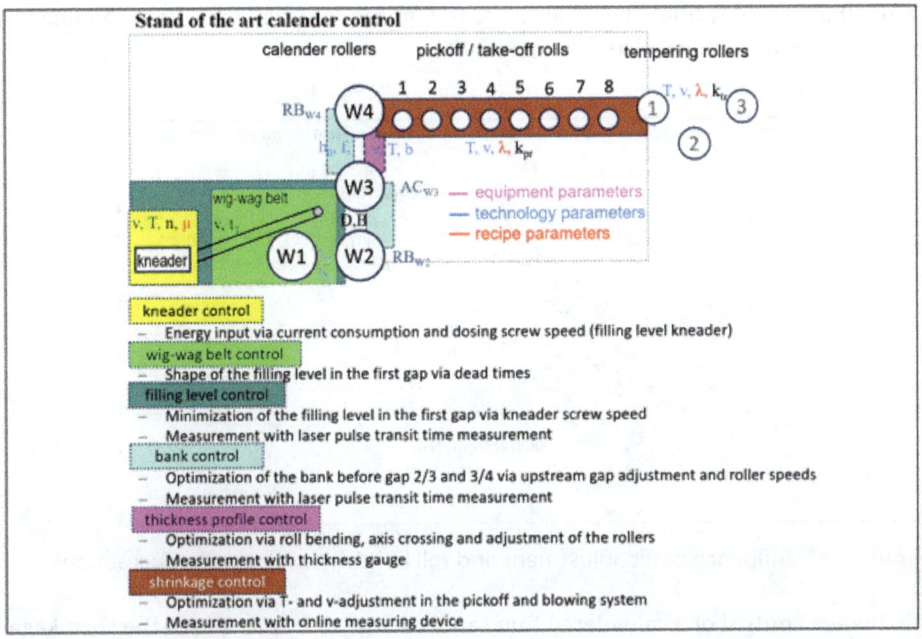

Figure 5.44 Modern calender controls

Energy input control can be carried out using the Buss kneader, which allows the energy input to be kept constant depending on the throughput and the wall temperature via the ratio of the speeds of the Stöbert hopper (intake hopper) and the main screw. This means that even if there is a short-term change in throughput due to a change in the thickness or width of the subsequent calender, a consistently good degree of plasticization of the polymeric material can be achieved.

If an output increase is necessary, the resulting reduction in the residence time of the material in the kneader or extruder is compensated for by a higher filling level of the fluxing unit, by increasing the revolution per minute (rpm) of the intake hopper. Subsequent conveyor belt controls allow the first nip to be filled evenly, considering material increases at the edges due to the wide flow effect between the calender rolls. For this purpose, the dwell time at the edges of the first nip at the stock guides is optimized.

Level control using the measuring options described under Section 5.2 can be carried out in order to control the fluxing unit. The aim is to optimally fill the material, which prevents the roll gap from running empty with subsequent possible roll damage. At the same time, it minimizes the dwell time due to the high level of material stabilization required. The banks of the subsequent calender gaps can also be regulated in a similar way.

Thickness cross-profile control, via the position of the rollers and optimization of the compensation options for roll bending and axis-crossing, is already offered in most thickness gauges. Here, pre-calculation of the optimal values makes sense, since control based on a compensation method or via an operator can only lead to local minima in the thickness deviation, while absolute minimization usually results in simultaneous optimization of both methods (Figure 5.45).

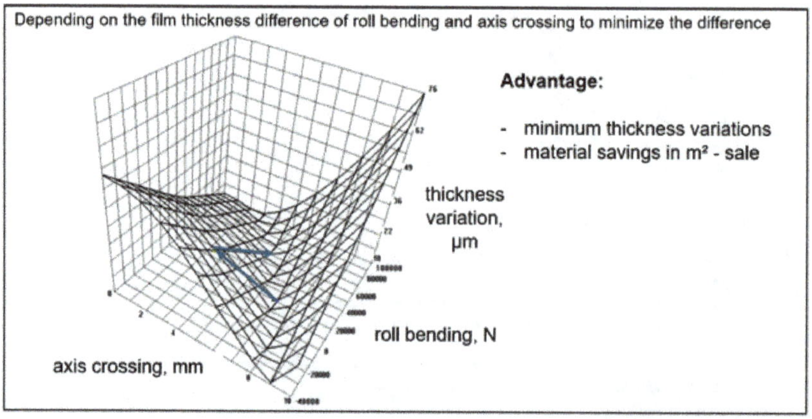

Figure 5.45 Influence of tilt adjustment and roll bending on thickness fluctuations

Shrinkage control of a calendered film can be achieved by measuring the shrinkage using polarizing filters, or pre-calculating the shrinkage depending on the rheological parameters, speeds, and temperature profiles (see Section 5.2). Instead of measurement, you can also use pre-calculation for control (see Figure 5.46 and Section 4.2).

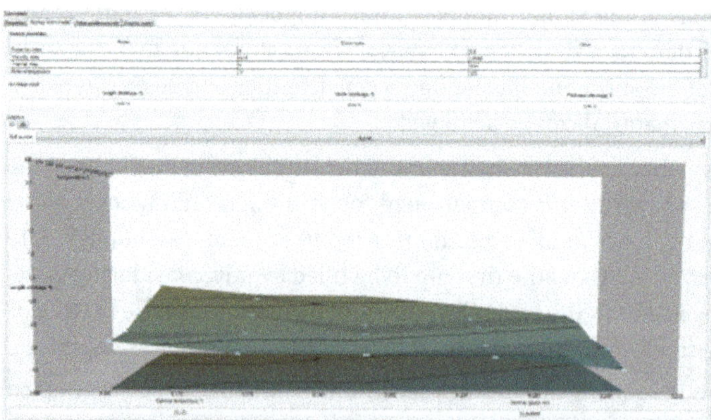

Figure 5.46 Calculated longitudinal shrinkage curve

References for Chapter 5

[1] Kohlert, C.: [lecture], Klöckner Pentaplast

[2] Schnabel, A.: Entwicklung eines Inline-Messverfahrens zur Bestimmung und Nutzung prozessrelevanter Materialeigenschaften an Kalander-Anlagen [Dissertation], TH Merseburg, 2010

[3] Mechanische Fertigung Schwickert [Company material], Oberahr, 2023

6 Properties of Polymeric Films

Christian Kohlert

6.1 Material and Processing Properties

Polymer films used in packaging are characterized by the lowest possible constant film thickness, which is sufficient for the necessary barrier properties, strength for product safety and transport, and printability for necessary product and safety information.

The following four figures show comparative properties of different polymer films. These properties are achieved either through the synthesis of the polymers or through the production of the recipes for further processing into films using calendering or extrusion. The comparisons are defined either by numerical values or by packaging experts.

PVC films, for example, can be tailored very well to customer requirements through their formulation, e.g. for UV protection, strength, surface modification, or antistatic treatment. PP and PS are also easily modified, while PET is more difficult to adapt to customer requirements.

Because of the manner of their production, PVC and PET have higher densities (1.3 g/cm^3) compared to PS and PP, meaning that the area yield for PS and PP is in principle up to 40% greater. However, this only has an impact on the area required, as the strength decreases significantly for the same thickness due to the smaller modulus of elasticity (Figure 6.1).

Figure 6.2 shows the transparency, maximum processing temperatures and tensile strengths of various polymer films. PET has the best transparency, while PP has advantages in terms of usable temperature, and PVC in terms of strength. This is why, for example, only PP films are used for microwave dishes, as other types of film change their shape, even when the dishes are heated at temperatures less than 100°C.

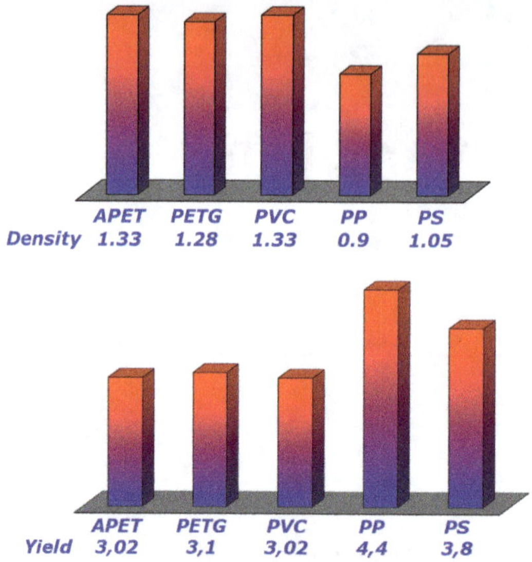

Figure 6.1
Density (in g/cm^3) and area yield (in m^2/kg for 250 µm) of various polymer films

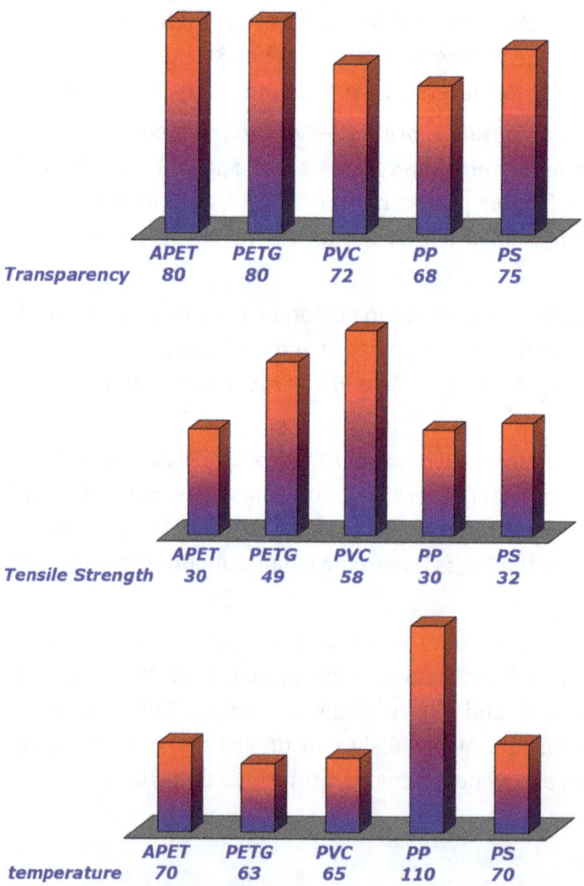

Figure 6.2
Transparency (in % for 250 µm), tensile strengths (in N/mm^2), and maximum operating temperatures (in °C) of various polymer films

Figure 6.3 shows the printability, thermoformability, and modifiability of various polymer films. Since printability depends heavily on the surface tension of the films, only PVC and PS can be printed without pretreatment, while other types of films must be pretreated using corona treatment. The surface tension required is over 40 mN/m (equivalent to the old unit dyn/cm).

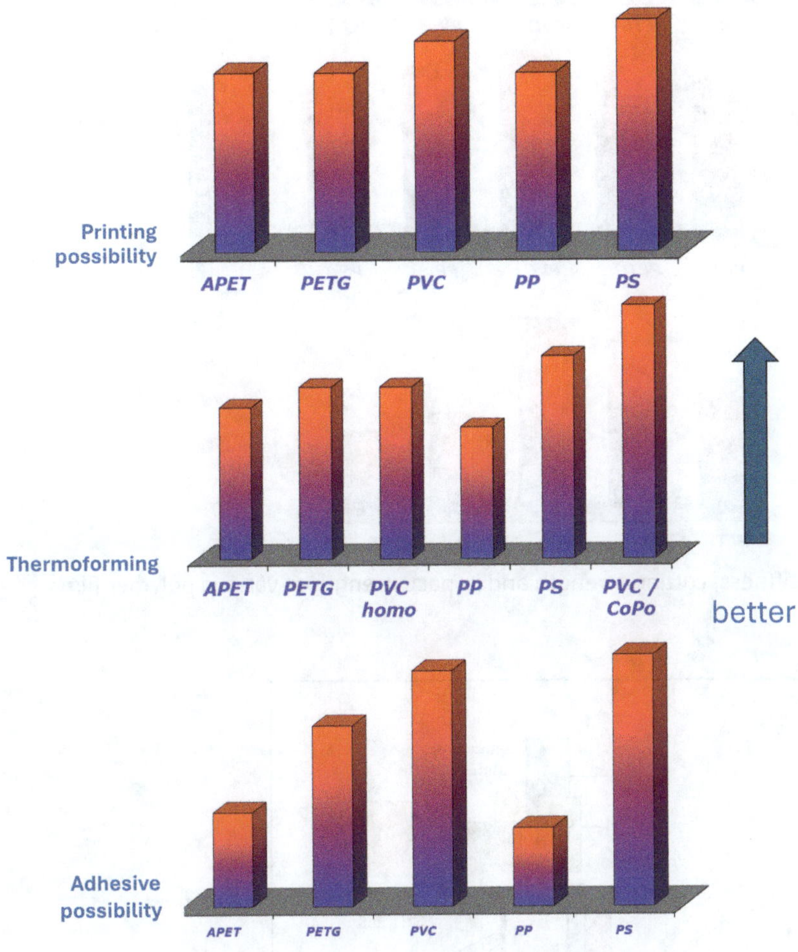

Figure 6.3 Printability, thermoformability, and modifiability of various polymer films

Figure 6.4 shows the stiffness, cutting strength, and impact resistance of various polymer films. PVC film is the leader in all these properties. PP is worse in terms of stiffness and PS in terms of impact resistance than the other types of film.

Figure 6.5 shows the oxygen and water vapor barrier requirements for various packaging goods. According to these requirements, the packaging material must be selected in terms of types and thickness. If monomaterials do not have sufficient barrier properties, multilayer films or special coatings are used.

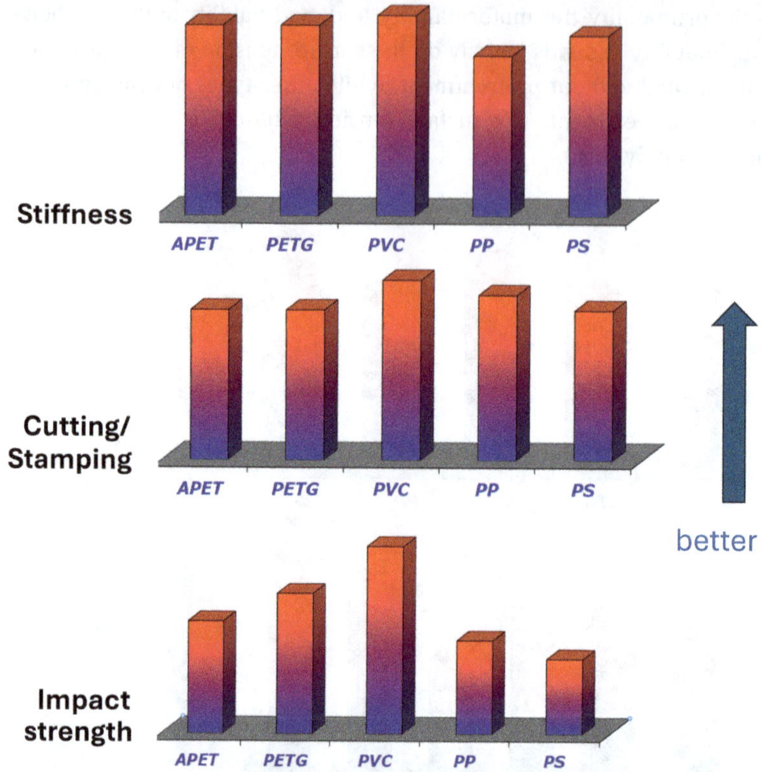

Figure 6.4 Stiffness, cutting strength and impact strength of various polymer films

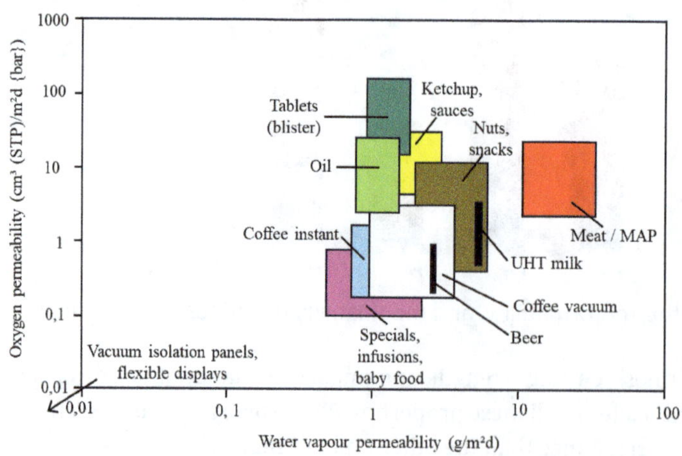

FIGURE 1: Barrier requirements of selected food and pharmaceuticals.

Figure 6.5 Barrier requirements for various packaged goods (Source: [1])

Figure 6.6 shows the barrier properties for oxygen and water vapor of various polymer films. EVOH and PVdC as multilayer additives or coatings on commercially available monofilms allow good improvements in the barrier properties of monofilms. These combinations allow the manufacture of packaging based on rigid films coupled with excellent barrier properties, especially for pharmaceutical packaging and long-lasting, cost-intensive consumer goods packaging.

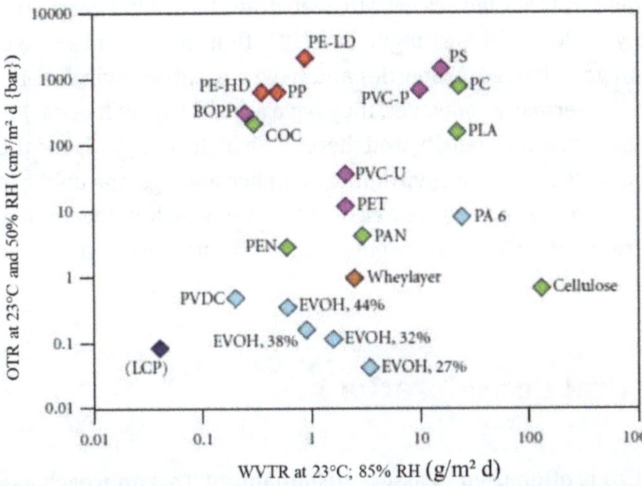

5: Barrier properties of whey-based layer versus other plastics commonly used in the packaging industry.

Figure 6.6 Barrier properties of various polymer films (Source: [1])

Figure 6.7 compares various material and processing properties of polymer films. Each type of polymer has its advantages and disadvantages, so that an assessment should be made regarding the acceptable material and processing properties.

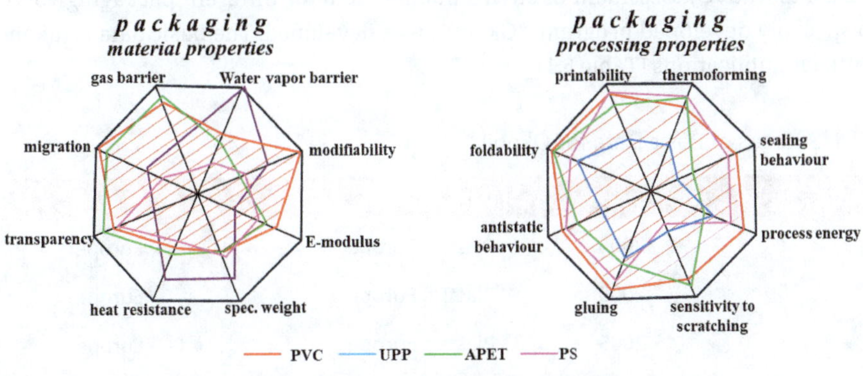

Figure 6.7 Comparison of material and processing properties of various polymer films (the further from the centre, the stronger the property)

The advantages of rigid PVC are modifiability, strength, and printability; the advantages of PET are transparency, ability to provide a gas barrier, and scratch resistance; the advantages of PP are ability to provide a water vapor barrier, heat resistance, and low specific gravity; and the advantages of PS are the ability to be deep-drawn, printability, and antistatic behavior.

In summary, it can be stated that, as already stated above, each type of polymer has specific properties that, in total, must meet the requirements of customers. Hard PVC films are characterized by the fact that they are easy to deep draw, have good barrier properties, and are easily recyclable and self-extinguishing. PET films have good strength, are highly transparent with good barrier properties and have a positive environmental image amongst the public in Germany ; however, they have a relatively high price. PP films have very good heat resistance, low density, and therefore a high area yield, and they are also easy to recycle and have a positive environmental image amongst the public in Germany. PS films have a low price and are very easy to thermoform, but due to their high level of monomer migration they have fewer possible uses for packaging goods containing fat.

6.2 Environmental Considerations

Life cycle analysis (LCA) is often used to assess sustainability. This approach assesses the environmental impact from the cradle (i.e. raw material extraction) to the grave (i.e. recycling or landfilling). A distinction is made between life cycle assessment, comparative life cycle assessment or a holistic approach.

To create a life cycle assessment, one starts from an objective, an inventory, an impact assessment, and a corresponding evaluation. Interpretation of the results should then lead to improved products, decision-making processes and often marketing benefits. For this comparative assessment of environmental data for different packaging variants, a specially developed program, "Ökobil", was developed. The basic data is taken from official publications (Table 6.1).

Table 6.1 Sources of Basic Data for Ökobil

Material	Inquiry date	Data source	Region
PS	2006	Plastics Europe	Europe
PET	2005	Plastics Europe	Europe
PVC	2005	Plastics Europe	Europe
PP	2005	Plastics Europe	Europe
HDPE	1995	Plastics Europe	Europe

Material	Inquiry date	Data source	Region
LDPE	1993	Plastics Europe	UK
PVdC	2005	Plastics Europe	Europe
Glass, aluminium, carton, paper, tin plate	1991, 1996	Federal Office for the Environment (FOEN)	Switzerland
PLA	2010	NatureWorks	USA
$CaCO_3$	2011	European Life Cycle Data 3	Europe

The metrics shown in Table 6.2 are considered within Ökobil for assessment of the environmental credentials of packaging materials. Using these, an "eco index" can be determined, that makes a statement about the relative values of environmental metrics for individual packaging materials. This evaluates the percentage shares of individual comparable packaging materials in relation to the total sum of the individualparameterss, each weighted by a predetermined factor.

Table 6.2 Metrics Considered within Ökobil for Assessment of the Environmental Credentials of Packaging Materials

Metric	Unit	Definition	Weighting factor (%)
Energy consumption	MJ/kg	Reference value for the total energy required (including oil consumption as energy) to produce and process 1 kg of a specific packaging material (e.g. film or cup)	30
Critical air volume	m^3/kg	Minimum volume per kg of packaging material to reduce the amount of harmful substances in the air to the legally permitted concentration	20
Critical volume of water	L/kg	Minimum volume per kg of packaging material to reduce the amount of pollutants in the water to the legally permitted concentration	20
Solid waste	cm^3/kg	Solid materials to be landfilled, considering their specific landfill volume	10
CO_2 fingerprint	mg/kg	Minimum volume per kg of packaging material to reduce the amount of pollutant CO_2 in the air to the legally permitted concentration	10
Crude oil consumption	g/kg	Total amount of petroleum used to produce 1 kg of packaging material	5
Water consumption	L/kg	Total amount of water used to produce 1 kg of packaging material.	5

Evaluation of the individual eco index is as follows (key figures can be changed in the program by the administrator):

$$\text{Eco index} = \left(\frac{\text{Energy consumption}}{\text{Total energy consumption}} \times 0.3\right) + \left(\frac{\text{Air consumption}}{\text{Total air consumption}} \times 0.2\right) + \left(\frac{\text{Crude oil consumption}}{\text{Total crude oil consumption}} \times 0.05\right) + \ldots$$

Figure 6.8 shows the opening image of this program. Clicking on "Start" takes you to the input options for two or more packages.

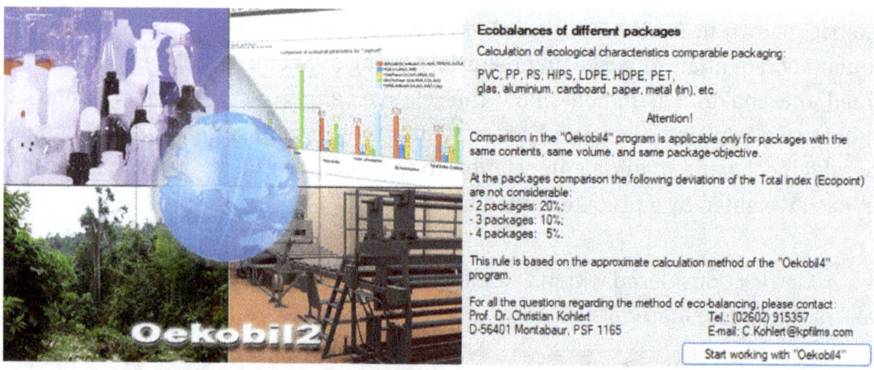

Figure 6.8 Opening image from Ökobil

Figure 6.9 shows an example of a comparative life cycle assessment for different pharmaceutical packaging materials, as calculated using Ökobil. Under the given conditions, the 5th packaging (PVC, paper, aluminium) performs best in terms of energy consumption and air pollution as well as landfill volume and CO_2 fingerprint. This is then also reflected in the total index. Variants 1–3 are less environmentally favourable.

Packaging	Aluminum	PVC, aluminum	PP, aluminum	PVC, aluminum	PVC, paper, aluminum
Capacity (pieces)	10	10	10	10	10
Bottom film (gramms)	Aluminum 1,3262	PVC 1,2075	PP 1,1984	PVC 0,8301	PVC 0,841
Top film (gramms)	Aluminum 0,8922	Aluminum 0,603	Aluminum 0,5989	Aluminum 0,5125	Paper, aluminum 0,23

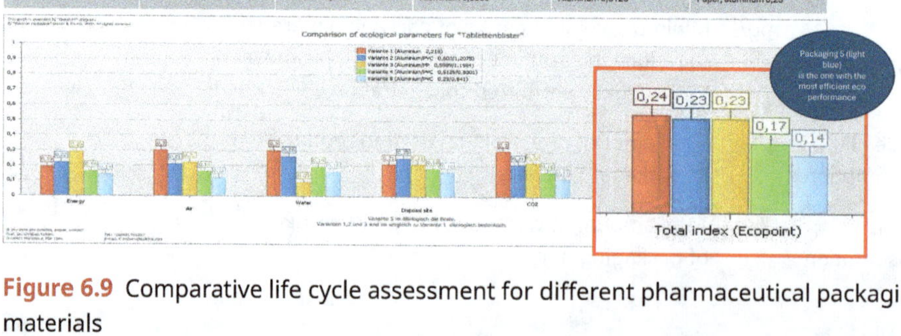

Figure 6.9 Comparative life cycle assessment for different pharmaceutical packaging materials

References for Chapter 6

[1] Langowski, H.-C. [lecture material]: Institut für Verfahrenstechnik und Verpackung Freisingen

7 Further Processing of Polymer Films

7.1 Introduction

Christian Kohlert

Polymer films are usually further processed before being converted to their end use, e.g. packaging. This can be done by lamination with other films that have better barrier or adhesion properties than the base film (Figure 7.1). Likewise, the base film may be coated to improve printability or antistatic properties, etc., or it may be surface-treated to provide properties that normal polymer films do not have.

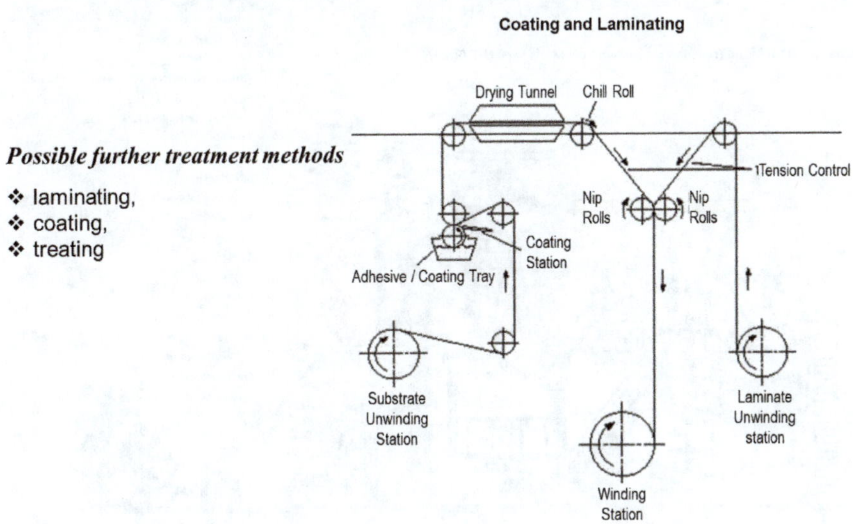

Figure 7.1 Lamination for pharmaceutical films

Lamination

Lamination is the process of combining different polymeric films with the aim of improving the film properties. For example, PE film is most often laminated onto PVC or PET film. This results in a multilayer film that combines the rigidity and barrier of the base film, e.g. barrier against oxygen and aromatics (PVC), with the adhesive strength and barrier against water vapor of the PE film.

This results in an excellent packaging film for a variety of foods where the barrier against oxygen is required (for example, in fatty products) and a barrier against water vapor is required (for example, in the pharmaceutical industry). Other innovative structures have also been developed (Figure 7.2).

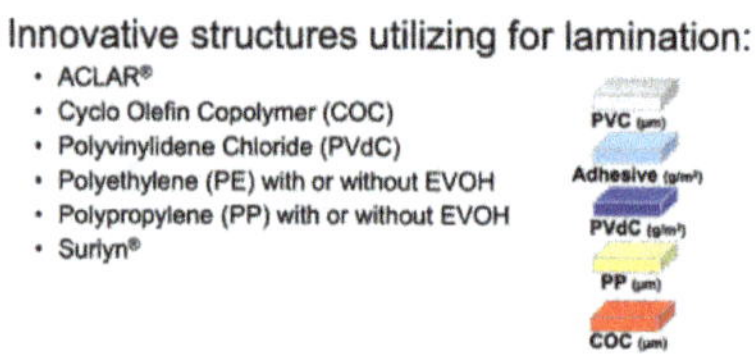

Figure 7.2 Multilayer films to increase the barrier properties by lamination

Lamination can be done with special laminating systems, but also directly in the calender downstream at the points where the film temperature is around 120°C. Multilayer films are mainly produced by blown film systems with 7–13 material extruders with total film thicknesses from 50 µm (Figure 7.3).

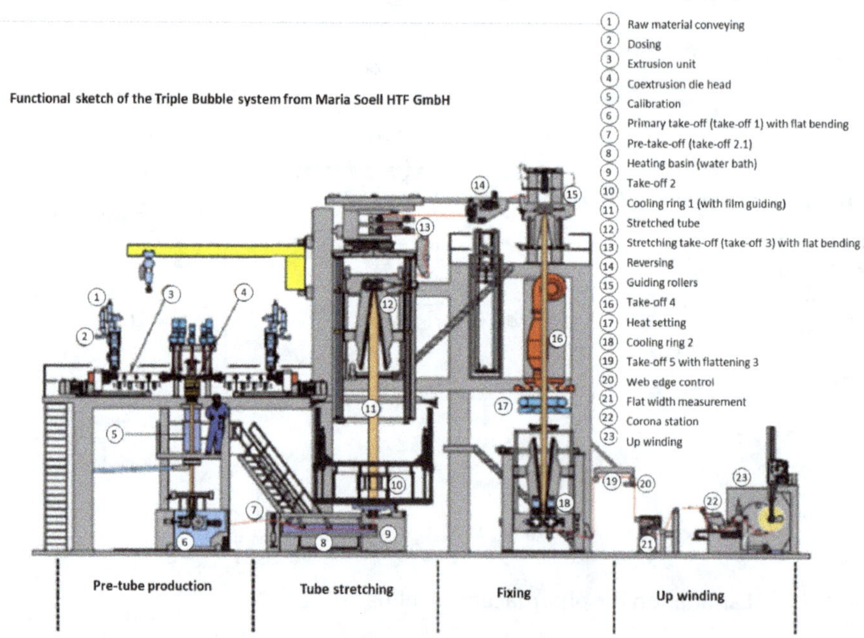

Figure 7.3 Multilayer films via blown film systems

Different polymeric layers can be used to produce multilayer films with different properties depending on the individual polymer layers (Figure 7.4). However, the associated recycling problem can also be solved using novel reactive extrusion processes (see Chapter 11).

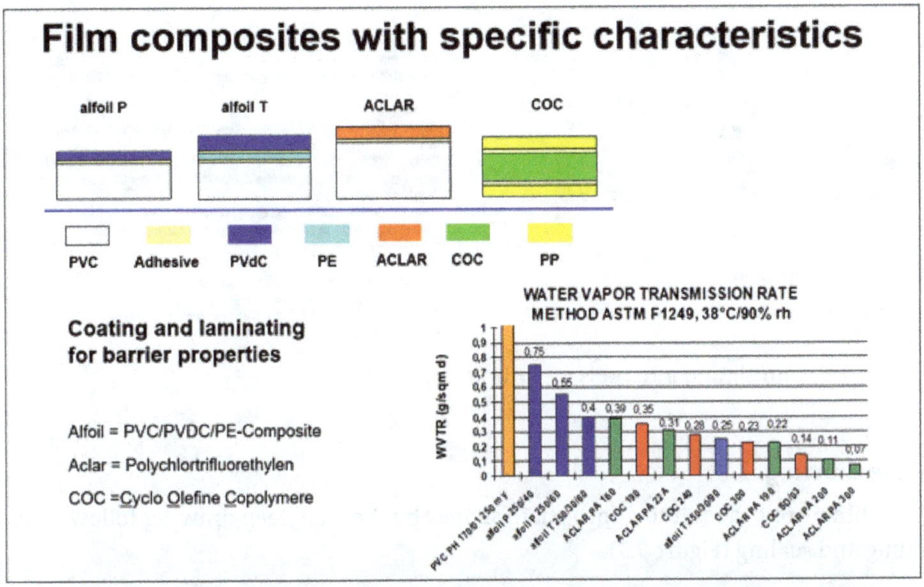

Figure 7.4 Change in water vapor permeability through multilayer films

Coating

For many applications, the water vapor barrier of PVC monofilms is not sufficient, so the water vapor permeability must be reduced using special PVdC coatings (Figure 7.5). This particularly affects pharmaceutical products for areas with high humidity.

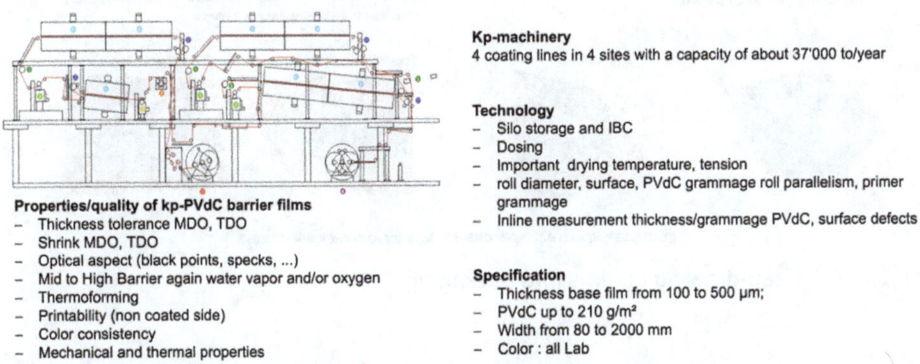

Figure 7.5 Coating with PVdC

Slitting

Since customer formats cannot usually be produced on large calendering systems due to the need to maximize throughput, slitter rewinders or format cutters are used to cut it into the required customer format (Figure 7.6). This is where anti-static equipment to prevent charging effects makes sense.

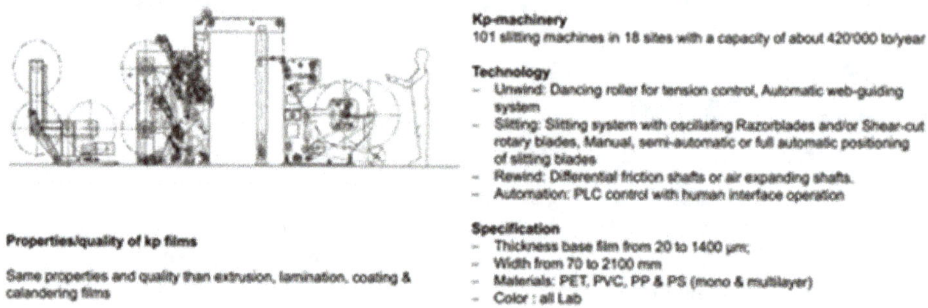

Figure 7.6 Film slitting into reels or cutting formats

Deep-drawing

Most films are drawn into appropriate shapes by thermal deep-drawing followed by filling and sealing (Figure 7.7).

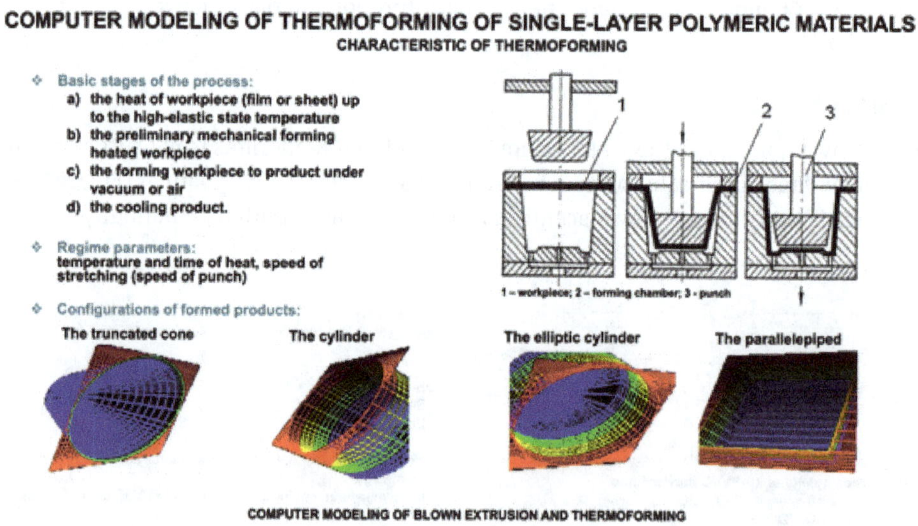

Figure 7.7 Deep-drawing of films into packaging

7.2 Surface Coatings From the Micro- to the Nano-Scale

Christian Kohlert

Coatings using squeegee rollers or spraying process are the state of the art, achieving surface properties corresponding to the coating substance, such as color, antistatic properties, conductivity, antimicrobial properties, etc. The type of coating and the thickness of the coating naturally result in a corresponding price. Moreover, novel surface coatings and treatments have resulted from the use of nanotechnology. "Nano" refers to particle sizes below 100 nm, in contrast to conventional coatings ranging from several µm to the mm scale.

Since light has a wavelength of around 500 ± 200 nm, individual particles in the range 100 nm and smaller are flooded with light in low concentrations. A nanocoating with a low concentration (1–5%) is not recognized as impairing transparency. Cellar films can therefore be given new properties without reducing the transparency desired in many packaging films. These nanoparticles can be applied to the formulation of monofilms, as a continuous coating on the surface, or as a non-continuous coating (treatment).

Nanoparticles are the smallest particles of molecular size, and can be smaller than fine dust and more dangerous than asbestos. Therefore, it is only possible to use them in systems bound to polymeric structures or in solution, which are manufactured by specialist companies using strict safety-relevant methods.

Since these are very thin layers, common coating methods such as squeegee blades or roller application cannot be implemented, as concentricity problems with the rollers would lead to strong thickness fluctuations in the µm range. Spray processes or so-called felt coatings are used, with subsequent (usually) IR drying making them technically ready for application (Figure 7.8).

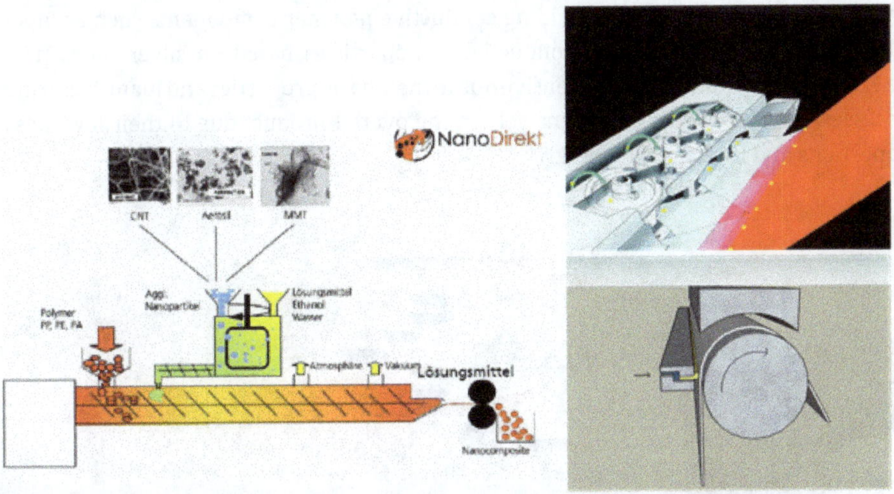

Figure 7.8 Incorporation and coating methods of nanoparticles (Source: [1])

Nanoparticles can be incorporated into the formulation, resulting in a volume effect, or they can be applied to the surface to achieve a surface effect. A distinction is made between the effects as "active" (acting on the filling material), "intelligent" (providing information about the condition of the filling material), and "special", i.e. providing new properties that polymer films normally do not have (Figure 7.9).

Active Effects	Intelligent Effects	Films with improved characteristics
▪ **Oxygen absorber** (1) ▪ **Anti microbial** ➢ Tangent (1) ➢ Out gasing (2) ▪ **Anti static / conductivity** (1) ▪ **Anti corrosion** ➢ Rust prevention (1) ➢ Rust removal (1) ▪ **Increase barrier properties** ➢ Against gases (2) ➢ Against water vapor (1) ▪ **Increase UV protection** ➢ Film (1) ➢ Product (2) ▪ **Film with aromatic properties** (2)	▪ **Indicator film irreversible** ➢ Time TI (2) ➢ Time / Temperature TTI (2) ➢ Humidity (2) ➢ PH-value (1) ➢ Temperature (2) ➢ UV-light (2) ▪ **Indicator film reversible** ➢ Temperature (2) ➢ UV-light (2) ▪ **Luminescent film** ➢ Self shiny (1) ➢ Activated by UV-light (1) ▪ **Dimming film** ➢ Electrical (1) ➢ Light (heliomatic) (2)	▪ **Surface haptics** (2) ▪ **Change printing characteristics** ➢ Increase surface tension (1) ➢ Decrease surface tension (2) ▪ **Scratch resistance** ➢ Microscratch (2) ➢ Surface rigidity (3) ▪ **Increase chemical resistance** (2) ▪ **Lotus effect** (3) ▪ **anticounterfeit** (2) (1) = practicable (2) = possible (3) = thinkable

Figure 7.9 Possible property changes of polymer films using nanotechnology. Ratings indicate: (1) ready to use; (2) possible, or in progress; (3) "nice to have", or ideas without any current possibility of implementation (Source: [1])

Electric conductivity

The often-desired increase in electrical surface conductivity for the use of films in photovoltaic systems is achieved by adding conductive polymer components such as laponite or Baytron® P, but also by conductive nanoparticles based on silver tubes (Figure 7.10). These nanotubes are expensive due to their basic properties and manufacturing processes, but can be used inexpensively in the overall product due to their low mass proportions and layer thicknesses.

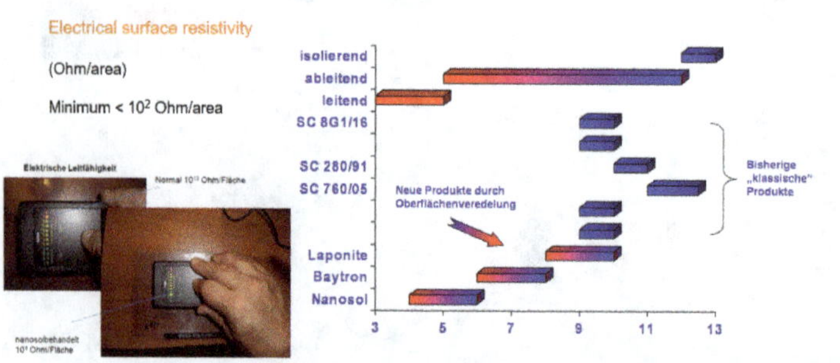

Figure 7.10 Nanotechnology to increase surface conductivity (Source: [1])

Surface tension

An increase in surface tension is often necessary to improve the printing properties of polymer surfaces, and this can be achieved using nanoparticles. It can also be decreased, and this finds application in anti-graffiti films (Figure 7.11).

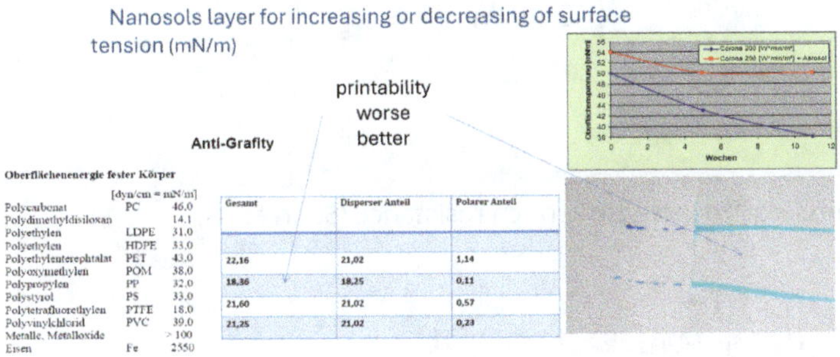

Figure 7.11 Nanotechnology to increase or decrease surface tension (Source: [1])

Color change

There are nanoparticles which, after initiation by UV rays, cause a color change depending on temperature and time (Figure 7.12). The color change occurs more quickly at higher temperatures (e.g. room temperature), than at lower temperatures (e.g. in cooling processes). Packaging films processed in this way would indicate an interruption in the necessary cooling process through a change in color.

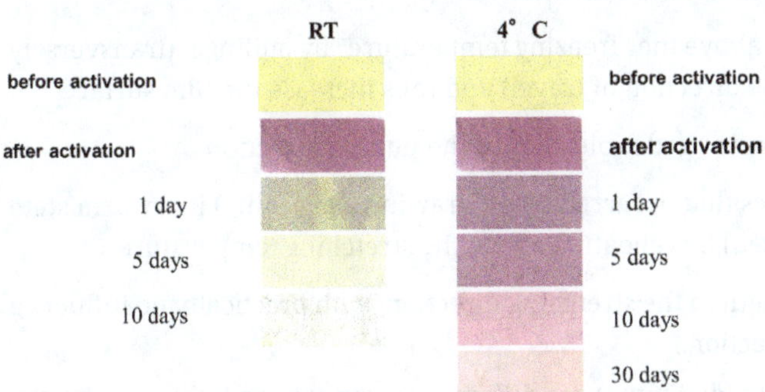

Figure 7.12 Nanotechnology for time-temperature indication (TTI) (Source: [1])

Scratch resistance

Special nanoparticles increase the scratch resistance of polymer films against vibration scratches and surface influences (Figure 7.13). These can be recorded using a measurement technology specifically developed for this purpose (see Section 5.3) and compared with other variants.

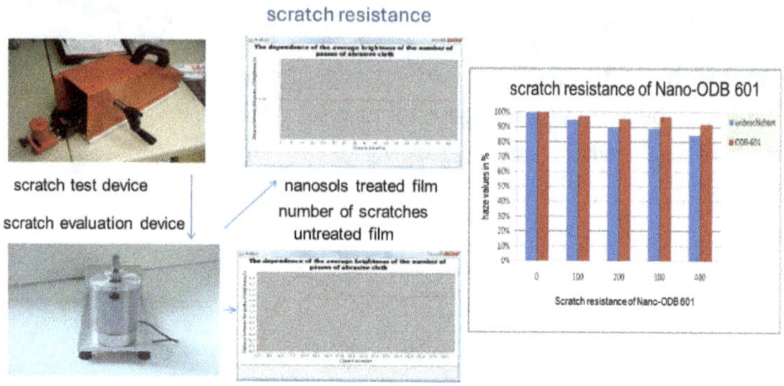

Figure 7.13 Nanotechnology to increase scratch resistance (Source: [1])

7.3 Stretching Films—MDO and TDO

Christian Kohlert, Helmut Welbers

7.3.1 Introduction

After calendering, films can be stretched in the machine direction (MDO) or transverse direction (TDO) (Figure 7.14). This means either that the films are removed at higher speeds during the cooling process after the calender (MDO), or that the width of the film is greatly expanded in a stretching frame also known as a stenter (TDO) (Figure 7.15). This "freezes" the molecular structures and the molecules are realigned. This process is called orientation. By this we mean:

- To deform a plastic film above the "freezing temperature" by pulling it (transversely and/or lengthways to the direction of travel) and thus increase the film surface
- Aligning the randomly arranged molecules in the pulling direction
- The dimensional changes that occurred when drawing (stretching) in a warm state can be reversed (shrinked) by reheating above the stretching temperature
- Increase in tensile strength in the stretching direction, with practically no influence in the non-stretched direction.

The stretch film is further divided into three different categories depending on the orientation of the macromolecules through the respective production process and thus the shrinkage behavior:

- Longitudinally stretched films—monoaxially stretched lengthways (MDO)
- Transversely stretched films—monoaxially stretched transversely (TDO)
- Biaxially stretched films—longitudinally and transversely stretched (MDO & TDO).

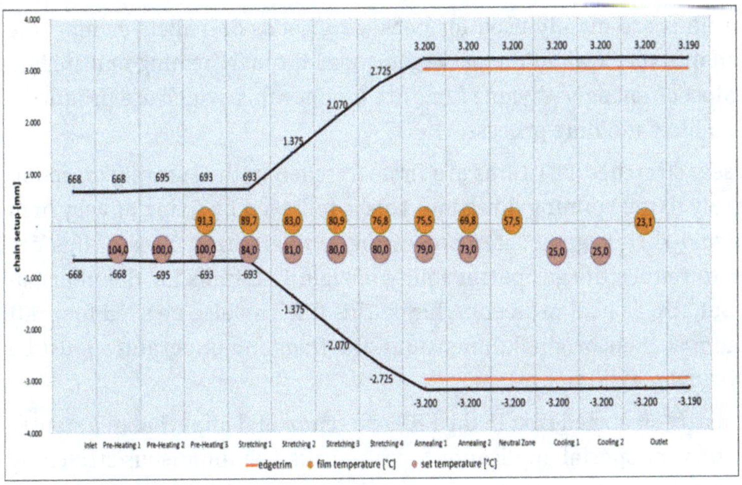

Properties/ quality of kp's stretched films

- High shrinkages realizable up to 80%
- Functional layers due to Co-extrusion
- Great printability due to Corona pre-treatment
- Excellent lay flat properties due to thickness control systems
- High optical quality due to melt filtration

Technology

- High standard regarding thickness control systems to ensure best lay flat properties
- Highly accurate heating systems powered by natural gas or steam
- Capability of adding high amounts of rework to support an environmental friendly production

Specification

- Majority of thicknesses from 35 to 75 µm
- Stretched widths from 2.0 m to 6.2 m
- Mainly clear high transparent films, colors possible

Figure 7.14 Longitudinal and transversely stretched films (Source: [1])

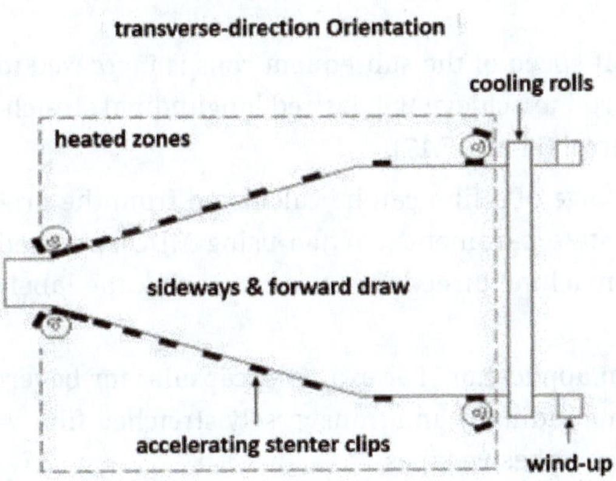

Figure 7.15
MDO process in the calendering process (Source: [1])

Biaxially stretched films are mainly used for packaging goods on pallets or machine parts, as the film adapts very well to the packaged goods through heating and shrinking on both sides. Most often, polyethylene films are used for this, which are produced, for example, using a blow-molding process.

Biaxially transversely stretched PET films are first stretched in the running direction and then transversely to the running direction. This is followed by fixing at very high temperatures to avoid shrinkage at later processing temperatures. These films are used, for example, to reduce the gas permeability of rigid PVC films for the pharmaceutical industry, but also as heat protection films. This film can also be used for craft work. To shrink them to their original dimensions, the fixation temperature must be exceeded (e.g. in the oven at home).

Normally these films are stretched first in the MDO direction and after that in a stenter in TDO direction. For very special applications also so-called simultaneous stretching is possible, but the machines are very expensive. The reason for this is that the clamps must be driven by single linear motors, to enable them to increase their longitudinal speed for MDO-stretching (Figure 7.16).

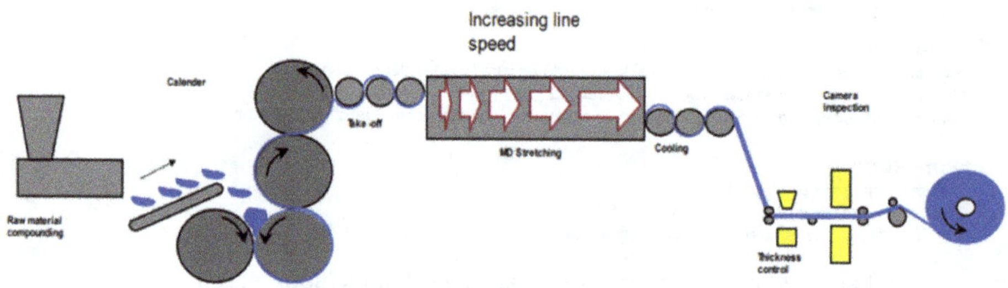

Figure 7.16 Biaxially stretching frame (or biaxial stenter) with longitudinal and transverse directions

In the calendering process, the take-off speed of the subsequent rolls is increased to 50–200 times that of the last calender roll to achieve the desired longitudinal stretching together with the temperature control (Figure 7.17).

Section 4.2.5 describes how the shrinkage of a film can be calculated from the process parameters and the rheological state parameters. When using MDO-stretched films, the stretching direction in the machine direction is used to shrink the labels (Figure 7.18).

Monoaxially stretched films overlap in application. For example, capsules for beverage bottles can be made from both longitudinally and transversely stretched film. A typical application for elongated films is adhesive tapes.

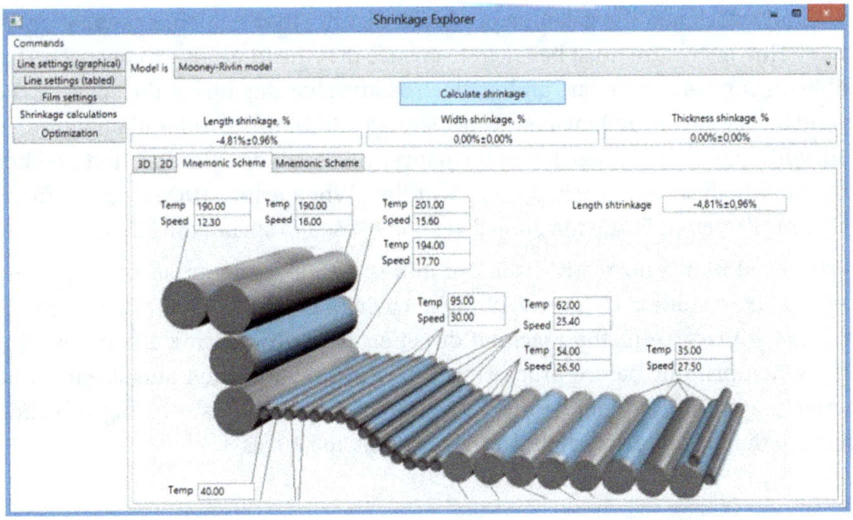

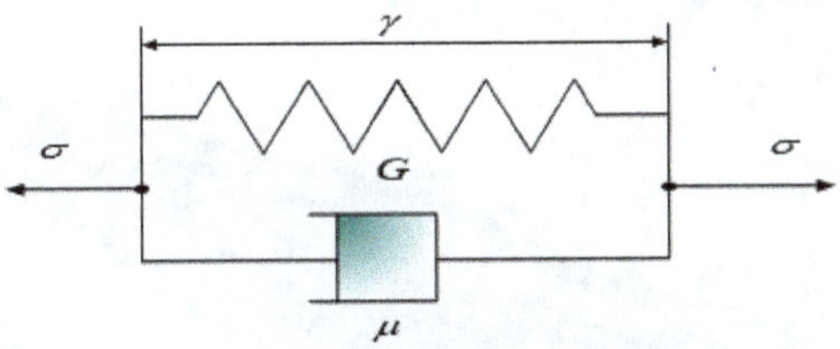

$$C \cdot S_l^{6} + A \cdot S_l^{4} + \left(B - C - 2^{n+1}\left(\frac{U_{i-1} \cdot n}{L(1-n)}\right)^{n} \cdot \left[1 - \frac{U_{i-1}}{Ui}\right]^{1-n} \cdot \frac{H_{i-1}W_{i-1}}{HW}\, \mu_0 e^{-b(T_{i+1}-T_i)} \right) \cdot S_l^{3} - A \cdot S_l^{2} - B = 0$$

Figure 7.17 Recalculation of the MDO shrinkage process in the calendering process (Source: [1])

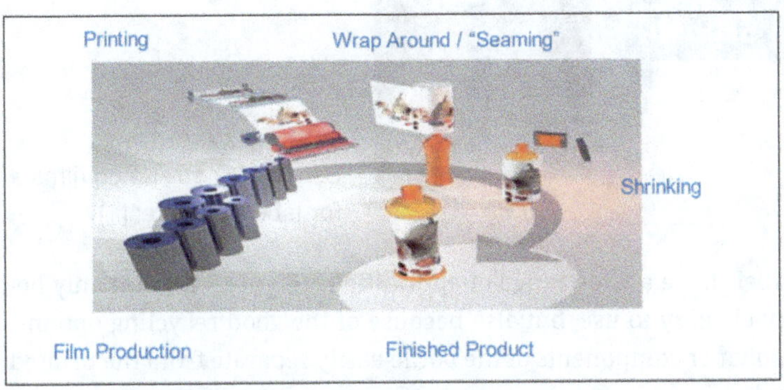

Figure 7.18 Use of MDO-stretched film for labels (Source: [1])

Longitudinally stretched films can be produced directly in the calender with a drafting system. However, care must be taken to ensure that the film cannot constrict. This is possible through lay-on rollers and/or electrostatic charging tips at the edges of the rollers and a very small gap between the rollers. In addition, the roller diameter must be small, with one disadvantage being the poorer thickness distribution, and another being the very high speeds reached when winding. When using MDO-stretched films, the stretching direction in the machine direction is used to shrink the labels.

A TDO-stretched film is normally produced in a spreader frame under certain speed and temperature conditions (Figure 7.19). When using TDO-stretched films, the stretching direction transverse to the machine direction is used to shrink the labels (Figure 7.20). When printing before processing, the stretching direction and size must be considered (Figure 7.21). When using TDO-stretched films, the stretching direction transverse to the machine direction is used to shrink the labels.

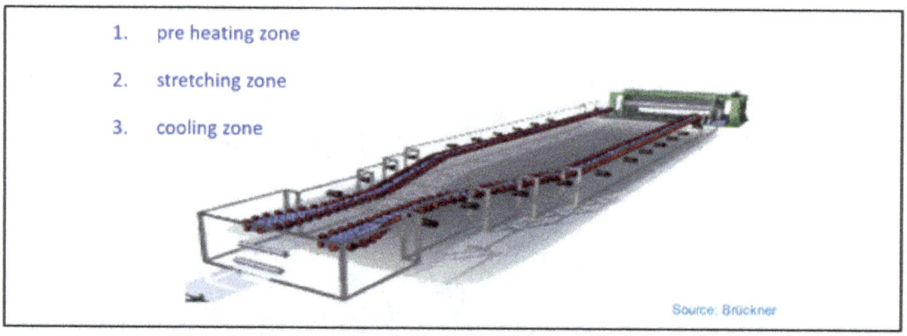

Figure 7.19 Zones of a TDO wide stretch frame (Source: [1])

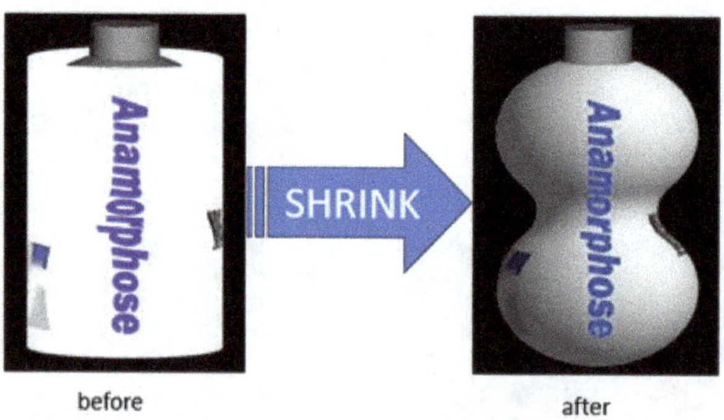

Figure 7.20 Use of TDO-stretched films for labels (Source: [1])

Stretched films as labels have a wide range of applications (Figure 7.22), not only because they are relatively easy to use, but also because of the good recycling options. When crushed, the polymer components of the bottle easily separate from the printed components, the labels.

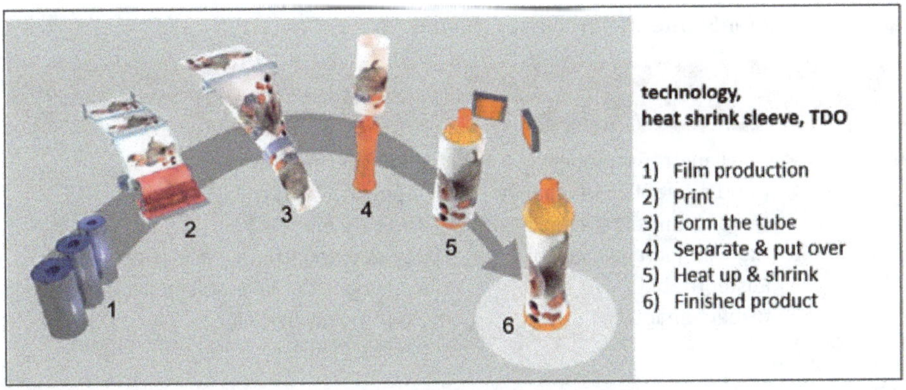

Figure 7.21 Printing process for stretched films (Source: [1])

Figure 7.22 Products made from shrink films (Source: [1])

7.3.2 Materials

Stretch films are mainly made from the thermoplastics listed in Table 7.1, all of which have their advantages and disadvantages.

Table 7.1 Films for Stentering Processes

Thermo-plastic	Advantages	Disadvantages
PVC	▪ Good all-round performance ▪ Good printability/glueability ▪ Good thickness distribution ▪ High transparency ▪ Good slip properties.	▪ High specific gravity (approx. 1.36 g/cm^3) ▪ High-shrink pasteurizable grade problematic.

Table 7.1 Films for Stentering Processes (*continued*)

Thermo-plastic	Advantages	Disadvantages
PET	▪ Excellent transparency ▪ Good printability/glueability ▪ Good chemical resistance ▪ High shrink ratio available ▪ Pasteurizable ▪ "Green" image.	▪ High specific gravity (approx. 1.36 g/cm^3) ▪ Increase in width ▪ Stress cracking (influence of solvents) ▪ Welding with THF problematic ▪ Poor slip properties ▪ More expensive.
OPP	▪ High transparency ▪ Excellent stretchability ▪ Low specific gravity (approx. 0.91 g/cm^3) ▪ Pasteurizable ▪ "Green" image.	▪ Corona treatment necessary ▪ Lower stiffness ▪ Poor thickness distribution ▪ Poor weldability.
OPS	▪ Excellent transparency ▪ Excellent stretchability (higher TD shrinkage possible) ▪ Low specific gravity (approx. 1.05 g/cm^3) ▪ Pasteurizable.	▪ Poor longitudinal shrinkage ▪ Poor thickness distribution ▪ Lower stiffness (therefore thicker films) ▪ More brittle than PVC ▪ High longitudinal shrinkage ▪ Only a few suppliers on the market (Sleever, Kasei, and Asahi Chemical).
COP	▪ High transparency ▪ Low specific gravity (approx. 1.0 g/cm^3) ▪ Pasteurizable.	▪ Expensive ▪ Sensitive against fatty acids (stress cracking) ▪ Corona treatment necessary ▪ Raw material very expensive ▪ No reports of weldability.

Transversely stretched films are discussed in more detail in this section. First, however, some additional information should be provided.

Before we take a closer look at transverse stretching itself, one difference needs to be mentioned. This difference lies in the heat source for heating the stretch film, which triggers the shrinkage of the stretch film:

- Dry heat—e.g. hot air or infrared heaters
- Moist heat—e.g. water vapor.

In processes that work with "dry" heat sources, the types of PVC used to produce the film are irrelevant. The film remains transparent. However, if the film is heated with moist heat, film made with S-PVC becomes milky due to water absorption.

This is related to the polymerization process. M-PVC must therefore be used for pasteurizable applications, as this type of PVC is not produced in an aqueous phase.

7.3.3 Areas of Application of Shrink Film

The areas of application for stretched film, which is called shrink film after stretching, are diverse. The film is used as:

- All-round labels (360° labels)
- Information purposes (advertising)
- Decorative items
- Splinter protection for glass bottles (the broken parts remain localized)
- Product protection
- Shape design freely selectable
- Thinner wall thicknesses (e.g. in glass containers)—lower costs, lower weight
- Bottle capsules (wine bottles)
- Multipacks
- Tamper-evident seals (detecting if packaging has been opened without permission)
- The applications also determine the type of base film in terms of shrinkage, film thickness, and recipe:
- Shrink film for bottle caps (thickness e.g. 60 µm; shrinkage 50%)
- Shrink film for labels (thickness e.g. 50 µm, shrinkage 60%)
- Shrink film for splinter protection (thickness e.g. 40 µm; shrinkage 70%)
- Shrink film for "steam" applications (thickness e.g. 50 µm; shrinkage 60%).

7.3.4 Capsule Film

In the past, lead capsules were used in valuable wines. However, lead is a toxic heavy metal and was therefore replaced by tin capsules from 1990 onwards. However, tin capsules are expensive and have been replaced by aluminium capsules. The aluminium capsules caused injuries because they formed a sharp edge when the bottle was opened. The solution was found in plastic capsules (Figure 7.23). These can also be designed to be visually sophisticated (printed) and thus visually enhance the wine. The purpose of the capsules on the wine bottles is, on the one hand, to protect the cork from infestation by cork moths and to reduce the evaporation of the wine by sealing it airtight.

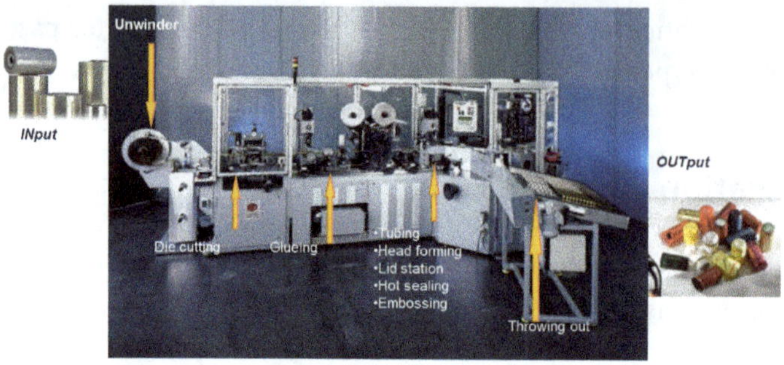

Figure 7.23 Capsule machine

Work steps for capsule production in the capsule machine

The production steps shown in Figure 7.24 dictate the following requirements placed on capsule films (which are also used for quality seals):

- They must be impact-resistant and must not splinter during the punching process
- They must be easily wettable by the adhesive used to form the capsule body
- They must be heat-sealable for bonding to the lid.

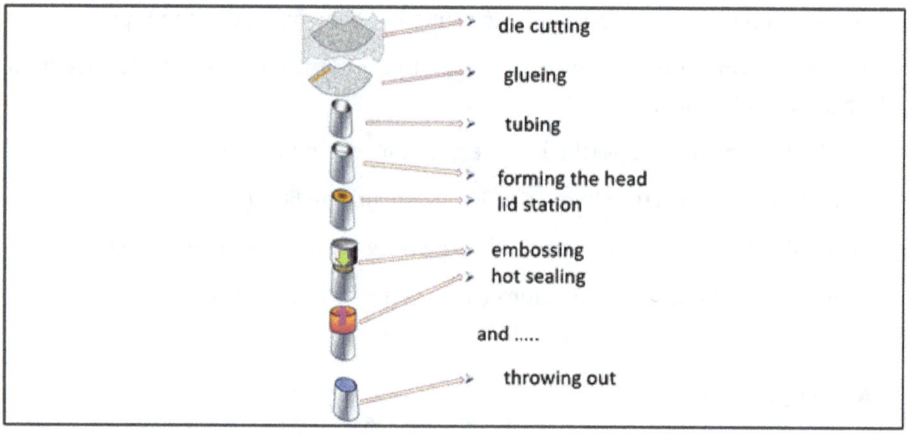

Figure 7.24 Production steps for capsule production

7.3.5 Sleeve Film (Label Film)

Sleeve films have a very wide range of applications. So-called sleeves are labels that completely (360°) enclose the product to be packaged or labelled. The customer's production steps are illustrated in Figure 7.25, and are usually:

- Printing
- Longitudinal cutting to desired width
- Apply glue to one edge, form it into a tube and wind it up again
- Load the unwinder of the "sleeve" machine with the rolls
- Cut the sleeve tube to the desired length, open it, and push it over the product with the support of rollers and air
- Heat the "sleeve" label in a shrink oven with hot air or steam. This means it hugs the product.

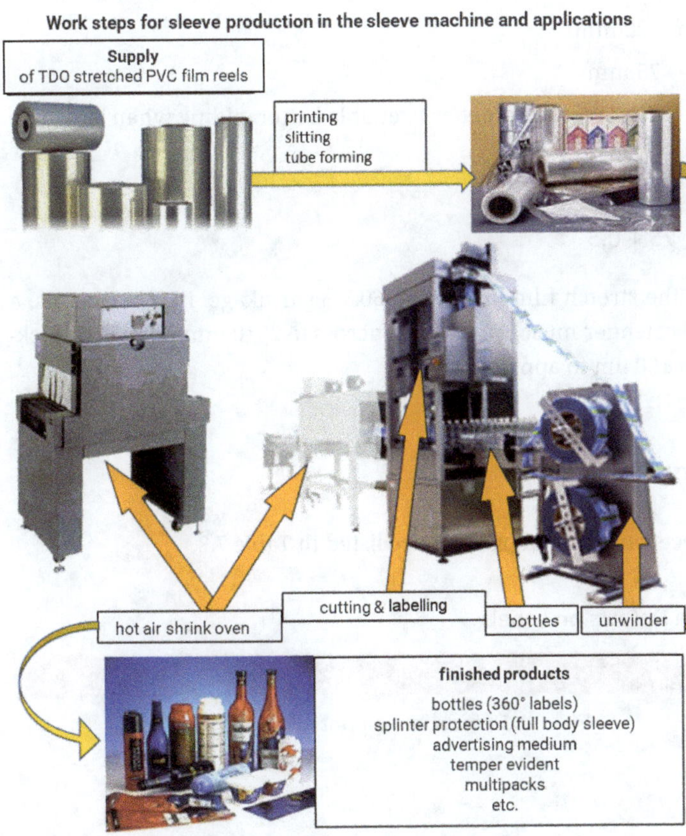

Figure 7.25 Work steps for sleeve production

7.3.6 Calculating the Required Shrinkage

To determine the required transverse shrinkage of the film, the largest and smallest circumference of the product to be wrapped must be considered. The easiest way to calculate this is for a bottle label (Figure 7.26), as illustrated by the following example.

Figure 7.26 Determination of the necessary shrinkage of a bottle capsule

The largest and smallest circumference (or diameter) of the bottle are measured:

d = smallest diameter = 30 mm

D = largest diameter = 75 mm

Then, 0.5 mm is added to the largest diameter, to enable better gliding when labelling. Therefore:

$$\text{Shrinkage} = 100 - \frac{30 \times 100}{75 + 0.5} = 60.3\%$$

Therefore, in this case the stretch film must have 60.3% shrinkage. In other words, a 1 m wide film from the calender must be stretched across to 2510 mm wide. The thickness then reduces from 100 µm to approx. 40 µm.

7.3.7 Effect of Film Defects

The effects of film defects on the final product are listed in Table 7.2.

Table 7.2 Effects of Film Defects on Labels

Film defect	Effects
Nibs	Defects in the print unprinted areas
Black dots	
Air streaks, bubbles	
Other foreign particles	

Film defect	Effects
Wavy edges	Folds Defects in print
Centre ballooning	
Slack edges	
Crosswise wrinkles	
Folds in the middle	
Color varies	Color of print-result varies from previous orders
Uneasy surface (flow lines, semolina)	
Shrink MDO too high (from calender)	Seam not straight, wavy
Shrink varies	Distorted print after reshrinking process
Shrink too low	Color of print-result varies from previous makes

7.3.8 Stretching Processes

During transverse stretching, the film is fed from the unwinder into the stretching frame through a take-off for web tension control (Figure 7.27). In the inlet, the film is fixed on both sides with clips that are arranged in a chain. The stretching frame consists of heating, stretching, tempering, and cooling zones. The take-off then takes place again, followed by the rewinder (see Section 7.3).

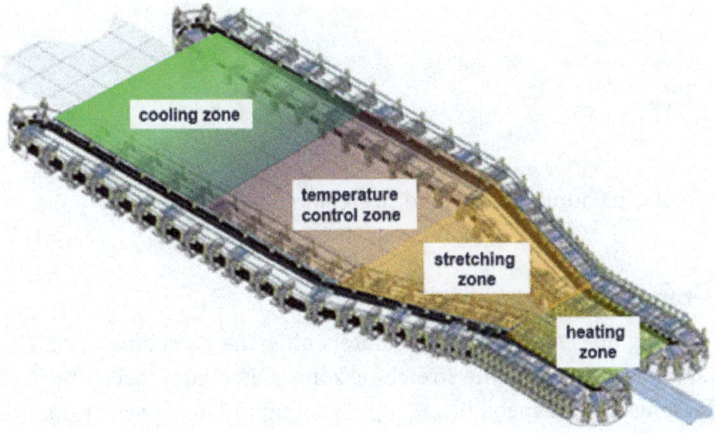

Figure 7.27 Schematic of stretching frame (Source: Lindauer Dornier GmbH brochure)

The temperatures in the stretching frame depend on the film thickness, the stretching ratio, and the speed. They are approximately:

- Heating zone: 110–90°C
- Stretching zone: 88–85°C
- Tempering zone: 80–75°C
- Cooling zone: 45–40°C.

Immediately after the stenter, a take-off unit, which is equipped with spreader rollers and cooling rollers, is required in front of the winder. The temperature of the cooling rolls may be a maximum of 20°C. The surfaces of the cooling rolls should be matt in the middle area and the edge zones should be approx. 200 mm high gloss.

Use of clips for transverse stretching

During transverse stretching, the macromolecules orient themselves in the direction of tension and are drawn across the width transversely to the direction of movement. This improves the tensile strength in the transverse direction, while it remains almost unchanged in the longitudinal direction (Figure 7.28).

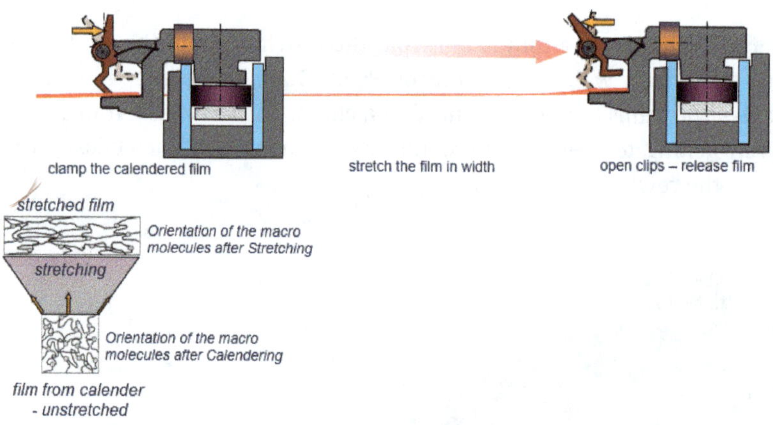

Figure 7.28 Influence of clips during transverse stretching

Influence of temperature control on shrinkage

It is important that cooling does not begin immediately after the stretching zone. The reason for this is that the web speed in the stretching zone is divided into a transverse speed ($v_{transverse}$) and a longitudinal speed ($v_{longitudinal}$). The longitudinal speed is smaller than the clip speed, leading to deformation of the web, especially at the edges. The shrink samples on the edges of the film are not rectangular but have the shape of a parallelogram (Figure 7.29). This results in poor appearance when the films are printed and then shrunk, and the print then appears completely distorted. A temperature control zone is therefore required to reorient the film.

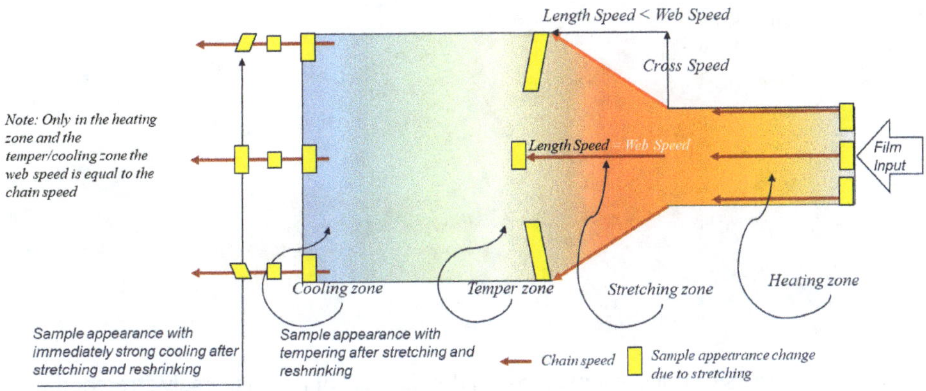

Figure 7.29 Zones of spreading

Effect of clipping on stretching

Formulae for calculating the clip (clamp) distances for stretching are shown in Table 7.3.

Table 7.3 Clip Spacing for Different Shrinkage Values

Shrink grade	Shrink rate approx. 50% ("Q5") [mm]	Shrink rate approx. 50% ("Q6") [mm]
Calender width	W	W
Web entrance	$W - 10$	$W - 10$
Heating zone 1	W	W
Heating zone 2	$W + 20$	$W + 20$
Heating zone 3	$W \times 1.09$	$W \times 1.20$
Stretching zone 1	$W \times 1.75$	$W \times 1.95$
Stretching zone 2	$W \times 2.1$	$W \times 2.54$
Tempering and cooling zone	$W \times 2.1$	$W \times 2.54$
Web outlet	$(W \times 2.1) - 20$	$(W \times 2.54) - 20$
Width at rewinder	$(W \times 2.1) - 250$[a]	$(W \times 2.54) - 285$[a]

[a] Width of the edge trim can vary.

To show the result of not taking corrective action, the effect of thicker film areas in the cross-profile is shown in Figure 7.30 and Figure 7.31.

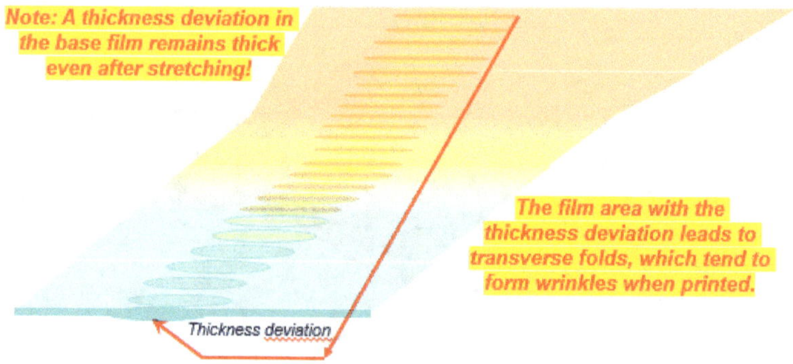

Figure 7.30 Effects of thicker film areas on the flatness of a transversely stretched film with thickness deviation without corrective measures after stretching

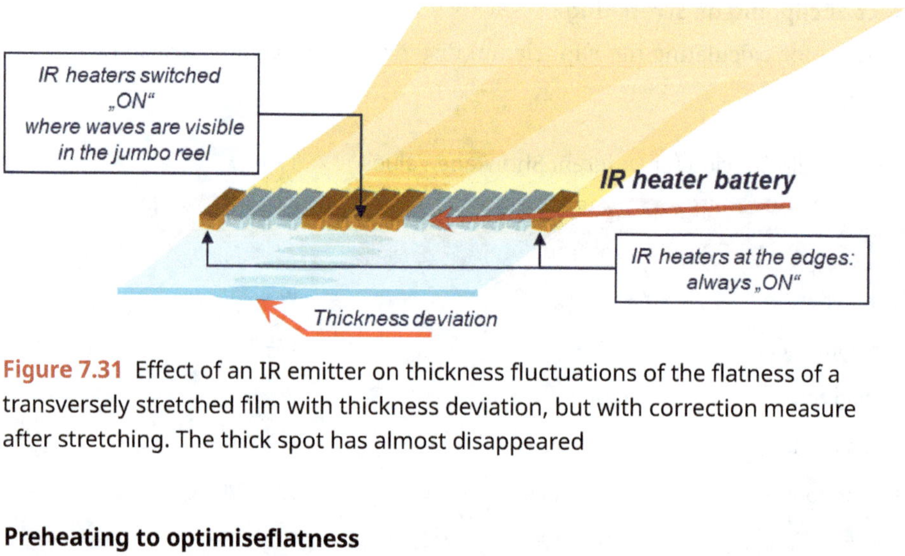

Figure 7.31 Effect of an IR emitter on thickness fluctuations of the flatness of a transversely stretched film with thickness deviation, but with correction measure after stretching. The thick spot has almost disappeared

Preheating to optimiseflatness

"Preheating" (e.g. with IR emitters) can help to minimize a flatness problem, in the following way. IR emitters partially heat the film before it enters the heating zones. In the heating zones, only the warmer parts of the film are stretched due to the preheating by the IR emitters. The result is that the thicker part of the film becomes thinner. The spread in the second heating zone can be increased to increase the expansion effect in the heating zone if necessary.

Measuring the thickness of the film at the entrance of the film into the heating zone does not help, because the thicker parts of the film were over-stretched during winding. When the film thickness is measured, no difference is seen, and the only sign of thickening is waviness.

The reason for heating the edges is that the heat flow goes into the warp and therefore more heat is needed at the edges to achieve the same elongation value (thickness) as the entire width of the film. Modern devices have a thickness measuring and control system (thickness gauge), allowing control the thickness by switching IR heaters or parts of the heating channel distribution on/off.

7.3.9 Considerations During Reel Unwinding and Rewinding

This section will briefly look at what needs to be considered when unwinding and rewinding the films.

Tension course during the unwinding process

A particularly gentle winding method is required when winding the generally thin films, some of which are less than 38 µm thick. Center winding is still possible with wide foils, as can be seen from Figure 7.32.

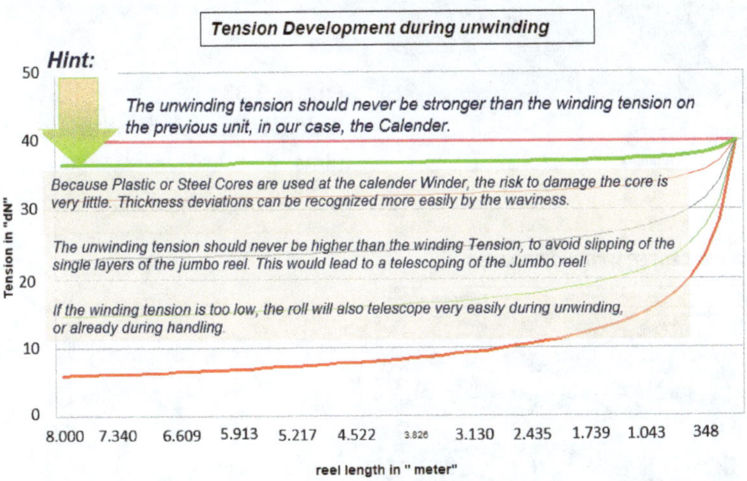

Figure 7.32 Tension during the unwinding process

This diagram shows the situation of the winding tension with a film thickness of 40 µm and a width of 2350 mm. It is easy to see that the longer the reel length, the more difficult the wrapping process. If the running length of the jumbo roll is 8000 m, the window for setting parameters on the winder becomes narrow, and the process technology limit has almost been reached for the center winder. The usual running length is 8000 m with an outside diameter of 1000 mm and a thickness of 100 µm.

If the roll width is further reduced in a subsequent work step, the situation becomes even more critical. Figure 7.33 and Figure 7.34 illustrate the situation on a winding machine with winding shafts (center winder). The possible roll length for a 40 µm film with a width of 596 mm is approx. 5000 m (diameter: 530 mm). This shows that a rewinder with only a central drive via winding shafts is only suitable for thin films up to a limited diameter, which is limited by the permissible winding tension. A circumferential winder has advantages for thin films; it allows a longer roll running length. This is because a circumferential winder drives the roll on the circumference, and the roll characteristics are determined by the contact pressure of the contact roll in conjunction with the cutting tension.

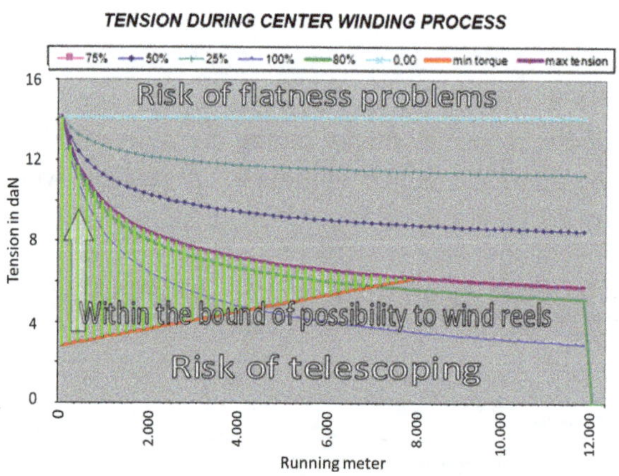

Figure 7.33
Risk of telescoping with wide films

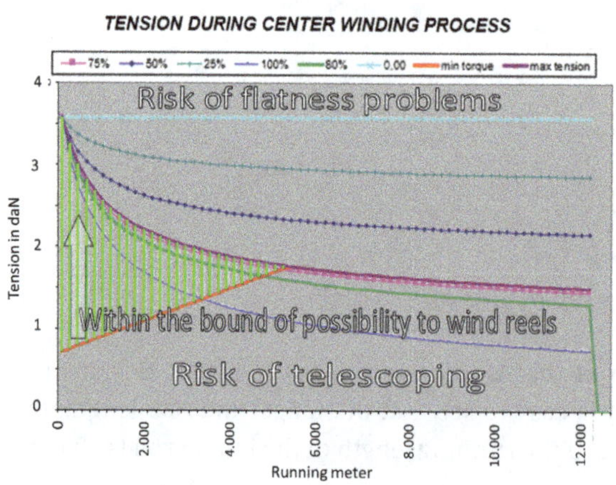

Figure 7.34
Telescoping risk with narrow film reels

7.4 Slitting and Winding

Helmut Welbers

7.4.1 Introduction

After production, the polymeric film must usually be cut to the width and length specified by the customer using the calendering or extrusion process. Of course, customer specifications must be carefully observed and adhered to. This work is carried out using so-called slitter rewinders. To maintain the qualities and tolerances of the finished product required by the customer, the work must be carried out with the appropriate care and quality.

The quality of the films produced on the calender or extruder is tested and approved according to the specification that states the customer's needs. A slitter rewinder cuts the larger parent reels into the appropriate width and length, but it must not be misused as a quality control for inferior film qualities to achieve effective operation with good reel quality.

To avoid misuse as quality control, cutting machines are now equipped with surface inspection systems, which enable errors to be localized, as is already usual with calenders or extruders. These reels with defects can then be individually reworked on so-called "doctor rewinders" to avoid unnecessary waste.

The appearance of the reels is the first impression that the customer receives when using them. This impression is important and immediately determines the competence of the supplier. Therefore, the reels must make a neat impression and have straight edges and clean winding. Customer satisfaction and trouble-free processing determine whether the customer comes back with a new order or whether they prefer another supplier.

7.4.2 Main Elements of a Slitter Rewinder

The main elements of a center winder are shown in Figure 7.35. A good result with the wound reels can only be achieved through a precisely coordinated interaction of the main elements. These main elements must be "separated" from each other by fixed points to be able to coordinate the tensile forces in the unwinding, cutting, and winding process.

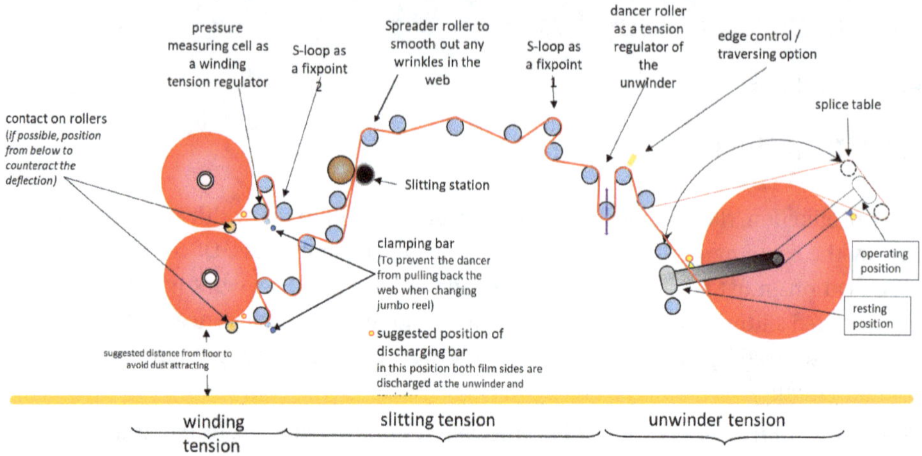

Figure 7.35 Main elements of a slitter rewinder

7.4.3 Unwinding Methods

The film web to be cut can be removed from the jumbo reel from "above" or "below". This method plays a minor role in the further slitting and winding process. The choice depends on the web routing in the machine and standardized specifications. As a rule, the winding takes place in the direction of rotation, as for the jumbo reel. This means that the reel outside of the jumbo reel is also the outside of the cut reel. Changing the outside is easily possible by changing the unwinding direction or the winding direction. However, changing the winding direction requires a conversion of the friction winding shafts or the use of friction elements that work in both directions. This could be shoulder friction, for example.

Winding tension in the unwinder

The unwinder picks up the jumbo reel to be cut. During unwinding, the web tension must always be such that a change in the web speed does not lead to an uncontrolled oscillation of the unwinding speed (Figure 7.36). To achieve this, the unwinders are equipped with a servo motor that ensures a targeted change in the tensile force and speed depending on the diameter. The tensile force must function as a braking force during unwinding to be able to build up web tension. This braking force generally increases as the diameter decreases. If the braking force is greater than the tensile force, the cutting machine stops; if it is too low, the film sags. In both extreme cases, a good cutting result is not possible.

Since the web speed remains the same during the cutting process, between the end of acceleration after starting and the beginning of deceleration before the end of the reel, the speed of the jumbo reel changes accordingly.

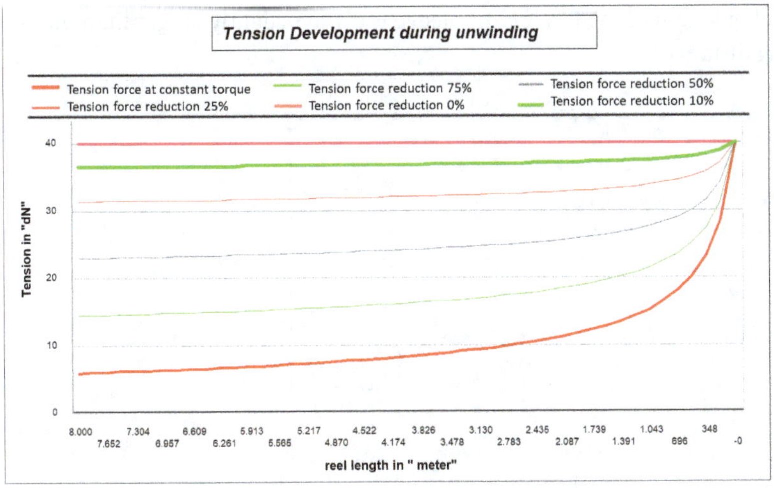

Figure 7.36 Course of the tensile force during unwinding

The change in speed is significant when there are only a few running meters left on the jumbo reel. This can be used to turn off the machine to avoid idling, and the associated time-consuming complete re-insertion of the film web.

Stopping the slitter rewinder due to the end of the reel in the unwinder

To avoid unintentional idling of the slitter rewinder, an interface must be created between the unwinder and the slitter rewinder drive. The diameter or speed of the reel in the unwinder can be used for this. As can be seen in Figure 7.37, the speed of the reel increases very quickly towards the end of the reel, and the diameter decreases very quickly. Considering the selected delay and the slitting speed, the remaining length on the core that is required until it comes to a standstill can be calculated.

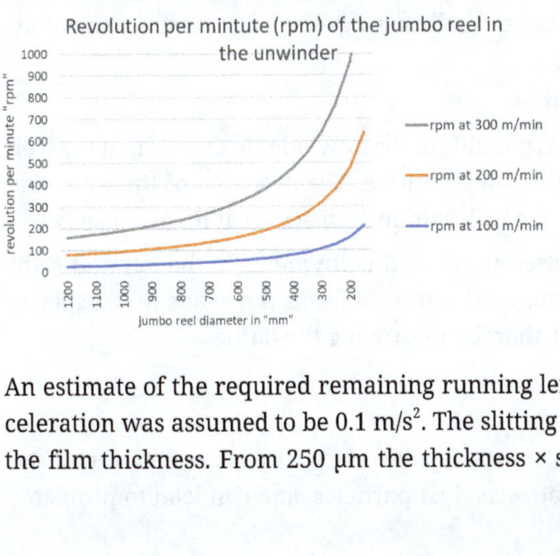

Figure 7.37 Jumbo reel speed in the unwinder

An estimate of the required remaining running length is shown in Figure 7.38. The deceleration was assumed to be 0.1 m/s^2. The slitting speed in m/min varies depending on the film thickness. From 250 µm the thickness × speed is constant. It should be noted that the remaining length of the film on the sleeve is always slightly longer than the required "braking distance".

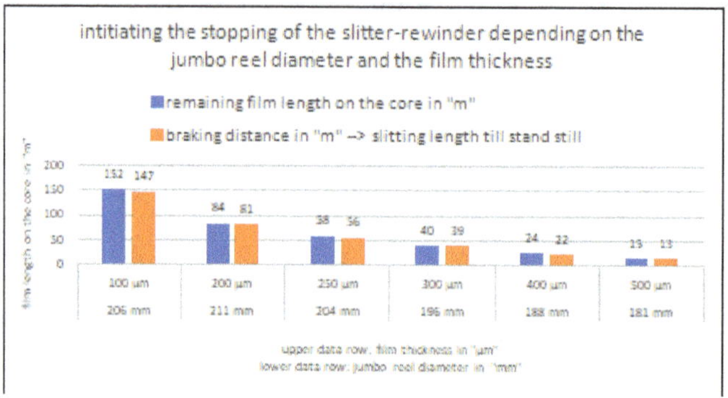

Figure 7.38 Stopping a slitter rewinder

Pressure cell or dancer?

Since it is not completely ruled out that a jumbo reel has a certain imbalance and therefore, or for another reason, does not run exactly true, the film distance between the unwinding point and the tension measuring point can change. The web can sag, and this leads to larger variations in tension. When using a tension cell, this sudden fluctuation can cause control problems and even increase the fluctuations. This difficulty can be avoided by using a dancer roller to measure and control tension. The dancer roller better compensates for fluctuations in the length of the material web because it can change its position, thus enabling length compensation.

Acceleration and deceleration can also be better compensated because braking or accelerating the jumbo reel require greater energy output due to its mass (particularly with a large diameter). The dancer device itself must have a low resulting mass and inertial force to enable rapid response. Of course, the ball bearings also must run very smoothly.

Edge control/oscillating movement

Side edge control is required as close as possible to the unwinder to ensure that the web is always guided over the rollers in the same position. The detection of the web edge serves to avoid disruptions caused by the web running, especially at higher speeds.

Scanning the web edge can also be used to move the unwinder to the left and right within certain limits. This can be helpful with thin films, as this leads to a displacement of thick areas (e.g. piston rings), thereby improving the flatness.

Ionizing bar

When non-conductive materials are unrolled, static electricity is created due to the separation of the layers. This easily attracts dust particles, and can lead to problems with films due to contamination. To avoid this, it is not advisable to increase the humidity using humidifiers until the desired effect occurs, as is common in the textile industry, for example.

Ionizing rods have proven successful in solving this problem with film webs. It is ideal if the bars can be arranged so that both sides are unloaded. During assembly, the information on distances must be observed. When the machine is at a standstill, the ionizing bars should be non-operational, to protect employees from electric shocks.

Splice table

A so-called splice table on the unwinder has proven to be very useful, as this auxiliary device makes it possible to produce a film splice without a large gap and with clean cut edges. It is ideal to have the splice table directly on the unwinder, arranged pivotally, and in a rest position during slitting so as not to disrupt the process. The middle part of the table should have strong suction, which creates a kind of vacuum and sucks the film halves so tightly that they cannot slip. A clean "butt-to-butt" splice point can then be made with adhesive tape. This device is also helpful when changing jumbo reels, and reduces machine downtime.

7.4.4 Cutting Methods

Slitting web-shaped materials can be done using various methods, each of which has its justification and advantages for certain materials. The most important and most common methods will be briefly mentioned below.

Scissor slitting is predominantly used for calendered films. Therefore, this method is discussed in more detail in Section 7.4.5.

Burst cutting with circular knife without counter-roller

Bursting cuts generally involve working without a counter-knife. The web must always be kept under a certain tension to achieve a smooth cut, which is why the two support rollers must not be too far apart. The cutting tool (round knife) must be driven and have a significantly higher peripheral speed than the material web to be cut. In addition, the circular knife must be very thin and sharp (Figure 7.39). Generally, this process is used for cutting cardboard, felt, rubber, foams, and carpets.

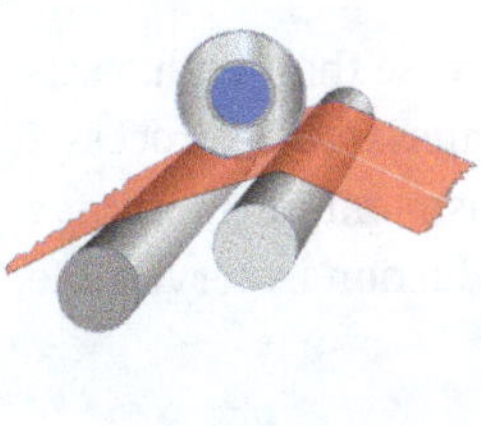

Figure 7.39
Circular knife for burst cutting without counter-roller

Razor blade cutting with counter-roller

A fixed razor blade can also be used instead of a rotating circular knife (Figure 7.40). However, this special method can only be used to a limited extent, e.g. for thin films. It enables narrow cutting widths—up to 4 mm—and is more cost-effective.

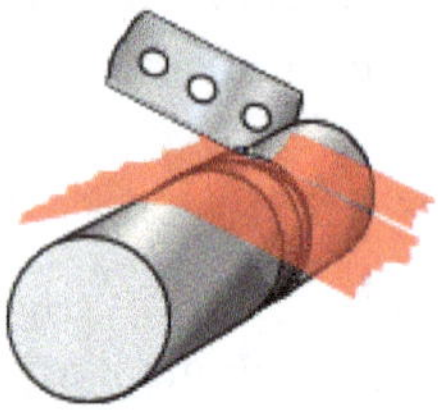

Figure 7.40
Razor blade cutting

With this cutting method, a razor blade is usually used, which dips into a counter-roller with a corresponding groove, but without touching it from the side. It is particularly suitable for very narrow widths from 2 mm, depending on the groove bushing used and its groove width. The "plunge angle" into the running web is crucial for the cutting quality.

This angle must be flat and adjusted to the material to be cut. Due to the narrow cutting widths, the diameters of the groove bushings must be the same. This can be done by grinding the "slit roller" equipped with the bushings. The method is suitable for very thin films made of plastic or metal, and of course, it is also possible to replace the razor blade with a circular blade.

Squeeze cutting

In squeeze cutting, the material to be cut is guided over a hardened shaft or bushing that does not have a groove (Figure 7.41). This glass-hard shaft or bushing serves as a counter-roller that absorbs the pressure of the squeezing knife on the material to be cut, prevents the material from "dodging", and thus enables a separating effect.

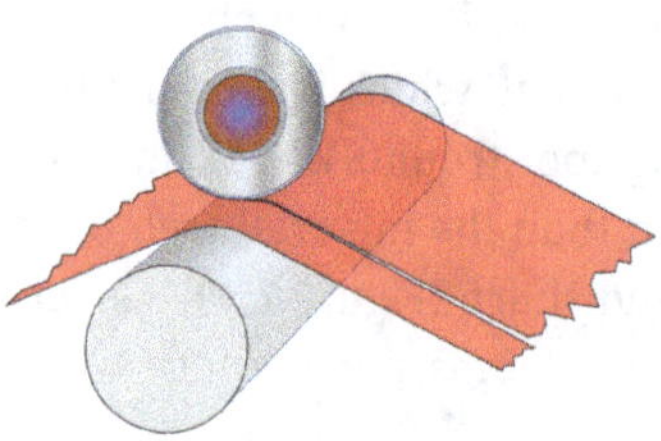

Figure 7.41
Squeeze cutting

Due to the "squeezing", the cutting edges are generally slightly worse than with a scissor cut. The quality of the cut edges depends very much on the material to be processed and its thickness. Another problem with this method is the adjustment of the contact pressure. If this cutting pressure is too high, this will lead to a reduction in service life;

if it is too low, then there is a risk that the material will not be cut through. If the cutting pressure is too low, it is also possible that the unavoidable, small manufacturing tolerances of the knife diameter and concentricity lead to vibrations that do not enable continuous cutting.

Nevertheless, this cutting method is well suited for cutting rubber, textiles, adhesive tapes, and paper. By using cutting cassettes, minimum cutting widths of 4 mm are possible. Knives with appropriate cutting shapes (e.g. zigzag) or profiles (e.g. perforation) can also be used. Zig-zag knives are used, for example, for cutting textiles to reduce "fraying" or profile knives for products that are subsequently separated by hand (e.g. streamers).

"Saw-cutting"

This method is rarely used. The application only takes place on the unwinding side to separate just one strip from a wider parent roll (Figure 7.42), which is to be used, for example, for a sample roll or a very small order. The special feature here is that the cutting knife must be completely flat on one side and slightly inclined towards the separated web to avoid streaks on the remaining parent roll side.

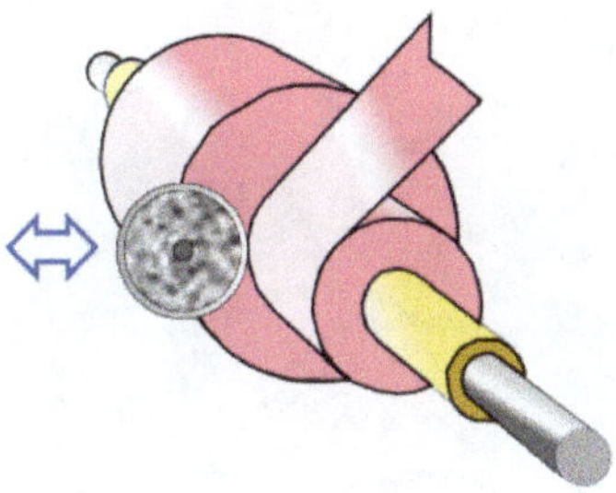

Figure 7.42
"Saw-cutting"

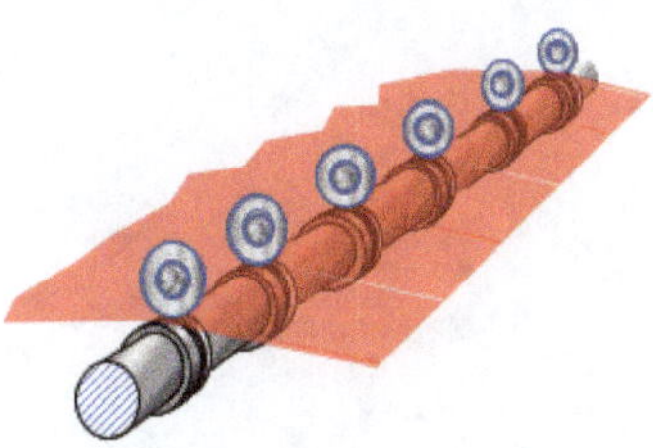

Figure 7.43
Scissor cutting

Scissor cutting

This cutting method is the usual cutting method for plastic films, such as PVC (hard or soft films), PET films, PS films, PE films (HD and LD) and is therefore discussed in more detail in Section 7.4.5 and the following sections. Some elements of scissor cutting are also important in the methods listed in the current section.

Other methods

Other cutting methods, which are only used for special purposes, are ultrasonic cutting, laser cutting, water jet cutting, and hot cutting. However, these are not important for cutting plastic films.

7.4.5 Scissor Cutting

This cutting method (Figure 7.44) is covered in its own section, because it is the usual method for cutting calendered jumbo reels into the customer's desired widths and lengths. The reason for this is the high speeds that are possible, combined with the very smooth cutting edges, generating very little cutting dust and avoiding so-called "angel hair". All of this assumes that the cutting knives (top and bottom knives) are of good quality, sharply ground, and not damaged.

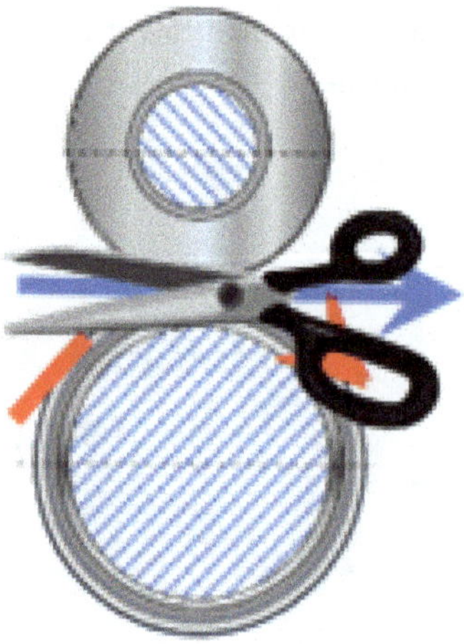

Figure 7.44
Scissor cutting

The most important criteria for a good cutting result are:

- The sharpness of the cutting blades
- The overlap of the knives
- The angle of attack
- The lateral contact force.

These criteria are discussed in the following sections.

The material of the knives depends on the material to be cut and must be determined together with the supplier. An HSS knife is recommended for "normal" applications.

With a well-adjusted scissor cut, the edges of the films are smooth, not frayed, not rough or broken, and have no burrs. Fluctuations in width are also avoided.

In scissor cutting, a distinction is made between the so-called "wrap cut" and the "tangential cut". With a wrap cut, the film is guided at a significant angle around the bottom knife shaft (as shown by the red arrow in Figure 7.44). With a tangential cut, the film theoretically only touches the bottom knife at one point, like a tangent (as shown by the blue arrow in Figure 7.44).

The wrap cut requires a continuously equipped shaft with female knifes, U-rings, and spacer bushings for the bottom knives. The circular knives can also be mounted on a continuous shaft, but individual knife holders are also possible. When cutting tangentially, the female knife shaft can be equipped with individual female knives, as a larger gap is possible without any problems. Here, individual knife holders are the rule for the male knives. This combination enables automatic knife positioning.

With single knife holders, "lubrication" of the male knives is also possible—and is also offered. However, great caution is required, especially with films for the food and pharmaceutical industries; the lubricant used must be "food-safe" and there must be no over-lubrication. Contamination of the film surface with the lubricant must be avoided at all costs.

7.4.6 Sharpness of the Slitting Blades

References for this section: [2–4].

A good cutting result requires sharp knives with the longest possible service life. Due to use, all knives are subject to wear and tear, and must be sharpened after a certain period. Regrinding the knives requires a lot of experience and precise work on a grinding machine certified for this work. The surface to be reground must be smooth (polished) and have a matt shine. No burrs should be present at the cutting angle. The area to be sharpened depends on the type of knife.

Top knife

Top knives are a dished type of male knife that need to be sharpened in diameter (arrow in Figure 7.45). The diameter decreases with each grinding process. If the diameter is so small that no overlap with the corresponding bottom knife can occur, the top knives must be replaced with new ones. When using the knives for wrap cutting on a top knife shaft, all top knives (in sets) must be ground to the same diameter to ensure an almost equal overlap. This is not relevant when using knife holders, as the overlap is adjusted individually.

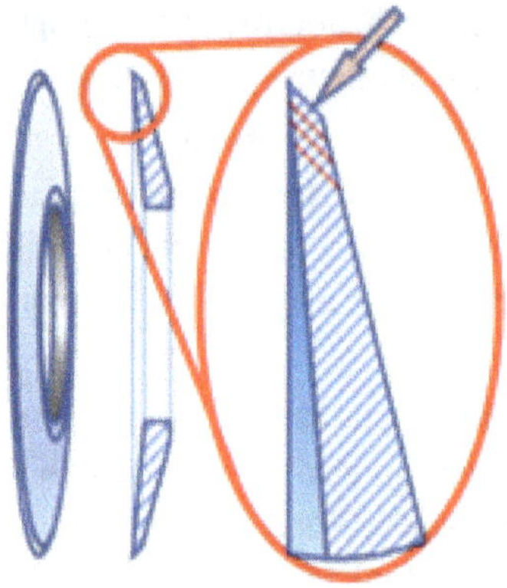

Figure 7.45
Top knife or male knife

Pot knife

Pot knives are a special type of top knife. They have a wide shoulder and cannot penetrate as deeply into the material to be cut (Figure 7.46). They are therefore also suitable, in a slightly modified form, as knives for "saw cuts". The regrinding takes place on the side surface, as for bottom knives (arrow in Figure 7.46).

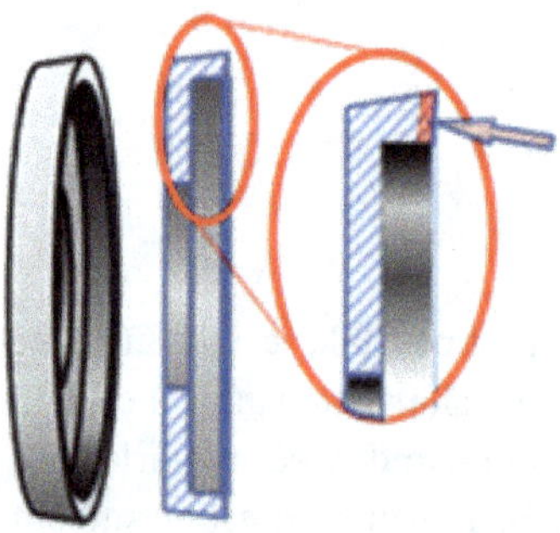

Figure 7.46
Pot knife

When using these knives, ensure there is a wider gap between the pot knife and the bottom knife (minimum 10 mm groove width, with a 300 µm thick film). The vertical pressure on the top knife must be sufficiently high to ensure that the knife overlap is maintained.

Pot knives are primarily used with knife holders; they are particularly suitable for thin films. The advantage is a longer service life and less heating due to the larger mass and the large cutting angle (approx. 80°).

Cutting point shift due to wear

During cutting, the cutting point shifts due to wear on the cutting tips on both the top knife and the bottom knife (Figure 7.47), which also leads to deterioration of the cutting edge. At the point when the knife wear has progressed to such an extent that the wear rings only touch at a tangent, a clean cut is no longer possible.

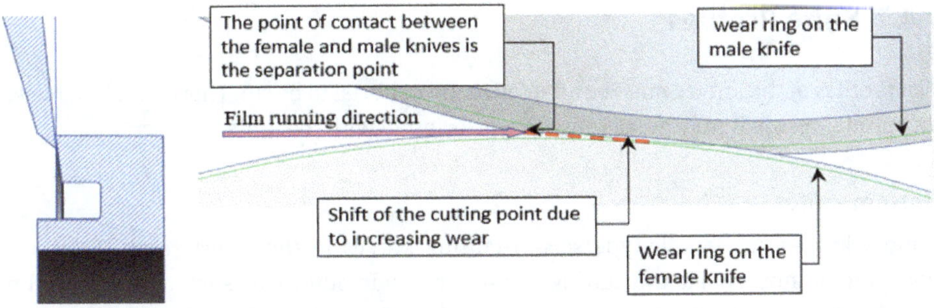

Figure 7.47 Cutting point shift

The wear area is shown in Figure 7.48. It can be clearly seen that the tip of the top knife is removed due to the abrasive (scraping) effect of the film. That is why the regrinding area is relatively high and the diameter reduction is therefore important.

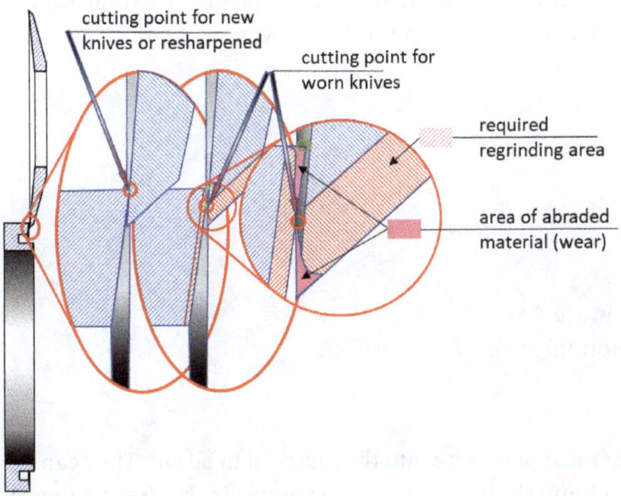

Figure 7.48 Wear area

Summary

- To get good cutting results, the knives must be well and smoothly ground with polished surfaces and without any burrs on the edges
- When cutting, the knives are subject to wear and tear, causing them to become "dull", which worsens the cutting result
- Top and bottom knives must touch each other at one point (the cutting point), and this is also necessary during the cutting process
- The position of the cutting point shifts during cutting due to knife wear.

7.4.7 Knife Profiles

The profiles of the knives must be adapted to the material to be cut. Therefore, to achieve an optimal result, it may make sense to work closely with the knife supplier.

Bottom knife

Bottom knives are female-type scissor-cutting knives in the truest sense. They are wide and do not have acute cutting angles. The wide outer ring surface supports the film during the cutting process and thus stabilizes its position. The edges must be broken and polished; they must not have any burrs, to avoid scratches in the film web. The wear is less than with the top knives, which are usually narrow dished knives and have a pointed cutting edge.

As shown in Figure 7.49, the regrinding takes place on the inner surface (see arrow) and does not change the diameter. When adjusting the roll width, the ground surface must therefore be used as a reference dimension for its adjustment. A clearance angle of 3° can be recommended as a standard for general cutting work.

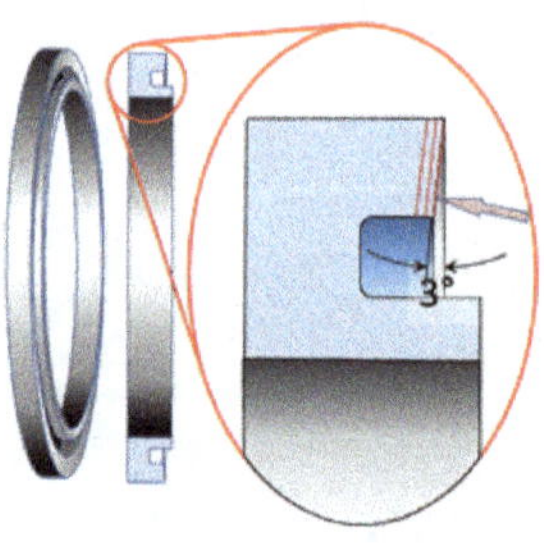

Figure 7.49 Bottom knife

Top knife

Top knives are the blades that must penetrate into the material to be cut. They can be arranged on a continuous top knife shaft and are moved accordingly on the support shaft when the width changes. This version is more cost-effective and can definitely make sense for narrow machines. Nowadays, individual knife holders are generally used. The advantages and disadvantages are discussed in Section 7.4.11.

Top knives can have different shapes and grinding angles. Figure 7.50 shows top knives with 45° and 10° grinding angles. In both cases these are dished knives, also known as circular blades, because they are very narrow. The cutting edges are very sensitive to any mechanical force and can be easily damaged. This is particularly the case with top knives with a very narrow cutting edge of only 0.3 mm to max. 1 mm, with a cutting angle of 10°, and which are only suitable for thin films (up to approx. 60 μm). Experience has shown that using these narrow cutting edges can prevent cutting burrs on thin films. This is due to the slight deformation of the film after the slitting process.

This is discussed in Section 7.4.6 in more detail. A cutting angle of 45° has proven to be well suited for normal cutting work on thicker films. The wear on these circular knives is higher than on the bottom knives. The type of regrinding is already described in Section 7.4.4 (Figure 7.45).

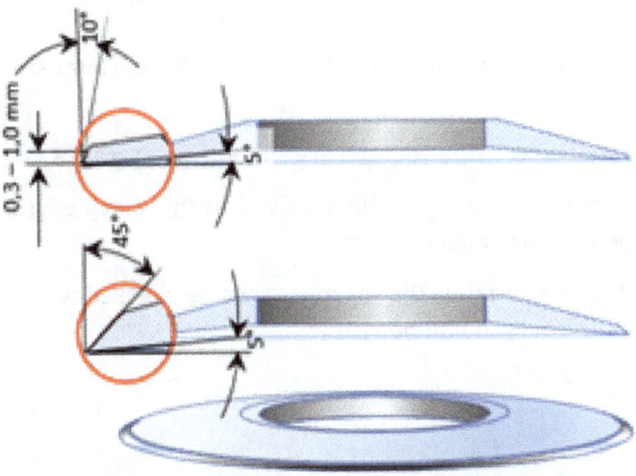

Figure 7.50
Top knife

Pot knife

Pot knives are a special type of male knife that are also suitable for thin films, up to around 300 µm. They are particularly suitable when using individual knife holders and for tangential cutting, as the distance between the bottom knives must be relatively large. The reason is explained in Section 7.4.11.

An advantage of these knives is the robust cutting angle—like a bottom knife—and the lower wear (Figure 7.51). The higher mass also means that the resulting heat can be dissipated better.

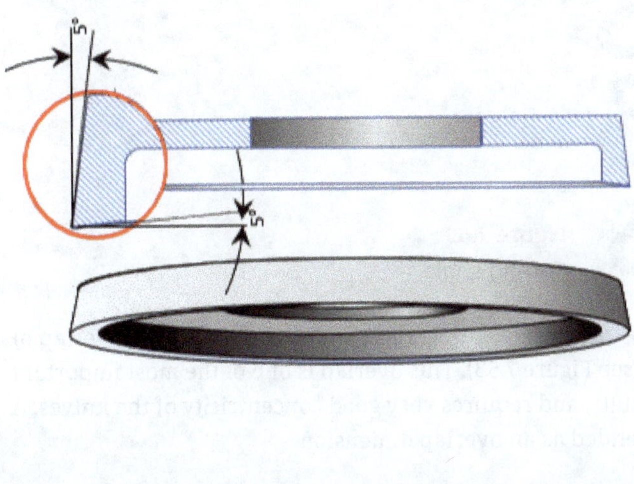

Figure 7.51
Pot knife

With thicker films, the vertical cutting pressure (not the contact pressure against the bottom knife) must be high to ensure that the film can be cut through safely. Even with thicker films, the wide surface allows the film to be guided better and the deformation is not as significant. This is discussed in more detail in Section 7.4.8.

Summary

- Knives are very sensitive to impact or falling, so they must be handled very carefully
- Magazines made by the manufacturer or by yourself are suitable for storage
- Caution must also be the top priority when handling knives, as the sharp edges, especially with dished knives, can easily lead to cuts
- The integrity of the ground surface and thus the sharpness can be checked with a flashlight; no reflective areas or burrs should be visible.

7.4.8 Overlap

References for this section: [2–4].

The overlap or immersion depth is an important criterion for the knife setting for the cut. Basically, the smaller the cutting angle, the better the cutting edge of the reel. This can be easily demonstrated using standard scissors (Figure 7.52). If the angle between the two legs of the scissors is acute (small), paper, e.g. 80 g/m^2, can be cut by pushing the scissors (assuming they are sharp), whereas this is not possible with a large angle.

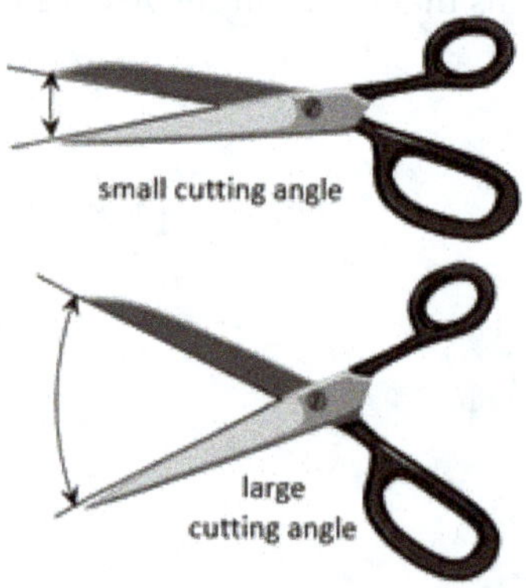

Figure 7.52 Cutting angle

When cutting with circular knives, the cutting angle is determined by the overlap of the top and bottom knives (see Figure 7.53). This overlap is one of the most important settings for good cutting results, and requires very good concentricity of the knives. A value of 0.7 mm is recommended as an overlap dimension.

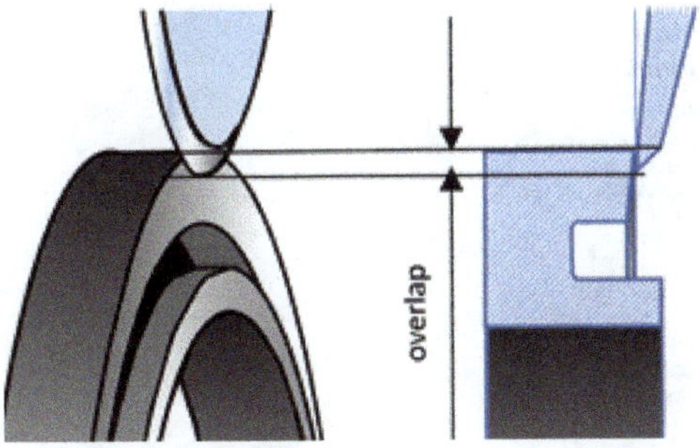

Figure 7.53
Overlap of the top and bottom knives

Figure 7.54 shows how the film deforms immediately after cutting depending on the overlap. It is easy to see from the illustration that the film deforms less with a smaller overlap than with a higher one. The greater deformation also results in a frayed and rough-cut edge of the film. Since the film webs cross each other on one side as they continue, this large deformation can lead to friction, which can result in the web being torn off.

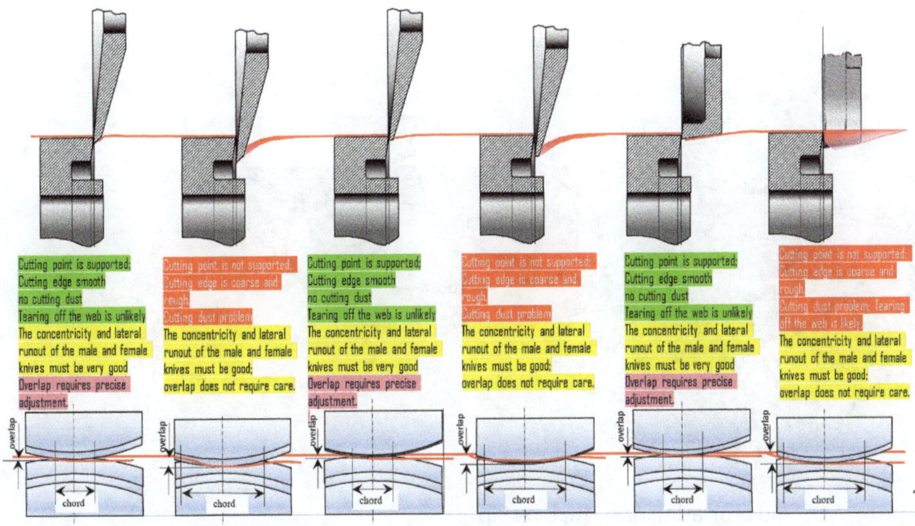

Figure 7.54 Deformation of the film due to overlap

Since an overlap of 0.7 mm is very difficult to estimate, a relation to the *width* of the overlap must be calculated. This width is the chord between the intersection points of the top and bottom knives. In Figure 7.55 one can clearly see that the length of the chord becomes longer with deeper overlap. The adjacent diagram (knife overlap) shows the result of the calculation. With a top knife 130 mm in diameter and a bottom knife 100 mm in diameter, the length of the chord is 12.6 mm. With a top knife 130 mm in diameter and a bottom knife 150 mm in diameter, the chord is 14 mm long. In both cases the overlap is 0.7 mm. The chord length in these dimensions can be better estimated than the 0.7 mm of overlap or immersion depth.

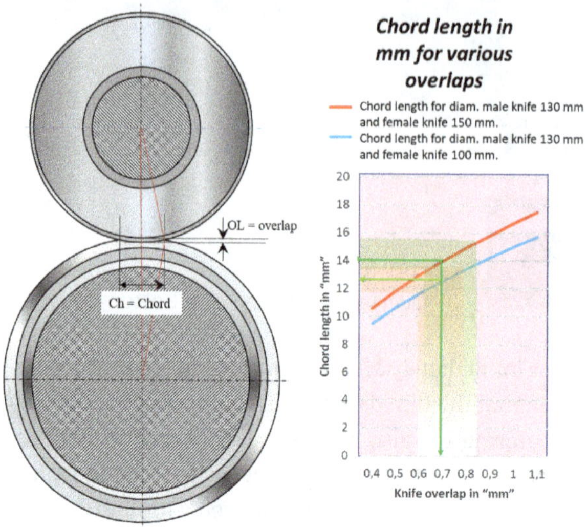

Figure 7.55 Chord length at different overlaps

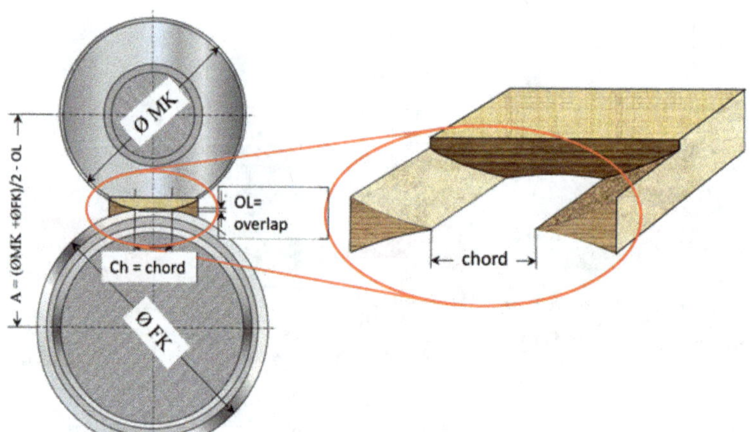

Figure 7.56 Aids for adjusting the overlap

The formula for calculating the chord length l_{chord} (Ch) is:

$$l_{chord} = \frac{d_{top}\ \arccos\left(\left(\frac{d_{top}}{2}\right)^2 + \left(\frac{d_{top}+d_{bottom}}{2} - l_{overlap}\right)^2 - \left(\frac{d_{bottom}}{2}\right)^2\right)}{d_{top}\left(\frac{d_{top}+d_{bottom}}{2} - l_{overlap}\right)}$$

where:

d_{top} = diameter of the top (male) knife

d_{bottom} = diameter of the bottom (female) knife

$l_{overlap}$ = overlap

An adjustment gauge is sketched in Figure 7.56, which helps to adjust or check the overlap of the cutting knives.

Summary

- The larger the overlap, the greater the deformation of the film by the top knife. This can lead to a rough-cut edge and tears in the film web (Figure 7.54)
- In Figure 7.54 it can also be seen that a concentricity error and the side runout must be very small with a small overlap
- Adjusting the overlap requires precise adjustment
- A good overlap is achieved when the length of the chord does not exceed approx. 13 mm
- Pot knives are better suited for tangential cutting, because the distance between the bottom knife must be relatively large.

7.4.9 Clearance Angle or Shear Angle

With a good pair of commercially available scissors, you can observe that the legs of the scissors are not exactly parallel, but rather cross each other slightly. This crossing is achieved in the scissor cutting of a rotary knife by setting the top knives at an angle to the bottom knives. This angle should be 0.25–0.5° for cutting PVC.

To illustrate the problem with the adjustment, this angle is shown greatly exaggerated in Figure 7.57.

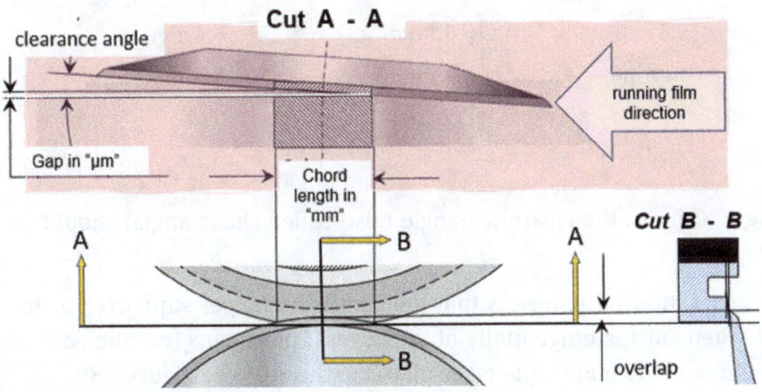

Figure 7.57 Clearance angle

For rotary knifes with a continuous top knife shaft, measuring the gap on the individual knives makes little sense. As can be seen from Figure 7.58, the gap at the end of the chord is approx. 50–100 µm. Therefore, on this type of machine, the axis crossing of the top knife shaft to the bottom knife shaft must be measured at the end of the top knife shaft. With single knife holders, this clearance angle can be precisely adjusted using an “angle vernier”.

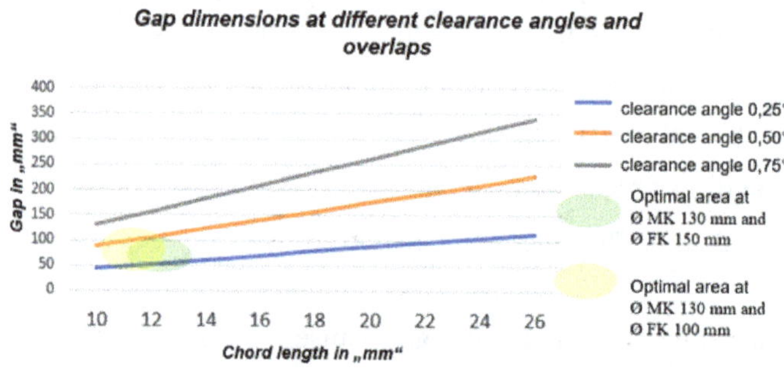

Figure 7.58 Gap dimensions at different angles of attack and overlaps

Table 7.4 shows the required axis crossing for different film widths and clearance angles. This value shows that, especially on wider machines and with tangential cutting, the film is not supported at the edges at the cutting point. Therefore, wider machines with a continuous top knife shaft should, if possible, only be operated with a wrap cut.

Table 7.4 Axis Crossing Values for Different Film Widths and Clearance Angles

Clearance angle	Film width		
	2200 mm	1400 mm	1000 mm
0.25°	4.8 mm	3.1 mm	2.2 mm
0.50°	9.6 mm	6.1 mm	4.4 mm
0.75°	14.4 mm	9.2 mm	6.5 mm

Summary

- When cutting PVC films, the clearance angle (also called shear angle) should be between 0.25° and 0.5°
- On wide machines, this value means that the film is no longer supported at the cutting point when cutting tangentially at the edges. This means that the contact between the film and the male knife takes place “in the air” (see Figure 7.59)
- When using a support shaft for top knives, the machine must not be too wide (max. approx. 1400 mm); this is one of the reasons for using single knife holders
- When using a support shaft for top knives, only a wrap cut makes sense because the cutting point is always supported due to the large wrap angle.

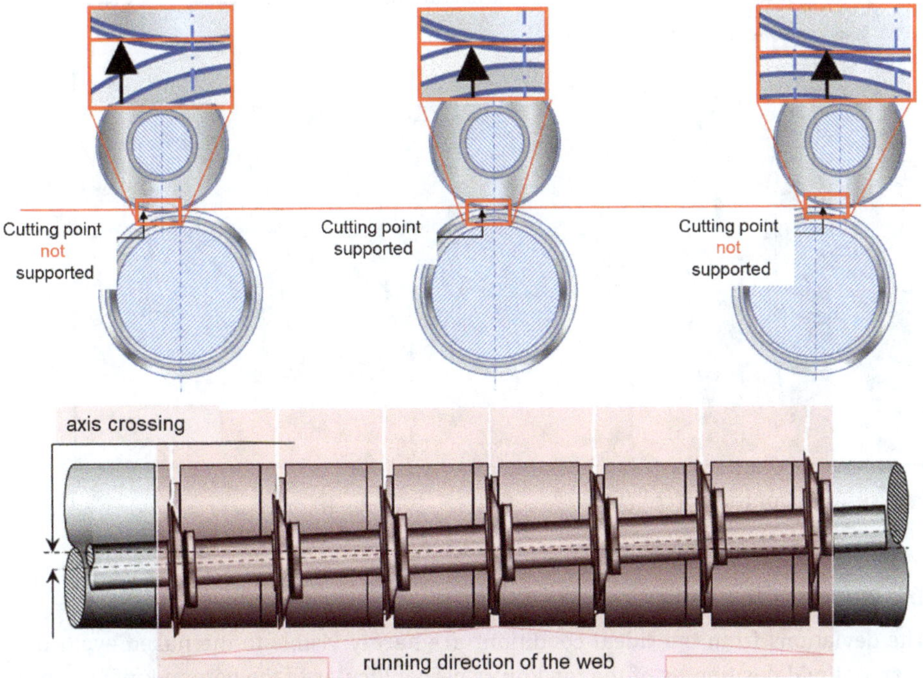

Figure 7.59 Clearance angle

7.4.10 Contact Force

References for this section: [2–4].

With standard scissors, cutting only works if the two legs are pressed against each other with a certain amount of force. This can be easily adjusted by loosening the screw that connects both legs.

When making a scissor cut in the cutting machine, the top knife must therefore be pressed against the bottom knife. This contact force must not be too low, but not too large either (Figure 7.60). If the pressure is too low, the cut will not be clean, and if the film is thicker, the material to be cut can push the top knife (male knife) away from the bottom knife (female knife). This is comparable to a connecting screw on conventional scissors that is too loose.

If the pressure is too high, wear is very high, and the knife's service life is significantly reduced. The cutting point is also lost, meaning that the cut becomes rough, and the web can tear. This is comparable to conventional scissors whose connecting screw is tightened too much.

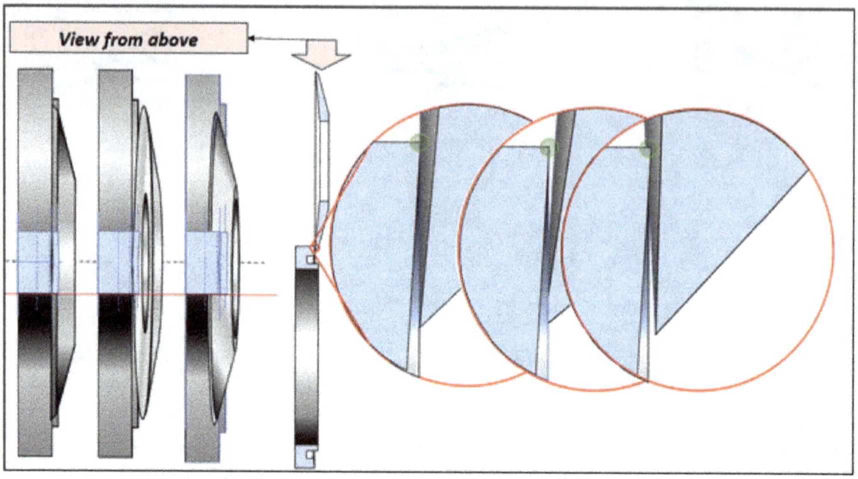

Figure 7.60 Contact force

Inclination of the knives

The deviations from the "ideal condition" are hardly visible to the naked eye. However, optimal positioning of the top knives perpendicular to the bottom knife is a prerequisite for good cutting results. To make this clearer, the following sketches show the inclination of the top knife to the bottom knife and the overlap in a greatly exaggerated manner.

Figure 7.61 shows the situation when the vertical axis of the top and bottom knives is at a right angle to the film web. Only then is the contact point between the top and bottom knives also the cutting point of the film. Only then is the cutting-edge smooth, without burrs, with very low levels of cutting dust, and with no issues from so-called "angel hair".

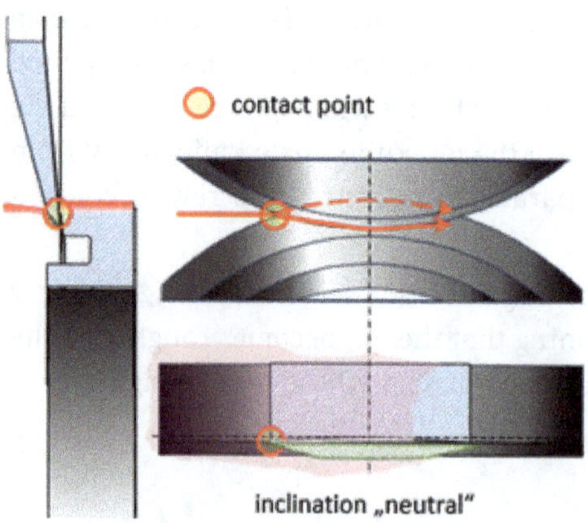

Figure 7.61 Blade inclination

Figure 7.62 and Figure 7.63 show the situation that arises when the axis is positive (axis to the film web; angle <90°; Figure 7.62) or negative (axis opposite the film web; angle >90°; Figure 7.63) is inclined. In both cases, the contact point of the film with the top and bottom knives is not the contact point of the top knife with the bottom knife. The contact point (cutting point) between the top and bottom knives is the vertex of the bottom knife with positive inclination, and the vertex of the top knife with negative inclination.

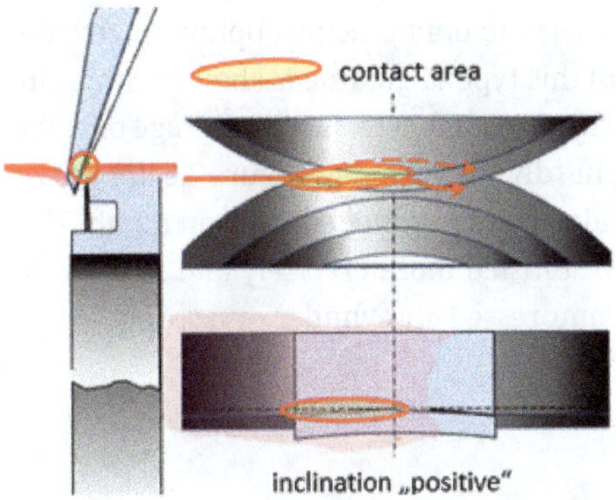

Figure 7.62
Positive inclination

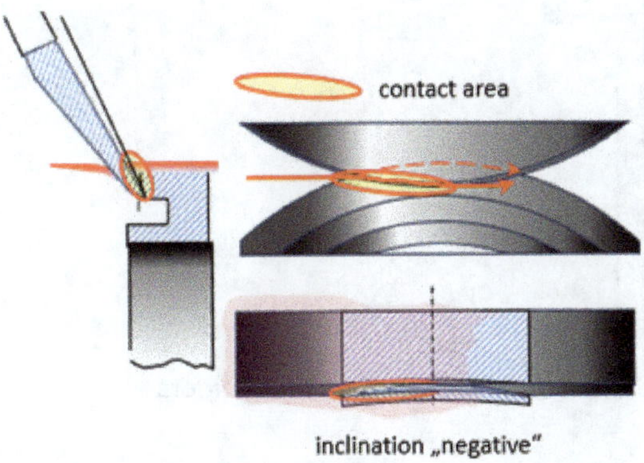

Figure 7.63
Negative inclination

As a result, the cutting point no longer exists, and a cutting area is created. In this area the film is "somehow" cut through but no longer cut smoothly. The "cutting edges" are rough, uneven and have a burr. The wound rolls show a rough side, cutting dust is created, and "angel hair" is possible. These very narrow stripes are created when the edge of the film touches the top knife (male knife) after it has been cut.

A positive or negative inclination is always associated with poorer cutting quality and must therefore be avoided. That's why adjusting the contact pressure is also very important.

Basic top knife holder and -guide

When holding and guiding the top knife, a distinction is made between a freely swinging guide on an adapter and a guide on a movable adapter. The free-swinging guide is mainly found on top knives with an adapter on a top knife shaft. To give the top knife a certain freedom of movement, with this type of guide either a plate spring or a spiral spring is arranged between the adapter and the top knife.

Figure 7.64(a) shows the freely swinging guide with an adapter on a continuous top knife shaft. This type of guide can cause a positive inclination—especially with high contact pressure. With a continuous top knife shaft, this is the only practical option to give the male knife a certain flexibility. In Figure 7.64(b) this type of guiding is shown with a single top knife holder. The same problems are to be expected here as with storage on a top knife shaft, which is why this construction is hardly in use today. Figure 7.64(c) shows the situation when using a single top knife holder with a sliding adapter on a hub. This arrangement allows the contact pressure to be adjusted more precisely and so to avoid inclination, using adjustable pressure from a compressed air cylinder or a spring.

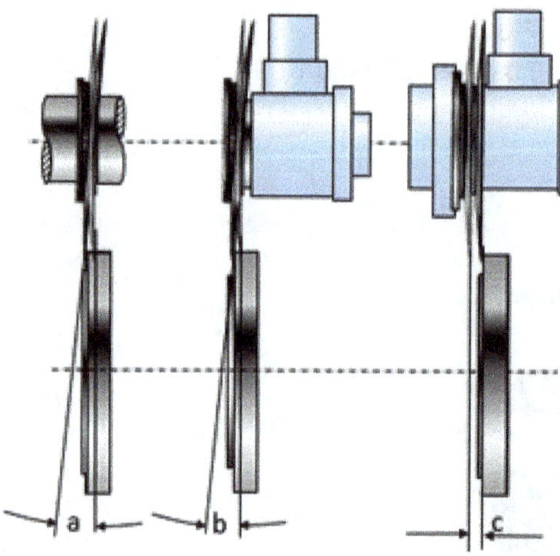

Figure 7.64 Various top knife holders

Summary

- The contact force between the top (male) and bottom (female) knives is the decisive parameter for a clean cut
- Cutting is not possible without contact force and the resulting contact between the top and bottom knives
- The lateral contact pressure must be chosen to be as low as possible so that "straight" cutting is possible

- However, the lateral contact pressure must be high enough so that the contact point between the knives is not lost during the cutting process
- The following applies as a guideline for cutting hard PVC films:
  - Approx. 10 N for thin films up to around 100 µm
  - Approx. 15–25 N for films between 100 and 750 µm
  - Approx. 30 N for thick hard PVC films from around 800 µm.

  These values are of great importance for the cutting protocol.

7.4.11 Single Knife Holder

References for this section: [2–4].

Single knife holders (Figure 7.65) have largely become established as the state of the art and are widely used today. They can be used for both wrap cuts and tangential cuts. This section will show the advantages and disadvantages compared to using a continuous top knife shaft.

Figure 7.65
Knife holder PSG1 (Source: [5])

Advantages

- Quick and easy change from one roll width to the next
- Automated width adjustment is possible (and the current state of the art)
- The overlap can be easily adjusted using a vernier
- A difference in diameter between the top knives can be compensated for

- The clearance angle (or shear angle) can be adjusted to the optimal value of 0.25° using an angle plate
- The contact force can be measured and adjusted precisely
- The possible linear guide (slidable adapter) ensures that inclination is avoided
- The setting values can be easily reproduced
- There are aids for adjusting the overlap and the contact force
- Replacing the top knives (male knives) is quick and easy
- It can be used in both wrap cuts and tangential cuts
- A cutting guard (hand guard) can be easily attached (and is standard).

Disadvantages

- Limited vertical pressure (depending on maximum compressed air pressure and cylinder area), which can cause problems when using a pot knife as the top knife with a thick film
- The peripheral speed of the top knives is lower than the peripheral speed of the bottom knives.
- Powered top knives would be very complex and impractical
- The acquisition costs are significantly higher compared to a solution with a continuous top knife shaft.

Summary

The comparison of the two possibilities—top knife on a continuous top shaft or single knife holder—clearly shows that using a single knife holder has advantages over the top knife shaft. Today, the single knife holders are available in a wide variety of designs for thinner and thicker materials, as well as various options for precise adjustment for cutting force, overlap, contact force and shear angle. Small cutting widths from approximately 20 mm for thinner films up to a minimum finished roll width of approx.150 mm for thicker products is standard. The most significant difference here is the vertical cutting pressure (knife pressure) and the horizontal contact pressure.

The only serious disadvantage that can be cited is the higher costs. However, these reduce with the operating time due to higher cutting performance and lower wear (regrinding costs). The lower peripheral speed does not play a significant role if the parameters are well adjusted.

7.4.12 Runout and Vibration

References for this section: [2–4].

The previous sections have mainly dealt with the cutting knives. Now we will take a closer look at the influences of cutting machines and cutting methods.

Good knives are of course important for a good cutting result, but the cutting machine itself must also contribute to a good result through vibration-free running and excellent concentricity.

Runout error

In Figure 7.66, the effect of a concentricity error in the top or bottom knife is exaggerated to make the problem clearer. A concentricity error in one of the two knives results in a fluctuating overlap. This affects the cut edge and the appearance of the side running surface of the finished roll. The fluctuations in the overlap become visible in so-called "annual rings".

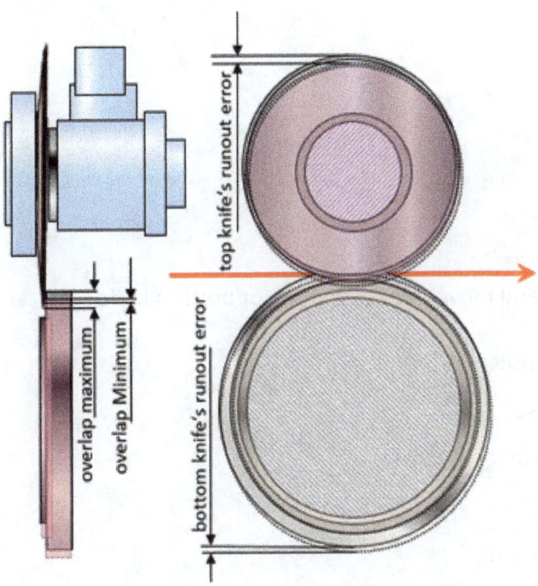

Figure 7.66
Effects of concentricity errors

The causes of this error are, for example:

- The shaft for holding the top or bottom knives is bent
- The play (fit) between the shaft and the knife or adapter bore is too large
- Impurities, e.g. if there is dust under the knives, adapters, or clamping rings
- The grinding machine does not work precisely; the hole is not in the center.

Vibration (lateral tremor)

Figure 7.67 shows, also exaggerated for better explanation, the situation when the top or bottom knife makes a lateral movement (e.g. a "wobble" or a "figure of eight"). This leads to changes in the lateral contact pressure and can also influence the angle of attack. This error results in a rough-cut surface of the rolls and can lead to web breaks. These problems mainly occur with continuous top knife shafts, but individual knife holders are not immune to them either.

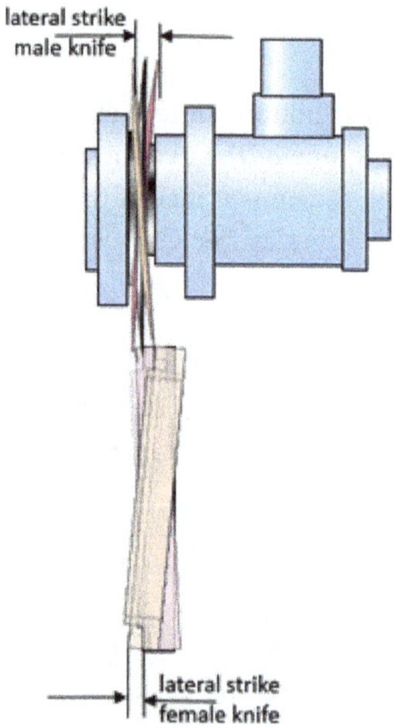

Figure 7.67
Lateral movement of the top or bottom knife

The causes of this error are, for example:

- The adapter on the shaft wobbles
- The adapters for the top knives are too narrow
- The hole is crooked
- The diameter of the adapters is worn
- The knives are poorly sharpened
- The storage in the individual knife holder is defective
- The bottom knife is bent
- The top knife is bent.

7.4.13 Slitting Geometry

When it comes to scissor cutting, a distinction is made between a wrap-cut and different variants of a tangential cut. The following sketches are intended to illustrate these different methods and show their advantages and disadvantages.

Wrap cut

During a wrap cut, the film wraps a significant distance around the bottom knife shaft relative to the bottom knives, both before and after the cut (Figure 7.68, point "a"). The parallel course after cutting over a longer distance can lead to the webs crossing over and thus to a rough cutting edge or, at higher speeds, to a web break. The bottom knife shaft must be fully equipped with spacers and bushings. When cutting, the deformation of the web caused by the top knife (male knife) is significant (point "b").

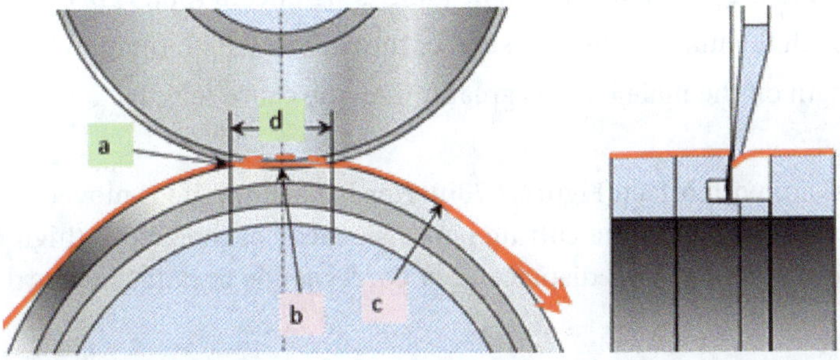

Figure 7.68 Wrap cut

The advantage of this cutting geometry is that the cutting point is always supported (point "a") and the contact length of the web with the knives is relatively short (point "b"). It corresponds exactly to the chord length and depends only on the overlap.

For slitter rewinders that are not very wide, this variant is justifiable if costs play a major role (see also Section 7.4.9). The top knives must be driven with a slight advance compared to the bottom knives.

True tangential cut

With a true tangential cut, the film web only touches the bottom knife shaft at one point. There are two variants here:

- The real tangential cut without offset (Figure 7.69a). The centers of both knives are exactly vertically above each other (at a 180° angle). However, this variant is not suitable for cutting PVC films.

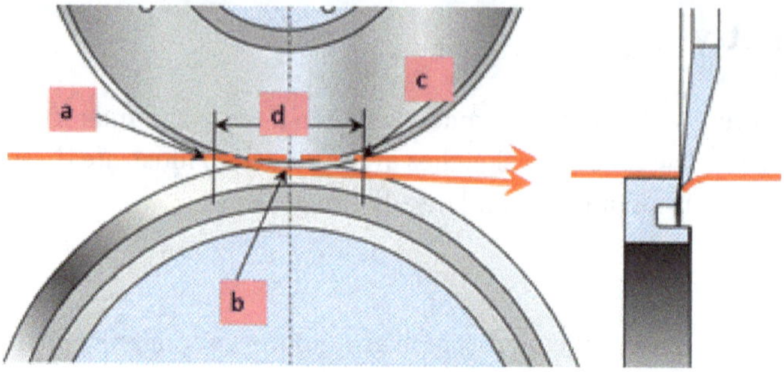

Figure 7.69 Tangential section without offset

The disadvantages are:

- The cutting point is not supported (point "a"). There is no scissor cut, but a burst cut. The cutting edge is rough, and a web tear is likely
- The deformation of the path along the male knife is significant (point "b")
- The web can touch the male knife at the end of the contact length (point "c")
- The contact length on the male knife is greater than the chord length (length "d").

- The real tangential cut with offset (Figure 7.70b). Here, the top knife is moved in the direction of web travel until the cutting point is exactly at the zenith (highest point) of the female knife. This displacement corresponds to half the chord length.

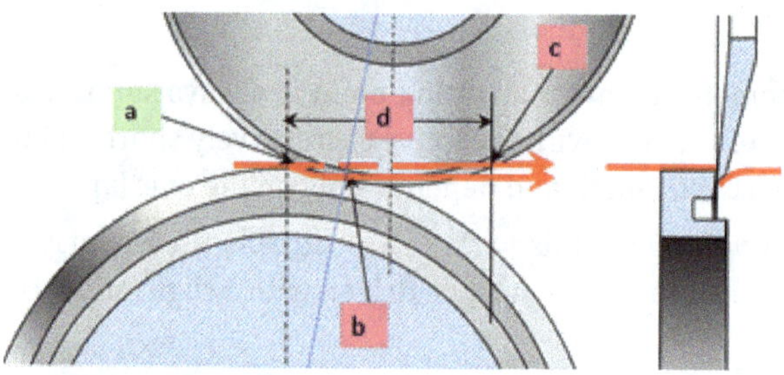

Figure 7.70 Tangent section with offset

This variant has the advantage that the cutting point is supported, and the overlap is relatively easy to adjust, since the contact point of both knives must always be at the zenith of the bottom knives (point "a"), this is relatively easy to estimate.

Otherwise, the disadvantages are the same as in the variant without offset:

- The deformation of the path along the top knife is significant (point "b")
- The web can touch the top knife at the end of the contact length (point "c")
- The contact length on the top knife is twice as long as the chord length (length "d").

Modified tangential cut without offset

This variant is best suited for a good scissor cut (Figure 7.71). The position of the rollers before and after the cut is decisive. The zenith of the guide rollers before and after the cut must be designed so that the film web touches the bottom knife just before the cutting point (point "a") and leaves the bottom knife at the same height after the cut (point "c"). The contact length of the film web corresponds to the chord length (length "d") and is therefore short. This variant is also well suited for the use of pot knives, which minimizes the deformation of the film. This is important for thin film to prevent one-sided "edge build-up" caused by the male knife.

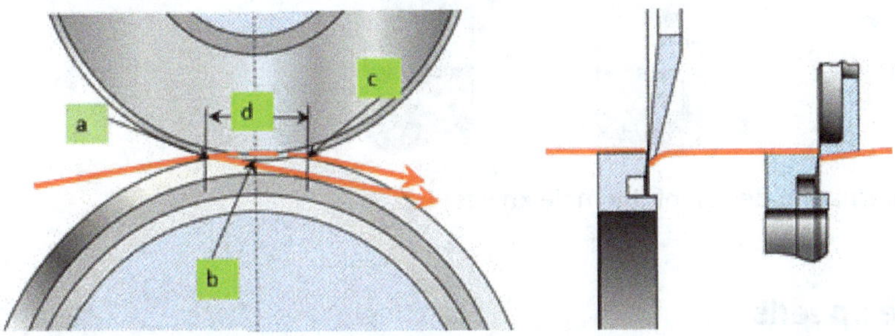

Figure 7.71 Modified tangential cut

Summary

Of the three variants discussed, the variant of the modified tangential cut without offset is best suited for slitting film stripes. All three options allow the use of single knife holders and enable automatic adjustment of the cutting widths for the reels. This variant is also very suitable for using pot knives. The "vertical" distance between the guide rollers "A" and "B" corresponds to the sum of the radii of both rollers plus the dimension of the overlap ($r_A + r_B + 0.7$), in mm (Figure 7.72).

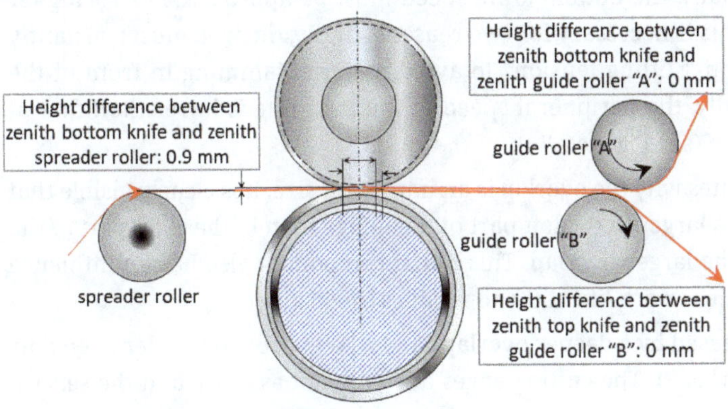

Figure 7.72 Slitting geometry

The pot knives allow for further improvement. If the male knives are used alternately in a right–left design (Figure 7.73), then the stripes can be guided to the winding shafts "without crossing" after the cut. This enables a further increase in web speed and thus better productivity.

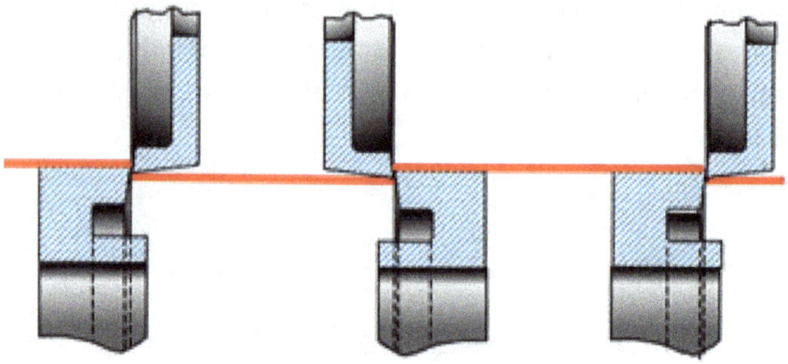

Figure 7.73 Alternating design of the male knives

7.4.14 Knife Speeds

Circumferential speed during wrap slitting

The peripheral speeds of the top and bottom knives must be in relation to the web speed to obtain a clean, smooth cut. Only with wrap slitting can the speed of the knives be the same as the web speed. If the top knives are on a common shaft, they are usually driven. The peripheral speed of the top knives can then even be greater than that of the bottom knives. But this is the only advantage, and it does not outweigh the other disadvantages.

Circumferential speed during tangential slitting

When slitting tangentially, the bottom knife speed must be approx. 3% to 4% higher than the web speed. This is necessary for two reasons: firstly, the web must constantly be kept under a certain "slitting tension" to avoid material jamming in front of the slitting section. Secondly, the peripheral speed of the top knife is lower than that of the driven bottom knife (see Figure 7.74).

Figure 7.74 also illustrates why the overlap is an important size. It is clearly visible that the radial component is larger in the left part of the image than in the right part of the image. This is due to the larger overlap. This radial component also has an influence on wear and thus on the service life of the top knives (male knives).

The cutting angle increased by a deeper overlap can be compared to a wider open pair of scissors (see Section 7.4.8). The cutting edges are then not as even, and the service life of the knives is significantly shorter because the wear on the top and bottom knives is higher.

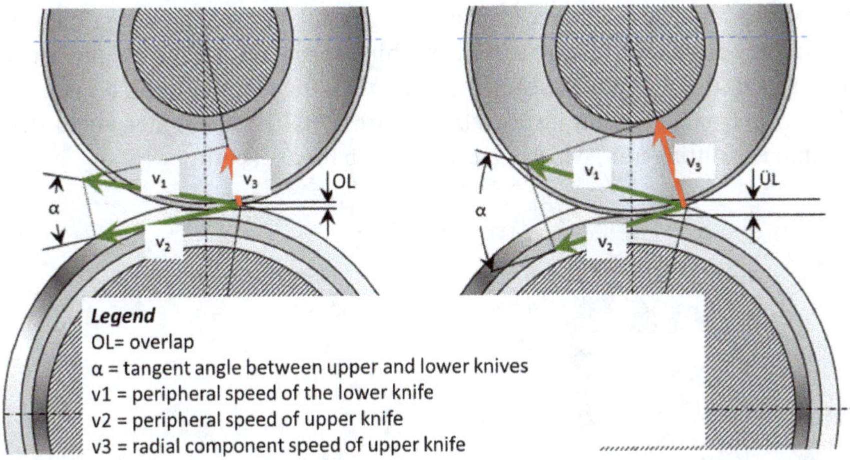

Figure 7.74 Circumferential speed

The knife diameters also influence the reduction in peripheral speed due to the radial component. Figure 7.75 shows a diagram of the difference in the top knife speed reduction (in %) for top knives with a diameter of 130 mm and bottom knives with a diameter of 100 mm (blue line) and 150 mm (orange line) at different overlap values. The larger the diameter of the bottom knife, the lower the radial component (v_3) and therefore the reduction in the top knife peripheral speed (v_2) is also smaller.

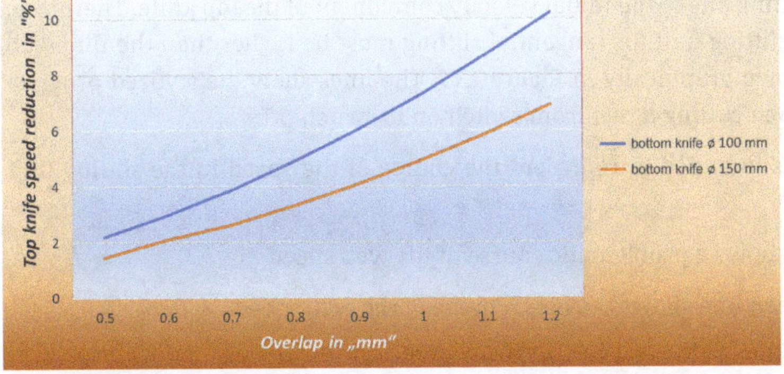

Figure 7.75 Top knife speed reduction

Slitting speed and web speed

References for this section: [2–4].

As already indicated before, the speed of the film web and the slitting speed must be in relation to one another to achieve a good slitting result. For example, the film web speed, commonly referred to as the cutting machine speed, must not be higher than the cutting speed. If the film web speed is higher than the cutting speed, a material jam will occur in front of the slitting point (see Figure 7.76). The film is then not drawn smoothly into the slitting unit, which leads to the web breaking off.

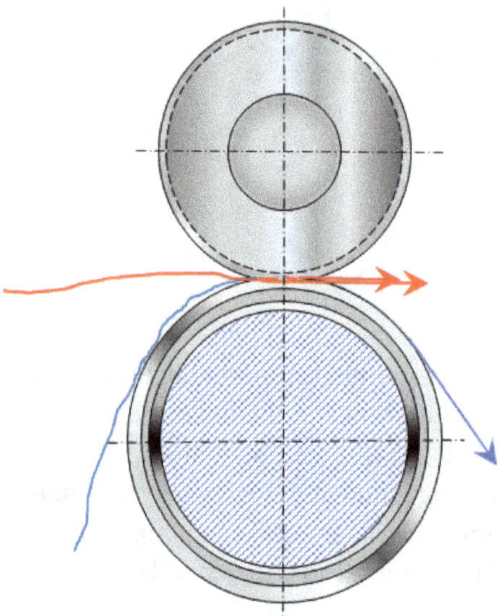

Figure 7.76
Slitting speed and web speed

Only if the film wraps the bottom knife shaft can the film web speed be at the same level as the slitting speed without the web breaking. However, this is not the case during tangential cutting due to the radial velocity component of the top knife. Therefore, the speed of the slitting unit for tangential slitting must be higher than the film web speed. This is shown graphically in Figure 7.77. The lines show the stylized progression of speeds when cutting a reel from switch-on to switch-off:

- The red line is intended to represent the course of the speed in the slitting unit (bottom knife)
- The blue line shows a problematic course of the web speed
- The dotted green line shows the ideal course of the web speed.

The blue line intersects the red line in three areas. In these areas the web speed is higher than the slitting speed, and this can lead to web breaks. The areas of speed increase (acceleration—area 1) and speed reduction (deceleration—area 3) are where the most tears off usually occur. However, breaks are also possible shortly after production speed is reached (area 2) if the acceleration of the web is not reduced shortly before the end of the acceleration phase to a value slightly lower than the slitting speed.

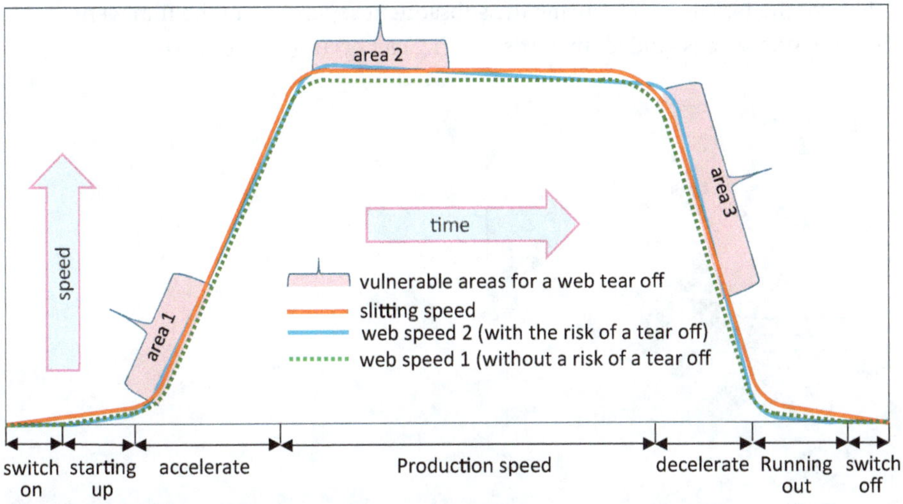

Figure 7.77 Speed progression from switch-on to switch-off

The situation with the green dotted line shows the idealized speed behavior for trouble-free work. As a rule of thumb, it can be stated that the speed of the bottom knife shaft during tangential cutting must always be around 3–4% higher than the web speed. However, these values only apply with an overlap of 0.7 mm. If the overlap is higher, the "bottom knife shaft advance" must also be higher.

Summary

- Only with wrap cutting may the cutting speed be greater than or equal to the path speed
- When cutting tangentially, the cutting speed must be slightly higher than the path speed
- In order to keep the wear on the knives within limits, the difference in speeds must not be too high, and the guideline value for hard PVC is 3–4%
- The film must not sag before cutting; it must always be under a certain tension.

7.4.15 Web Guidance Immediately After Cutting

Another risk of web tears exists immediately after cutting. The film web, which is guided along the male knife, lies below the web that runs over the female knife. If both webs are guided to a common roller, contact between the two webs can easily occur. In the worst-case scenario, the film web edge of the male knife gets slightly under the film web edge of the female knife. During the subsequent separation of the film webs, if the underlying film web is guided upwards, tearing can occur (Figure 7.78).

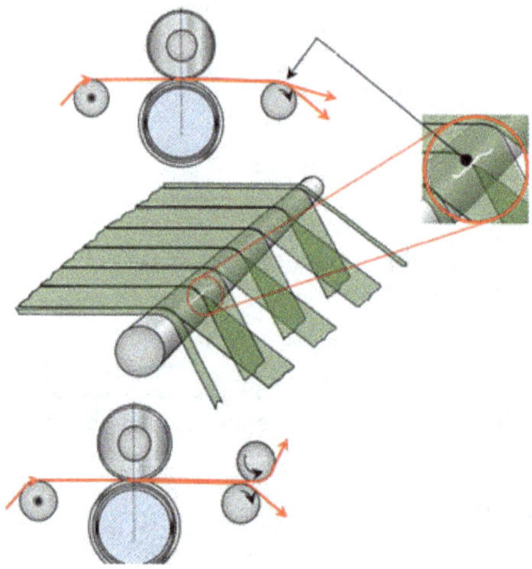

Figure 7.78
Web guidance

To avoid this source of interference, it is recommended to install a second deflection roller, so that the two film webs are guided separately immediately after the cut. The edge trims are particularly at risk, and are further discussed below.

Handling of the edge trims

The edge trims can become a problem at higher speeds because they are narrow and have a more pronounced tendency to get under the adjacent film web when they run parallel for a long time. Here too, separating the webs immediately after cutting offers advantages.

If the edge trims are fed directly after cutting to the recoiler for winding or to the suction system for shredding (Figure 7.79), without support, the edge trim will tear off sooner or later. Continuing the edge trim over a longer distance like the film web only helps to a limited extent. There is a real problem that the edge trim touches the film web when it is separated and then tears off when it gets under or over the film web. The edge trims must therefore be supported at the cutting point (Figure 7.80). The state of the art today is no longer to wind the edge trims on a reel, but to vacuum them and feed them directly to a shredder.

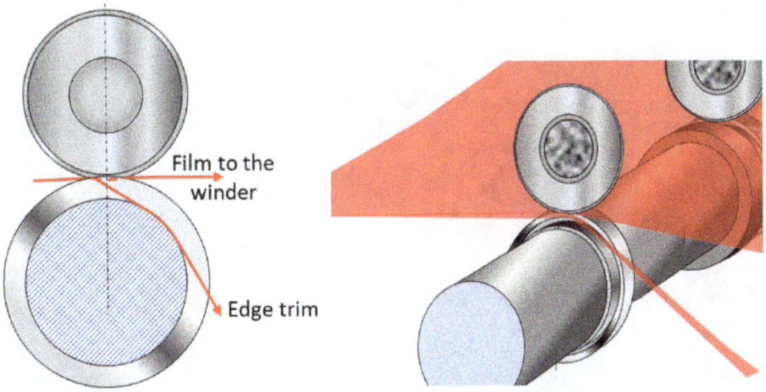

Figure 7.79 Edge trim guidance without support

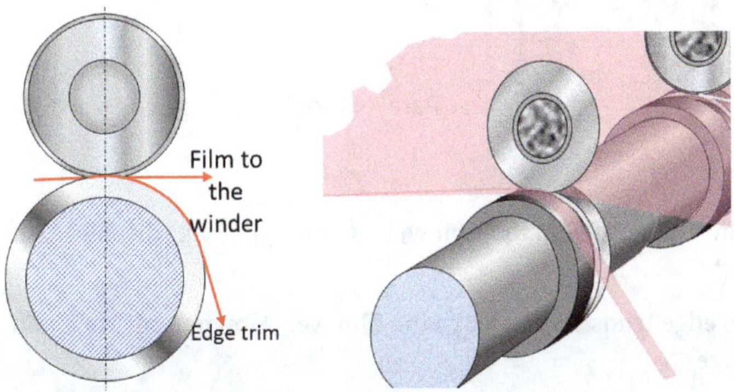

Figure 7.80 Edge trim guidance via the bottom knife

When vacuuming the edge trims, it is important to ensure that the air flow is adjusted so that no pulsating tensile stress can build up in the edge trim at the cutting point. This would be counterproductive, as a "tear-off" would then become possible again. The direct shredding of the edge trims is also a measure to be able to feed the granulated edge trims back into the production process as clean regrind.

The issue of edge trims on the calender should not be underestimated either. If the edge trim is deflected too much directly in the cut (Figure 7.81), greater tensions arise in the "kink", which can lead to the edge trim tearing off and to a rough edge of the film that may have tears down it. It doesn't matter whether the cut is made with circular knives in a scissor cut or with razor blades in a bursting cut.

Another problem with the calender can occur if the edge trims are guided over a distance that is too long, parallel to and at the same height as the web, over several rollers. Since the edges of the film cross-profile are thicker, the cut edge trims tend to orient themselves somewhat towards the center (Figure 7.82). If the edge trims brush against the film web when the web guide is being separated, this will result in edge waviness or even tearing.

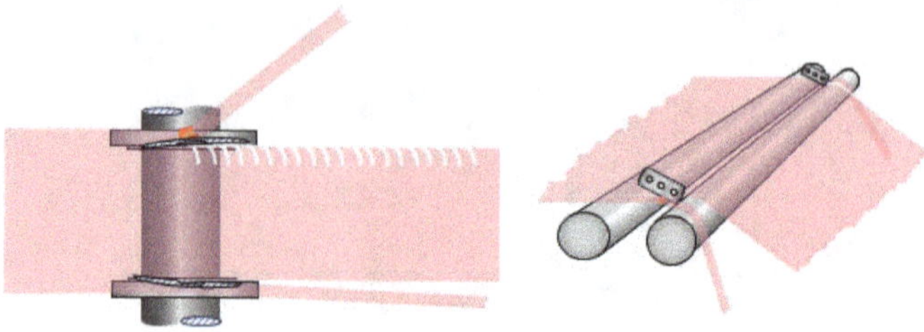

Figure 7.81 Deflection of the edge trim on the calender

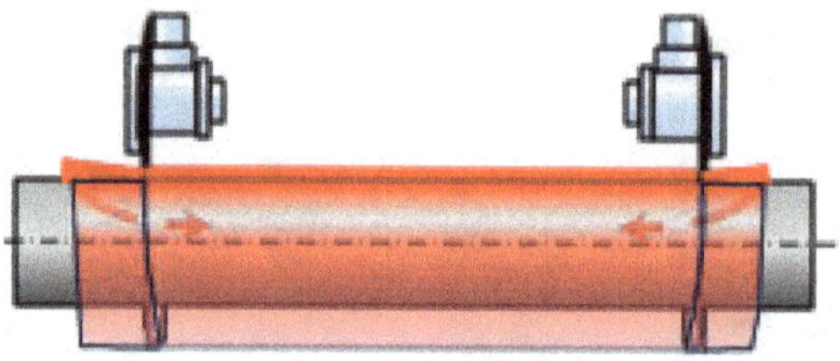

Figure 7.82
Parallel running of the edge trims

Summary

- The edge trims should run parallel to the film web over a significant distance, but not at the same height
- When separating, the edge trims should leave the film web “downwards” to avoid grazing at the web edge
- The tension acting on the edge trims (generated by the reel or the exhaust fan) must not extend to the cutting point
- The edge trim must be supported at the cutting point.

7.4.16 Slitting and Winding of PVdC-Coated Film

Mono-PVC films have a relatively high water-vapor permeability and are therefore only suitable to a limited extent for products that are sensitive to moisture. To solve this problem, PVC films can be coated with PVdC (poly(vinylidene chloride)). The PVdC forms a barrier against water vapor, protecting the product if it is sensitive to moisture, as is the case with some medications and foods. This is particularly important in climatic zones with high humidity, especially during monsoon periods.

Figure 7.83 shows a cross-section through a 250 μm thick PVC film, coated with 40 g/m^2 PVdC. The PVdC particles are very finely dispersed in a liquid; this dispersion is applied to the PVC film and then dried. The liquid evaporates and the PVdC remains on the film.

Figure 7.83 PVdC coating

Since the PVdC layer has a crystalline structure, it is significantly more brittle at low temperatures than the PVC base film. Tests have shown that the temperature range around 35°C for cutting is a good compromise to achieve an acceptable speed with a good cutting edge (Figure 7.84).

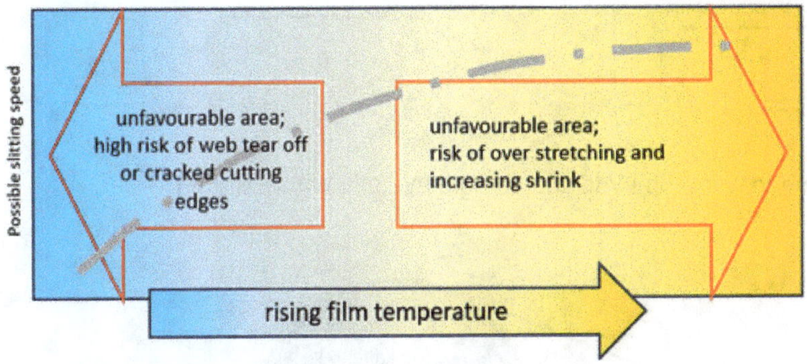

Figure 7.84 Optimal area for cutting PVdC-coated film

Since the problem of tear down and tear off depends heavily on the film temperature, difficulties arise particularly in winter when the PVdC-coated film cannot be stored in temperature-controlled rooms. To get a feeling for the length of time it takes for a coated PVdC reel with an outer diameter of 800 mm to warm through (film thickness 250 µm; PVdC coating 40 g/m^2), a series of measurements was carried out. This was carried out in a heated container at a temperature of approx. 45°C.

The diagram (Figure 7.85) shows the result. A reel with a diameter of 800 mm should be tempered for approximately 60 h (2.5 days) to safely reach approximately 35°C, but even then, there is still a difference between the film layers on the core and the outer film layers.

Another thing to keep in mind when cutting PVC/PVdC coated film is that the PVdC layer must always face the bottom (female) knife (Figure 7.86). The reason for this is that excessive deformation of the PVdC layer must be avoided.

The problem of temperature can be avoided if an opportunity to heat the coated side of the film is created as close as possible to the cutting section, for example using an infrared emitter. However, it is important to ensure that the heat supply is switched off immediately and that the heater moves away from the film when the slitter rewinder is stopped. If this is not considered, the film can shrink or even be thermally damaged.

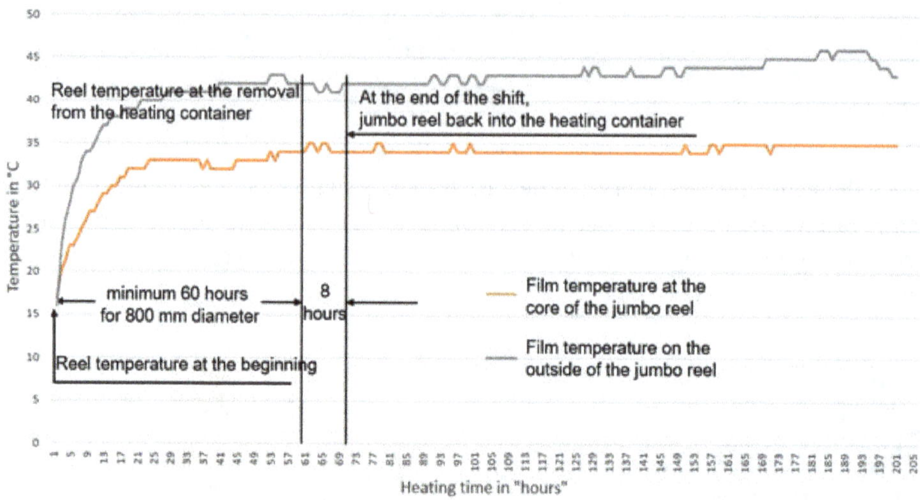

Figure 7.85 Time to warm up PVdC-coated rigid PVC film before slitting

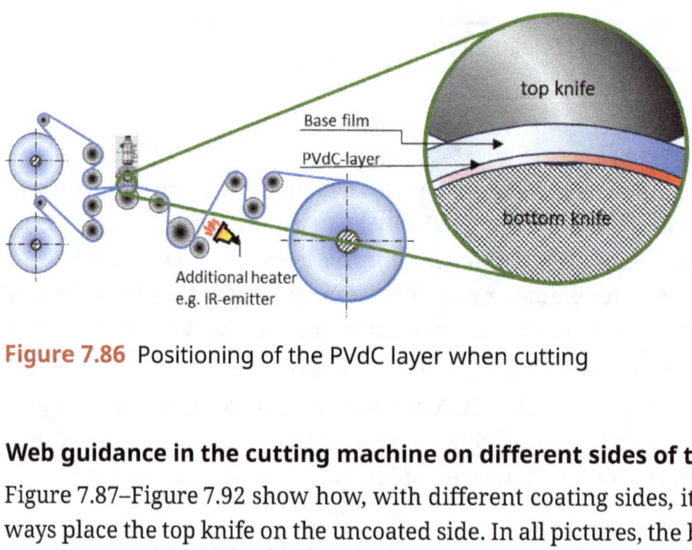

Figure 7.86 Positioning of the PVdC layer when cutting

Web guidance in the cutting machine on different sides of the coating

Figure 7.87–Figure 7.92 show how, with different coating sides, it is still possible to always place the top knife on the uncoated side. In all pictures, the blue side is the coated side, and the green side is the uncoated side.

a) Figure 7.87:

   Jumbo reel: Outside coated (blue side).

   Customer request: Outside of the roll coated (blue side).

   This arrangement should be our standard when cutting coated films. The direction of rotation of the jumbo reel is the same as the direction of rotation of the cut reel.

   Advantage: The coated side is guided over the bottom knife.

   Disadvantage: None (standard).

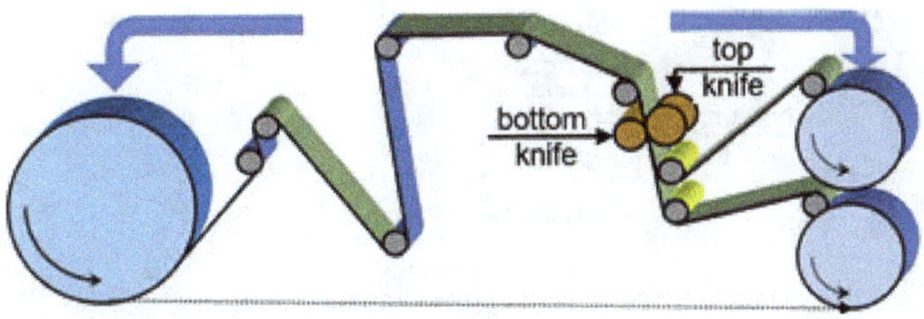

Figure 7.87 Cutting of PVdC coating (example a)

b) Figure 7.88:

Jumbo reel: Coated outside (blue side).

Customer request: Outside of the roll uncoated (green side).

With this web guide, only the direction of rotation of the unwinder must be changed; the direction of rotation of the winding shafts remains the same.

Advantage: Easy to handle, with no additional downtime for the cutting machine.

Disadvantage: The uncoated side of the film is guided over the bottom knife; the top knife cuts on the coated side. The deformation of the coating is high, and this can lead to a rough, partially torn cutting edge or even to the tearing of individual cut webs. If the cutting speed is greatly reduced and the film web is preheated by an additional heater (see Figure 7.88), cutting is possible. This solution therefore makes sense for small, individual orders.

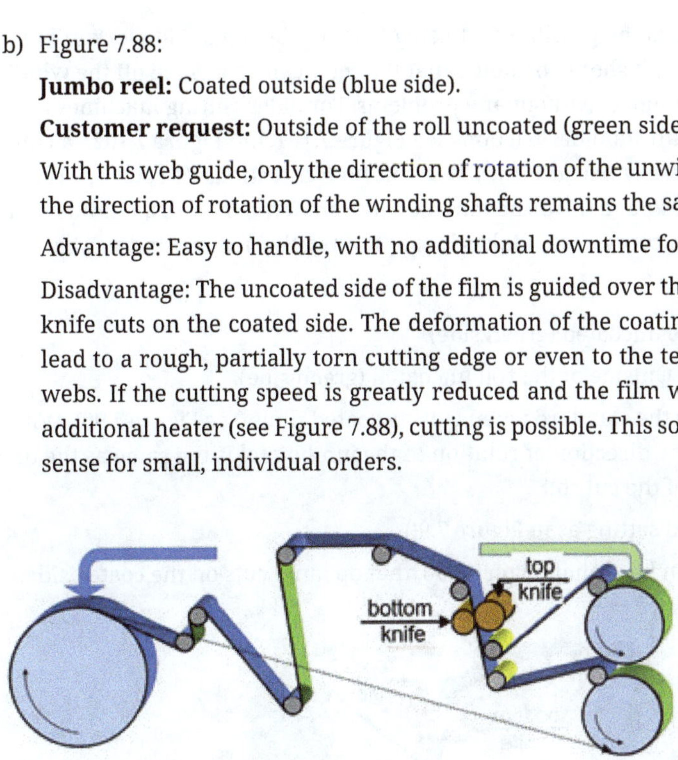

Figure 7.88 Cutting of PVdC coating (example b)

c) Figure 7.89:

Jumbo reel: Coated outside (blue side).

Customer request: Outside of the roll uncoated (green side).

With this web guide, the direction of rotation of the rewinder must be changed. The direction of rotation of the unwinder remains as standard.

Advantage: The coated side is guided over the bottom knife.

Disadvantage: A second set of friction winding shafts is required, or an order change takes a very long time, as dismantling the friction winding shafts and turning the friction elements is very time-consuming.

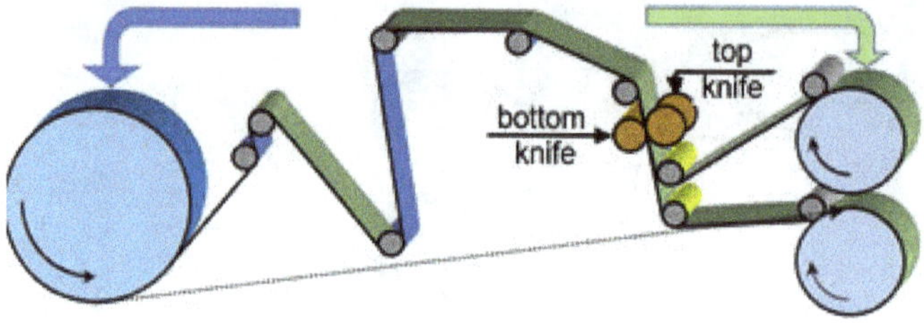

Figure 7.89 Cutting of PVdC coating (example c)

However, there is also the possibility of using friction elements that work in both directions of rotation. It should be noted that the reels can be pushed off the winding shaft onto the mandrel without any problems. For older cutting machines that are still operated with shoulder friction (see Figure 7.101 and Figure 7.102), a conversion is not necessary because they work in both directions of rotation. However, these machines have the disadvantage that removing the finished rolls and reloading them with cores always takes longer than with friction winding shafts.

d) Figure 7.90:

Jumbo reel: Outside uncoated (green side).

Customer request: Outside of the roll uncoated (green side).

This arrangement is the standard when cutting coated films (see Figure 7.90). With this arrangement, the direction of rotation of the jumbo reel is the same as the direction of rotation of the cut roll.

Advantage: Standard setting as in Figure 7.90.

Disadvantage: As can be seen in Figure 7.90, the top knife cuts on the coated side.

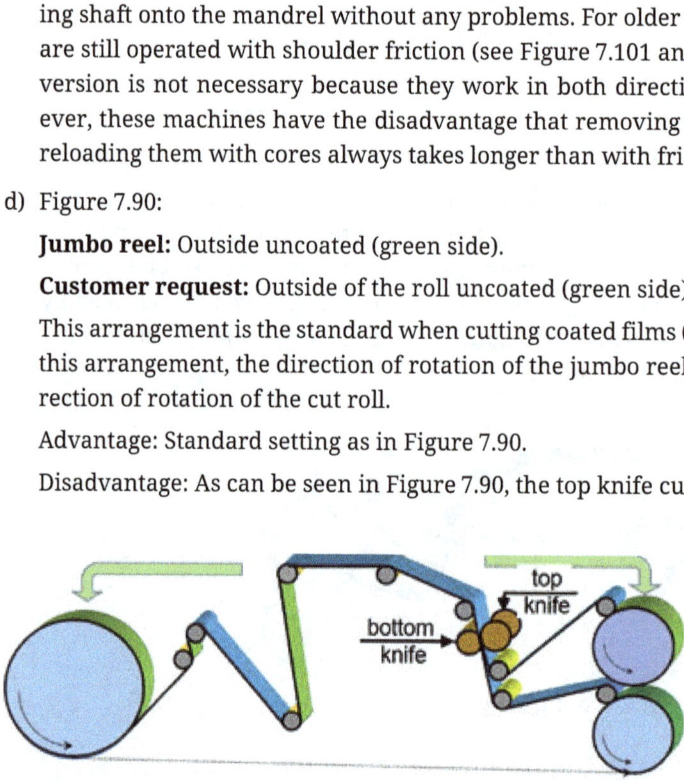

Figure 7.90 Cutting of PVdC coating (example d)

e) Figure 7.91:

Jumbo reel: Outside uncoated (green side).

Customer request: Outside of the roll coated (blue side).

With this web guide, only the direction of rotation of the unwinder must be changed; the direction of rotation of the winding shafts remains the same.

Advantage: Easy to handle, with no additional downtime for the cutting machine. The coated side is guided over the bottom knife.

Disadvantage: A change in the direction of rotation of the unwinder, but no further disadvantages.

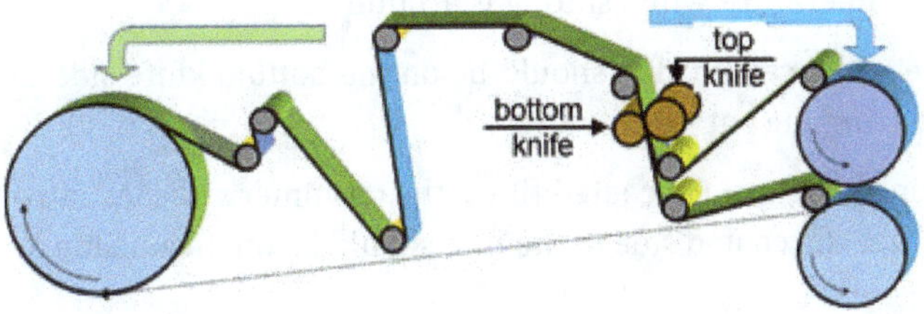

Figure 7.91 Cutting of PVdC coating (example e)

f) Figure 7.92:

Jumbo reel: Outside uncoated (green side).

Customer request: Outside of the roll uncoated (green side).

With this web guide, the direction of rotation of the rewinder and the unwinder must be changed.

Advantage: The coated side is guided over the bottom knife.

Disadvantage: A second set of friction winding shafts is required, or an order change takes a very long time, as dismantling the friction winding shafts and turning the friction elements is very time-consuming. But there is also the possibility of using friction elements that work in both directions of rotation. This is the situation as already described in Figure 7.92.

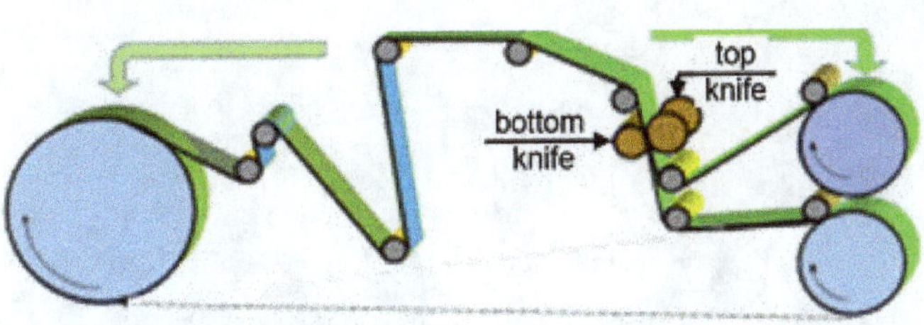

Figure 7.92 Cutting of PVdC coating (example f)

Summary

- The crystalline structure of the PVdC coating and the amorphous structure of the PVC film lead to a more brittle behavior of coated films compared to mono films
- Due to this brittleness, the temperatures of the film composite must not be too low, but also not too high, to avoid stretching of the film when cutting due to the required tensile stresses
- The recommended film temperature when slitting is around 35°C
- During the slitting process, the coated side should be on the bottom knife side to minimize deformation during the cut
- When planning the slitting program for coated films, the customer's wishes must be considered as to whether the coated side of the film should be on the outside or inside of the reel.

7.4.17 Winding Methods

Center winder

- The principle of center winders is the central drive of the winding shafts. A distinction is made between two types of implementations:
- Winder with two winding shafts, the so-called duplex winder (Figure 7.93)
- Winder with two reversible winders, on each of which two winding shafts are arranged, the so-called turret winder (Figure 7.94).

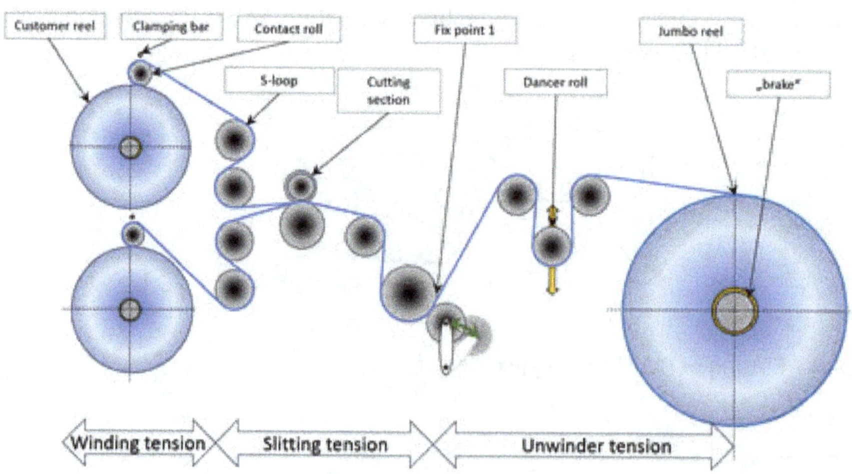

Figure 7.93 Center winder with two winding shafts

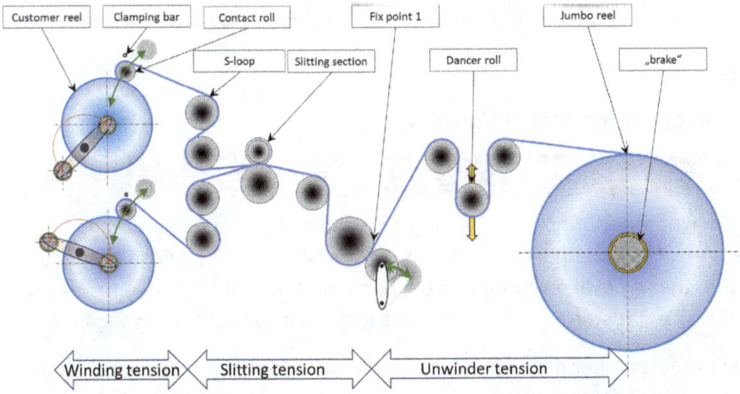

Figure 7.94 Turret winder

Both methods require continuous winding shafts that can accommodate the individual reels, and which allow the speed of the individual rolls to be balanced through friction. This is necessary because the films can have the slightest, unavoidable thickness deviations, which lead to different speeds and thus affect the winding tension. This would result in one roll being wound too hard and another too soft. These special elements work based on friction and can compensate for these differences (for more details, see Section 7.4.18).

Figure 7.95 shows a comparison of the working heights between the center winder with two winding shafts (duplex winder) and the center winder with two turning winders (turret winder) and two winding shafts each. With the turret winder, the working height of the second winding shaft increases by the outer diameter of the reels and the required safety distance between the outer diameter of the reel. This leads to a slightly smaller maximum possible reel diameter when the winder position is horizontal. However, this can be avoided by tilting it by (for example) 45°, and this even offers an advantage (green line).

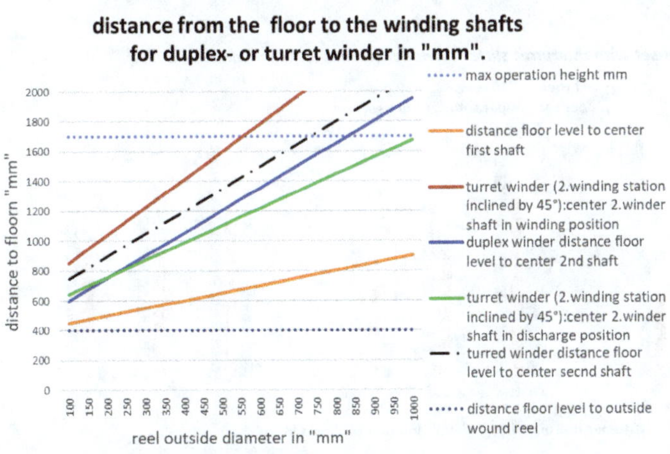

Figure 7.95 Working heights

Table 7.5 compares the advantages and disadvantages of these two center winder variants.

Table 7.5 Comparison of Center Winder Variants

Turret winder	Duplex winder
Very short machine downtimes during rewind change over time. Only around 40 s is required to turn into the removal position, cut and stick the stripes, and restart.	Longer machine downtimes when changing blocks. The duration of the standstill depends on the number of reels; it is needed to remove the reels and re-equip the winding shafts.
The downtime of the slitter rewinder depends on the slitting speed, but this only applies to small diameters (approx. 300 mm).	The downtime of the slitter rewinder is independent of the slitting speed and the reel diameter of the customer roll.
Higher throughput due to lower downtime (see Figure 7.96).	Lower throughput due to longer standstills (see Figure 7.96).
If the reversible winder is arranged horizontally, diameters of 900 mm are possible[a].	Roll diameters of up to 1000 mm are possible[a].
Investment costs are higher.	Investment costs are lower.
Maintenance effort and costs are higher.	Maintenance effort and costs are lower.
Personnel costs are higher (see Figure 7.97).	Personnel costs are lower (see Figure 7.97).

[a] Rated as neutral because the load on the winding shaft determines the maximum diameter.

In Figure 7.96, an estimate of the time savings between a duplex and a turret winder has been made. The time saved essentially corresponds to the set-up time when changing sets. From the diagram, the diameter of the customer's roll, the thickness of the film (as this influences the slitting speed), and the number of rolls per set greatly influence the time savings.

Figure 7.96 Time savings between duplex and turret slitter rewinder

The number of rolls per set has the greatest influence on time savings, followed by the diameter of the customer roll. The narrower the customer rolls usually are and the larger the diameter of the rolls, the more worthwhile a turret winder is. At the same time, however, there is an increased need for personnel—see Figure 7.97.

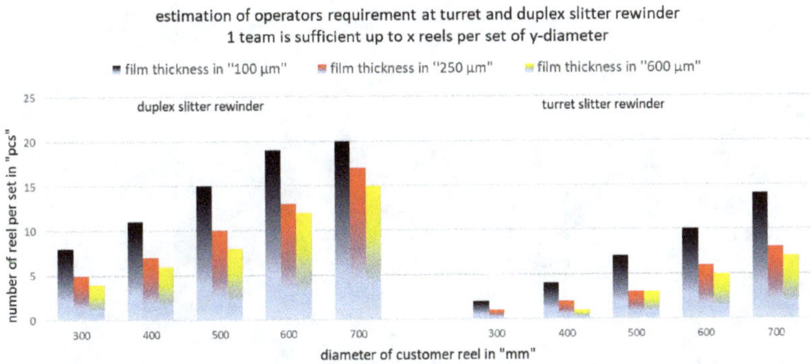

Figure 7.97 Personnel requirements for duplex and turret slitter rewinders

Referring to Figure 7.97, it should be noted that in order not to extend the downtime of the cutting machine due to reel handling, additional personnel are required for a turret slitter rewinder with six or more reels per set, a roll diameter of 600 mm and a film thickness of 250 µm. With a duplex cutting machine, this is only the case from 13 numbers of reels per set onwards. This is offset by a time saving of approx. 50% with the turret winder compared to a duplex winder. A turret slitter rewinder has clear advantages here, because two duplex slitter rewinders would be required to produce the same production quantity at the same time. The exact ratios are also determined by the slitting speed.

Surface winder

A distinction is also made between two types of surface or circumferential winders:

- Surface winder without an additional central drive via a friction winding shaft
- Surface winder with an additional central drive via a friction winding shaft.
  - With a real surface winder, the slitted reels are not driven centrally via a winding shaft, but rather by a driven contact roller (Figure 7.98). This contact roller is provided with a low-wear friction lining (e.g. a special rubber mixture), which enables the rollers to be driven almost slip-free.

The individual reels are pivoted individually against a contact roller that serves as a drive roll and, using adjustable pressure, the film is pressed against the contact roll and wrapped around the core. This type of reel winding is particularly suitable for thin films, as the winding hardness is not generated by the winding tension via a central drive, but rather by tension between the fixed point before slitting and the pressure of the individual reels against the contact roll. This pressure must be regulated depending on the current reel diameter. The slitting tension is constant over the entire length of the reel and does not affect the winding hardness of the reel. This is regulated by the contact pressure of the reel against the contact roll. In principle, the film web is guided through two "rollers". After this "transport point", the web has practically no tensile load, but it is wound up and does not simply fall to the ground.

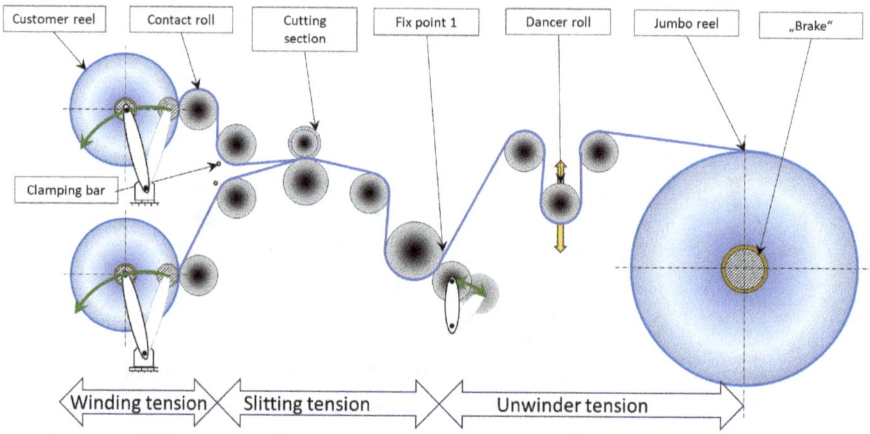

Figure 7.98 Surface winder without central drive

This type of winding is also well suited for larger diameters with thinner film. However, the films must have a good thickness distribution across the direction of travel (flatness). This type of winding also makes it possible to wind rolls with different diameters next to each other.

Surface winder with additional central drive

This winding method essentially combines a center winder (duplex winder) with a surface winder. There is therefore the option of winding with a gap between the reels and the contact roll or with direct contact between the reels and the contact roll.

Lay-on rolls

When slitting, especially at higher speeds, straight-edge winding is made more difficult by an air cushion that forms between the reel and the film web. This is further exacerbated if the electrostatic charge on the web is not eliminated or is insufficiently eliminated.

Wrapping without a lay-on roll

When winding without a lay-on roll (Figure 7.99), the air cushion that is formed by the rotational movement of the reel and the linear movement of the web has a negative effect on the fixation of the film web. With a small diameter, the "angle" between the reel and the film web is blunter than with larger diameters. The "pointier" inlet allows air to be introduced between the layers, which can potentially lead to the top layer shifting. The reel then tends to telescope, especially at higher speeds.

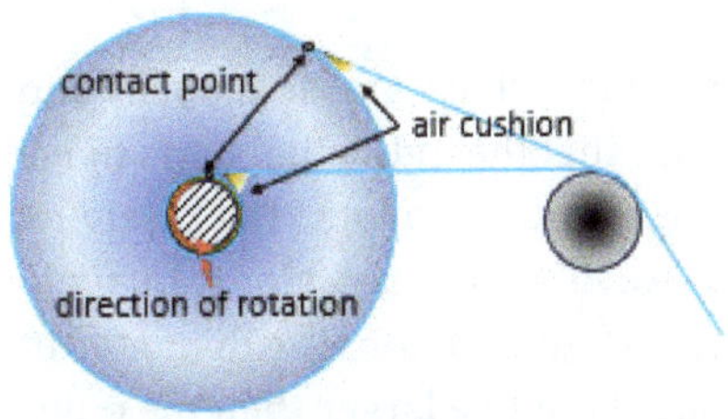

Figure 7.99
Winding without a lay-on roll

Wrapping with a lay-on roll

When using a lay-on roll or contact roll, the web is held in position immediately in front of the contact point, and contact with the roll prevents the layers from moving sideways (telescoping). The lay-on roll or contact roll must exert a certain amount of pressure on the reel to prevent air from being introduced. However, the pressure must not be too high, otherwise it can lead to wrinkles.

In the center winder (Figure 7.100 left), this roll is called the lay-on roll. It doesn't need to be powered. Arranging the lay-on roll from below is advantageous in terms of process technology, as it then reduces the deflection of the winding shaft and ensures contact between the middle reels and the lay-on roll, as this then acts as a kind of "support roll". In the surface winder, the driven roll forms the contact roll, and prevents air from being introduced into the reels.

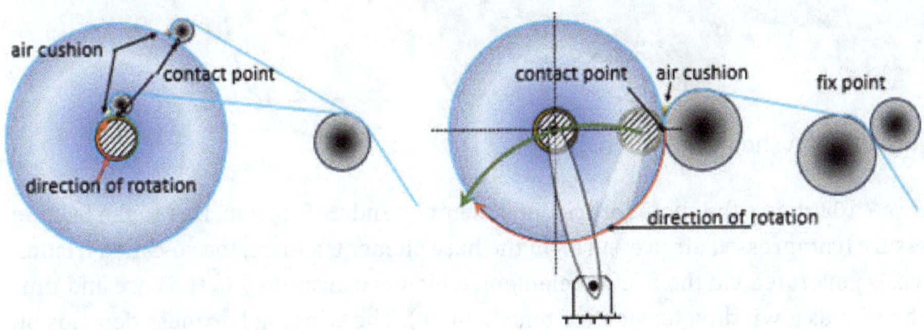

Figure 7.100 Winding with (left) a lay-on roll and (right) a contact roller

7.4.18 Frictions

"Frictions" are auxiliary elements that enable different speeds on a winding shaft during winding. They must be able to wind several lengths of material on a common, centrally driven shaft without allowing the tensile forces to stretch the material. Depending on the application, a distinction is made between a shoulder friction and a friction winding shaft.

Shoulder friction

Figure 7.101 shows the elements of a shoulder friction, which are assembled and mounted on a common shaft. It is important to ensure that the individual parts move smoothly among each other, and that the friction disk is not worn out. The receiving element for the sleeve should have a way for the sleeve to be taken along, to avoid slipping between the sleeve and the receiving element. In addition, the diameter of the collar that is in direct contact with the sleeve should not be larger than the outer diameter of the sleeve. This is important because the friction can heat up, and the resulting heat would cause the film layers to "weld" together near the sleeve.

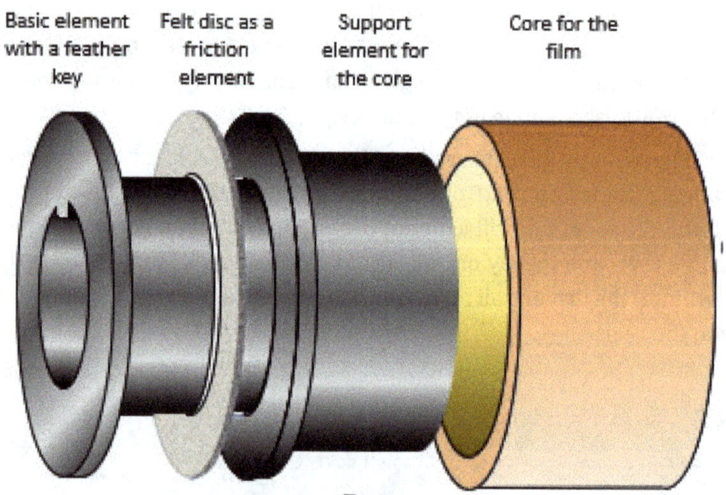

Figure 7.101 A shoulder friction

Figure 7.102 shows the shoulder friction assembly, and its function. Due to the friction pressure (compressed air pressure) on the base element, a force, the so-called friction force, is generated via the friction element, which is transmitted to the core and thus to the film as a winding tension (or tensile force). The winding hardness depends on the friction force F and the effective diameter D_e of the friction element. In this picture you can also see that the collar of the core receiving element is smaller than the outer diameter of the core.

Figure 7.103 shows a fully equipped winding shaft with shoulder frictions and fully wound reels. The distance between the reels must correspond exactly to the reel width, and the gaps between the frictions must be filled with spacers. In addition, care must be taken to ensure that all elements slide easily on the winding shaft, otherwise the friction pressure is not the same over the entire length. This would lead to different winding hardnesses of the reels and/or a telescoping of individual reels. It is recommended to spray the winding shafts at regular intervals after cleaning with a food-safe, silicone-free lubricant and release agent (e.g. "Florin SLX").

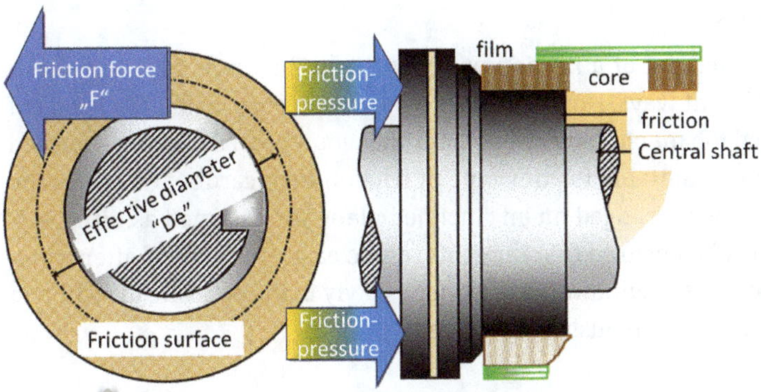

Figure 7.102 An assembled shoulder friction

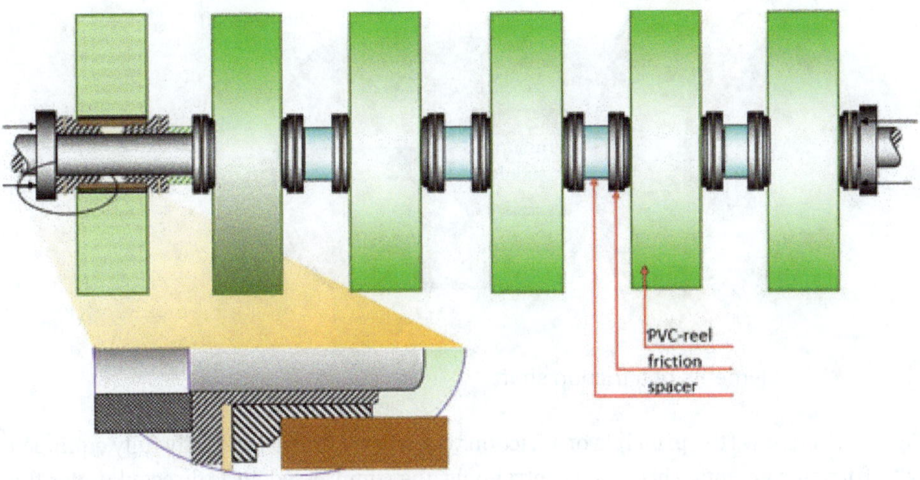

Figure 7.103 Fully equipped winding shaft

Compared to friction winding shafts, shoulder friction shafts have the disadvantage that they require time-consuming maintenance and longer machine downtimes when changing sets, meaning that they tend not to be the first choice. On the other hand, they have the advantage that they can be used in both directions of rotation without modification.

Core material

The cores to be wound with film can be made from special cardboard or plastic. When using shoulder frictions, the risk of paper dust forming when using cardboard tubes is particularly high. This can get between the layers of the film, which would make the film unusable for the customer. Plastic sleeves are therefore preferable.

Friction winding shafts

Figure 7.104 shows an element of a friction winding shaft. The drivers for the sleeve can be designed in different ways. However, it is important to ensure that the inside of the core is not damaged too much, so as not to block the wound reel from being pushed off. Plastic cores are more suitable for this type of friction than cardboard cores. The driving elements are either arranged on an "inclined plane" or designed as tilting elements. However, it must be ensured that the reels can be easily pushed off after winding. This works very well if the winding shafts rotate slowly in the opposite direction to the winding direction during the push-off process.

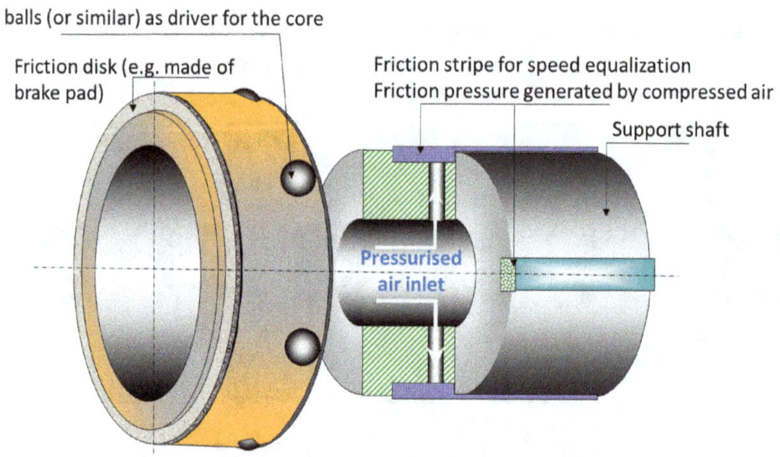

Figure 7.104 Elements of a friction shaft

Figure 7.105 shows the principle of a friction winding shaft. This is usually fully equipped with friction elements. Through a central hole, the compressed air is directed under the friction stripes via many radial holes and presses them against the core support elements. This pressure creates the tensile force to wind up the film. The driving elements transfer this friction force to the cores. These must not slip radially or axially on the friction elements, to ensure a clean, straight-edged winding pattern. After stopping and glueing with adhesive tape, the friction winding shaft must slowly rotate in the opposite direction of rotation for winding. This releases the drivers and allows the rolls to be pushed off.

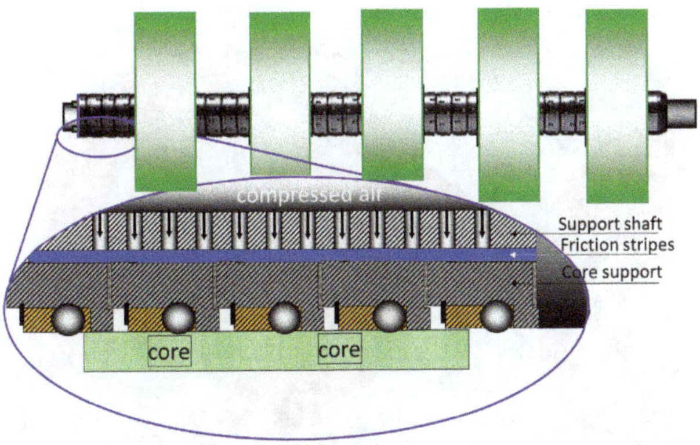

Figure 7.105 Principle of a friction shaft

Since there are no spacers between the reels with friction winding shafts, the reels can be pushed together directly and pressed onto a mandrel. Since this loosening is necessary to remove the reels, the friction shaft should rotate slowly in the opposite direction. Friction winding shafts with standard rings can only be used in one direction of rotation. To change the direction of rotation, these friction elements must be rotated, or a winding shaft must be used for each direction of rotation. However, there are also so-called "bi-rings" that can be used in both directions of rotation. With the friction elements acting on both sides, "turning backwards" can lead to problems and must therefore be coordinated with the supplier of the friction elements in advance.

When re-equipping with cores, the exact positioning of the cores is signalled by a laser beam. This enables quick and accurate loading of new cores, minimizing winder downtime.

Friction pressure

To wind film onto a core, it is necessary to bend it around the core and wind it layer by layer by turning it. This requires forces that are generated by compressed air and friction in the case of center winders, and by pressing against a contact roller in the case of surface winders. This pressure is called friction pressure and is specified in bar or daN/cm^2 (daN = decanewton = 10 N). In order to obtain a well-wound roll, it must not be too high, but also not too low. This friction pressure is decisive for the hardness of the wound roll.

Friction force in the center winder

In Figure 7.106 and Figure 7.107 the situation for the center winder is shown. In the left part of each figure the roll is shown in partial section to better show the function.

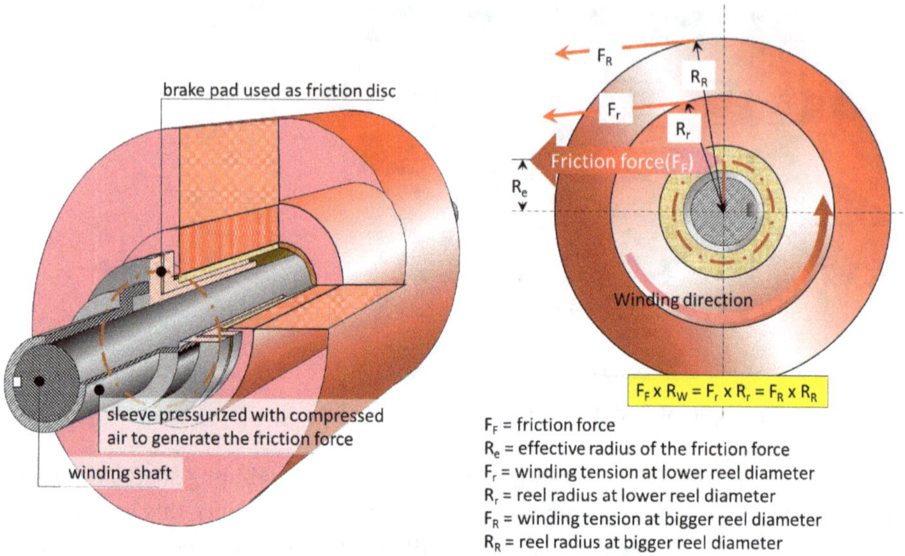

Figure 7.106 Winding with compressed air and shoulder friction

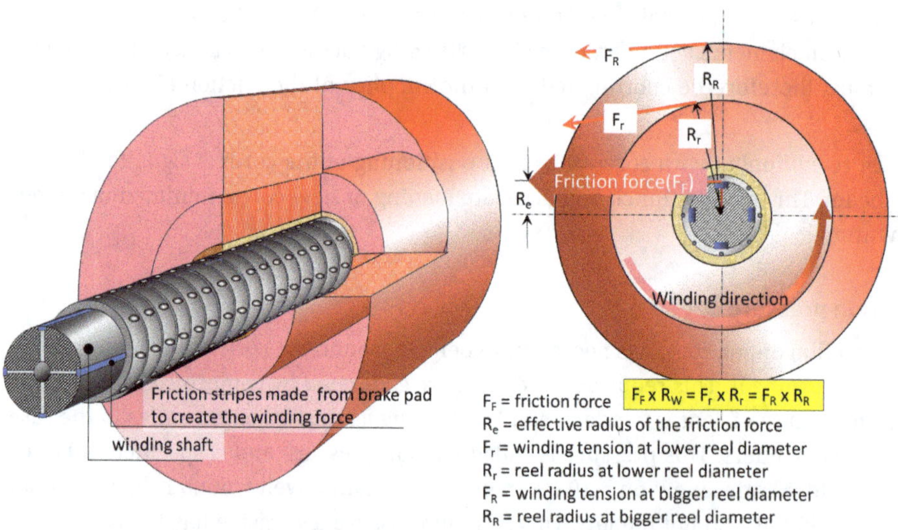

Figure 7.107 Winding with compressed air and a friction shaft

With shoulder frictions, a movable sleeve is pressed against the frictions with axially acting compressed air. As already described in this section, a friction has a base element, on which the sleeve (pressurized with compressed air) exerts pressure. The base element in turn presses on the friction disc, and this presses on the support element for the core. The pressure on the friction disc acts like a brake on a car; it transfers the force to the receiving element and thus to the sleeve through sliding friction. The friction force generated in this way is the friction force for winding. The winding shaft always rotates slightly faster than the sleeve, which is why the "sliding friction factor" must always be considered when calculating the winding tension. Over the course of use, the sliding friction factor changes slightly due to the continuous sliding between the base element and the receiving element. It is therefore not possible to specify exact values for the friction pressure, but the information can only serve as a guideline. The machine operator therefore must fine-tune the air pressure according to the winding result.

In the right part of each figure, the direction of action of the various forces is shown by arrows and the associated diameters. For orientation purposes, a formula is also given for how friction force is calculated and how it acts as torque in interaction with the associated roller diameters. When winding with a friction winding shaft, the situation is like shoulder friction, but only the friction force generated by sliding friction is transmitted via the friction strips that are embedded in the support shaft. With the friction winding shaft, the friction force acts directly on the diameter, while with shouldefriction it acts on the diameter, which divides the friction disk into two equal surfaces.

Calculation of friction force (for reference only)

For a shoulder friction:

$$D_{\mathrm{W}} = \sqrt{\frac{D_{\mathrm{a}}^2 - D_{\mathrm{i}}^2}{2}}\ [\mathrm{mm}]\,; F_{\mathrm{f}} = \left(D_{\mathrm{a}}^2 - d_{\mathrm{i}}^2\right)\frac{\pi}{4}\ [\mathrm{mm}^2]\,;\ F_{\mathrm{k}} = F_{\mathrm{f}}\,p\mu\,[\mathrm{daN}]$$

For a friction winding shaft:

$$D_{\mathrm{W}} = \frac{D_{\mathrm{f}}}{2}\ [\mathrm{mm}]\,;\quad F_{\mathrm{f}} = aB_{\mathrm{f}}L_{\mathrm{h}}\ [\mathrm{mm}^2]\,;\quad F_{\mathrm{k}} = F_{\mathrm{f}}\,p\mu\,[\mathrm{daN}]$$

where:

D_a = outside diameter of the friction disk

D_e = effective diameter

D_f = diameter of the friction winding shaft

D_i = inner diameter of the friction disk

F_f = friction surface

F_k = friction force

p = air pressure

μ = sliding friction factor.

B_f = width of friction stripes for friction winding shaft

a = number of friction stripes

L_h = length of the sleeve (width of the roll)

Friction force on the surface winder

In the circumferential winder without a central auxiliary drive, the reel is driven exclusively by the friction between the reel and the contact roll (Figure 7.108). Telecopying of the roll is avoided by the contact pressure of the contact roll against the reel. The winding tension is generated between the contact roll and the fixed point after the cut. This tension is reduced to almost 0 dN at the contact point. This is the reason why this winding method makes it possible to produce bigger roll diameters even with thin films. The disadvantage, however, is that very thick films cannot be wound without an additional central auxiliary drive because the required bending force cannot be achieved. This would only be possible with a very large core diameter.

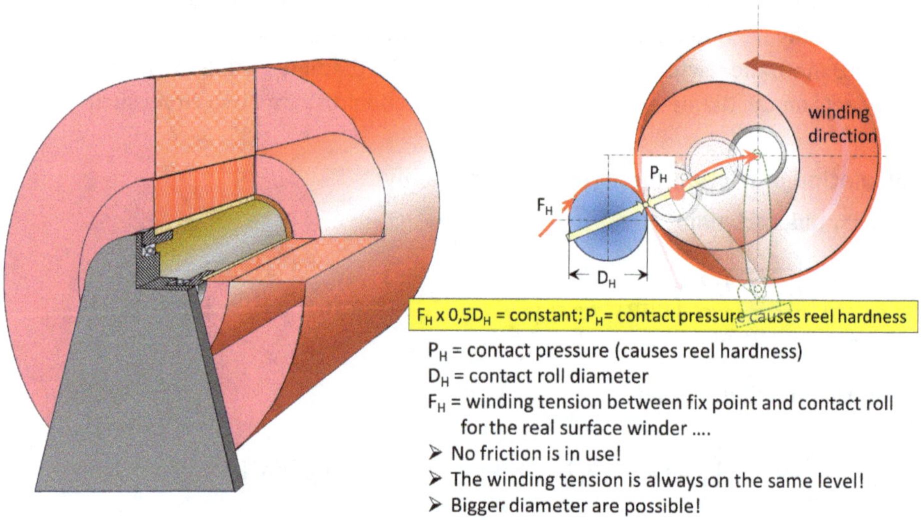

Figure 7.108 Real surface winder without a central auxiliary drive

With the surface winder with auxiliary drive (Figure 7.109), the required bending force for thicker films is generated by the tensile force of the center drive, as with the normal center winder. This means that the advantages of the surface winder can also be realized with thicker films.

Since surface winders are usually designed for bigger reels, the individual stations can be equipped with an "auxiliary drive" to use the full advantages of the center winder.

For reel slitter rewinders that have a friction winding shaft as a central auxiliary drive, it is recommended to arrange the contact roll below the reels to compensate for the deflection and achieve contact between all reels and the contact roll. If not all reels are in contact with the contact roll, uniform winding quality cannot be ensured for all reels. As a rule, surface winding of calendered films should be limited to thinner films and wider rolls for cost reasons.

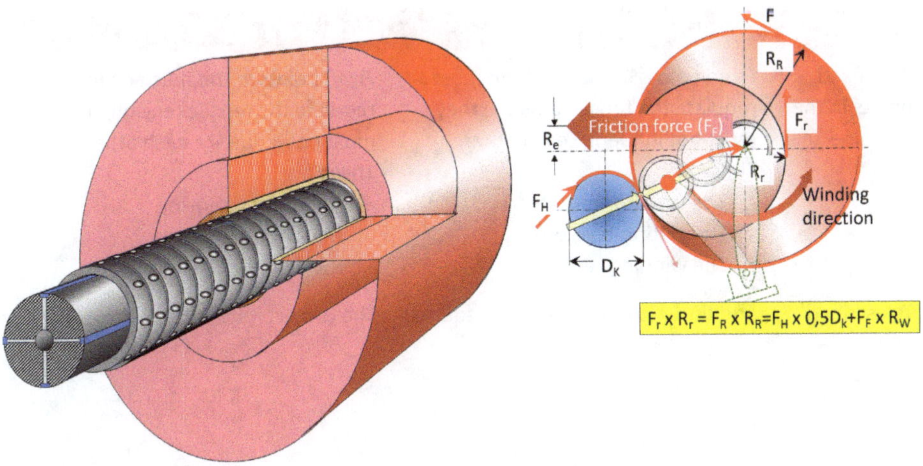

Figure 7.109 Surface winder with central auxiliary drive

For "normal" hard PVC film thicknesses, no additional central auxiliary drive is required. Thick films from approx. 800 µm are usually not wound up, as the resulting tendency to curl has a detrimental effect. Films of this thickness are usually used as sheets.

Summary

Marking	Center winder	Surface winder
Flatness/ high edges	Good thickness distribution (flatness) is required.	Good thickness distribution across the web (flatness) is very important.
Friction force	The friction force is transferred from a central drive via a support shaft to the friction element and from there to the reel. The friction force must be controlled via the pressure of the compressed air to adjust the hardness of the wound reels. Mechanical influences can affect the winding result.	With the surface winder, winding occurs without friction; web tension is transferred between a fixed point after cutting and the contact roll to the reel through the contact pressure. The hardness of the reel is regulated by the contact pressure between the reel and the contact roll.
Web tension or winding tension	The tensile force or winding tension is generated by the drive motor of the winding shaft and regulated via the friction pressure. To avoid overstretching of the thicker areas, the winding tension must be reduced as the diameter increases.	This web tension is greatly reduced at the contact point between the contact roll and the reel, so that it is wound up without any tension. Since the remaining tensile tension after the contact point is low, the film is not over-stretched.

Marking	Center winder	Surface winder
Web fixed points in the machine	Fixed points in the machine must regulate the tensile tension; they must not influence each other. These are: ▪ The unwind tension ▪ The slitting tensions ▪ The winding tension.	Fixed points in the machine must regulate the tensile tension; these must not influence each other. These are: ▪ The unwind tension ▪ The slitting tensions ▪ The winding tension.
Telescoped reels	Additional lay-up rollers are required.	The contact roll serves as a lay-up roll.

7.4.19 Winding Tension and Speed When Winding

When speed changes occur due to acceleration and deceleration, the tensile stresses within the web must not lead to overstretching or loosening due to fluctuations. Loose (sagging webs) can lead to web tears, especially around the winding tension.

This risk is greater with center winders than with surface winders because the reels are driven by the motors of the winding shafts. The control of the machine speed must be precisely coordinated with the speed control of the winding drives. With the surface winder (without a central auxiliary drive), the speed of the reels automatically adapts to the circumferential speed of the contact roll.

Reel rotational speed (rpm) when winding at a constant speed

It must be noted that with each revolution the outer side of the material to be wrapped is longer than the inner side by the factor πs, where s = material thickness. For a film 50 µm thick, the outer side is approximately 0.157 mm longer than the inner side; for a 500 µm film this is already 1.57 mm. This elongation of the material is the same for every layer, regardless of the diameter (Figure 7.110).

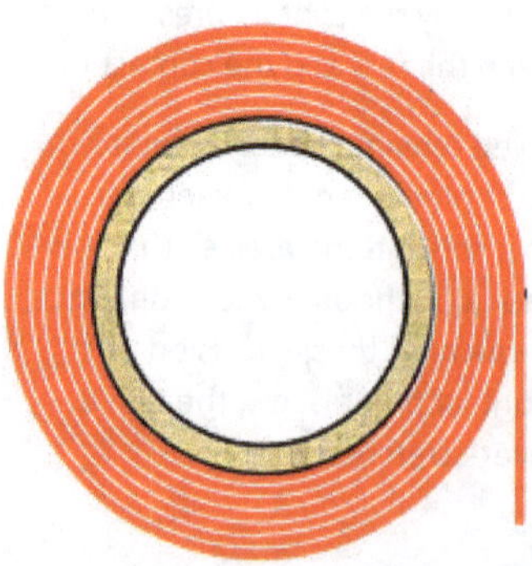

Figure 7.110
Wound reel length

This consideration leads to a formula for calculating the length ("running meters") of the reel, L:

$$L = \left(D_a^2 - d_i^2\right)\frac{\pi}{4s}$$

where:

D_a = reel outside diameter

d_i = outer core diameter

s = film thickness

From this, the outside diameter of the roll can be calculated by changing the formula for known or desired length:

$$D_a = \sqrt{d_i^2 + \frac{4Ls}{\pi}}$$

In addition, the speed of the winding shaft can be calculated if the speed and acceleration or deceleration are known; this is shown in Figure 7.111.

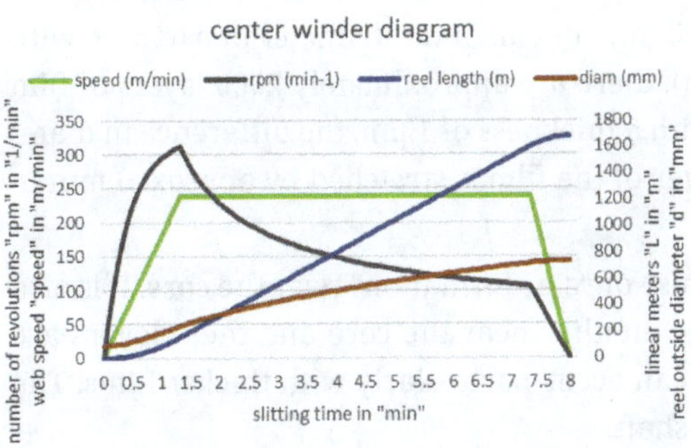

Figure 7.111 Determination of the speed of the winding shaft on a center winder

For simplicity, switch on, starting, running out, and switch off are not shown in Figure 7.111. It is clearly visible that when the speed of the film web remains constant, the reel length increases continuously (blue curve, L). The reel diameter increases less as the size increases, as more linear meters are wound per revolution (dark grey curve, d). The curve does not start at "0", but at the size of the core outside diameter. Of course, this linearity is not present when accelerating and decelerating (braking). When the acceleration phase begins, the speed increases sharply, but then levels off again before reaching the final speed (red curve, n). This is due to the rapidly increasing diameter. When the constant final speed is reached after acceleration, the rotational speed (rpm) of the winding shaft further reduces as the diameter increases.

This part of the curve shown follows the geometric shape of a hyperbola, which is why this type of winder is also called a "hyperbola winder". A hyperbola occurs when the product of two variables is constant. In this case, the speed v is equal to the product of distance s and time t ($v = st$), and is constant. The green curve, v, shows the web speed with acceleration at the beginning and deceleration at the end of the winding process.

From Figure 7.111 it can also be deduced that the center winder requires very precise and quick-reacting control, so that there is no risk of tearing when accelerating. If the speed deviates too much from the corresponding web speed, either the web becomes loose, and the film layers are not wound up with straight edges (speed too low), or friction causes excessive heating (speed too high). Both lead to problems.

As the real surface winder (surface winder without auxiliary drive—Figure 7.108) works without a winding shaft, these control problems do not occur. The reels are driven by the contact roll which, except for the acceleration and deceleration phase, runs at a constant speed and constant rotational speed (rpm), and thus drives the roll. The web tension and contact pressure can also be controlled without the influence of the winding shaft speed.

Winding tension at the center winder

When using a center winder, special attention must be paid to the winding tension. If the winding tension is chosen too high, there is a risk that the film will partially stretch. This leads to flatness errors. This risk exists primarily with thin films. If we look at a 100 µm thick film, with a diameter of 600 mm and an outer diameter of a 6″ core with a wall thickness of 15 mm (D_a = 182 mm), there are approximately 2000 layers of film on top of each other. For a piston ring with a thickness of 1 µm, the difference in diameter is theoretically 2 mm. The outer layer of the film is stretched by approx. 6 mm in this area. This inevitably leads to folds.

The pressure can also cause a core collapse or “star formation” near the core. It is also possible that some layers slide over one another near the core and these layers are displaced laterally (Figure 7.112a). This can occur particularly with thicker films. The core can also be moved on the winding shaft.

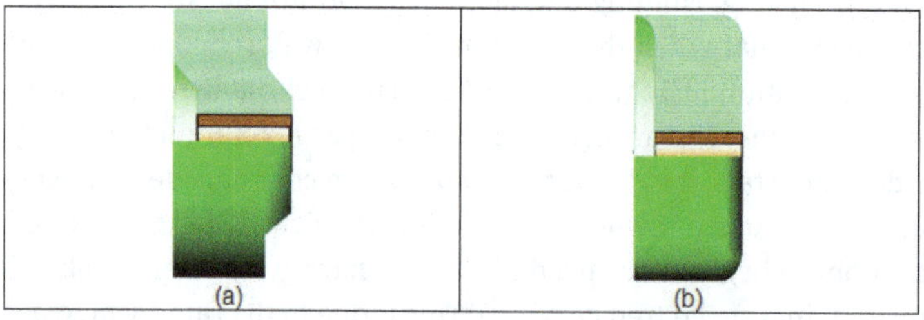

Figure 7.112 Telescoping with incorrect winding tension

If the winding tension is too low, then the layers will shift laterally (telescope) once the roll reaches a certain diameter (Figure 7.112b). The greater danger here is that this only happens at the customer's premises because a higher unwind tension is set on the unwinder than when winding in the slitter rewinder. In extreme cases, these reels can telescope during handling.

Figure 7.113–Figure 7.115 show the course of the winding tension at different friction pressure settings.

Wrapping with constant friction pressure

If winding is carried out with constant friction pressure (Figure 7.113), the torque remains the same throughout the entire winding process. This leads to a sharp drop in winding tension and so the reels will telescope once they reach a certain diameter. At the beginning of the reel, the winding tension can be so high that core collapses or star formation occur, and the flatness of thinner films is adversely affected. This setting is not suitable for producing good quality reels.

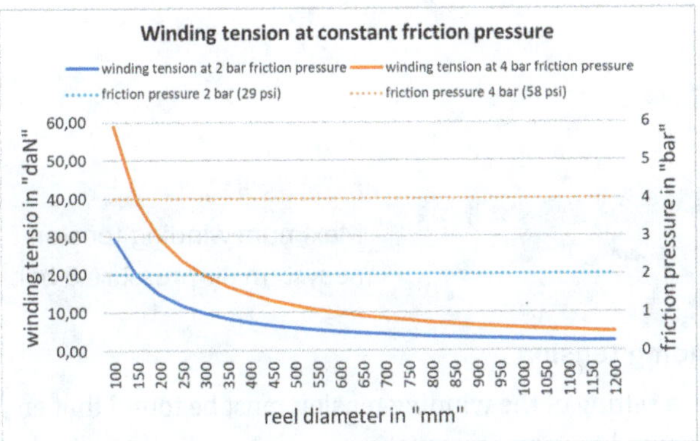

Figure 7.113 Winding with constant friction pressure (Note: 1 bar ≈ 14.5 psi)

Winding with constant winding tension

If the reels are wound with a constant winding tension (Figure 7.114), then the friction pressure must also increase as the diameter increases. If the initial values are chosen too high, the maximum possible system pressure of the compressed air will be reached very quickly (red line). A further increase in pressure is no longer possible; further winding would then correspond to winding at constant friction pressure but at maximum system pressure, which inevitably leads to unusable reels. Setting the initial pressure to the lowest possible friction pressure does not help either, because the reels would telescope right at the beginning of the winding process.

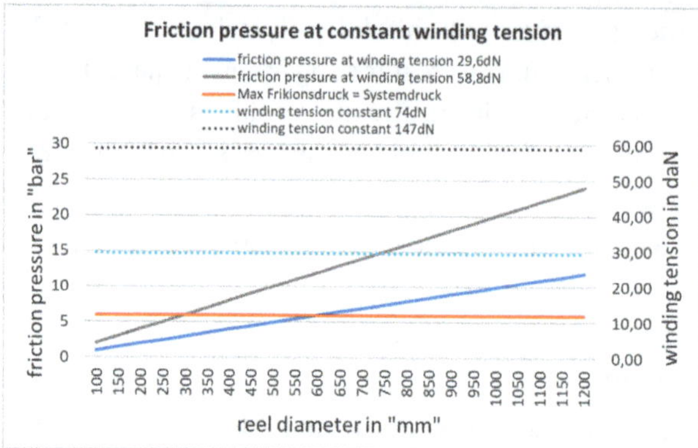

Figure 7.114 Winding with constant winding tension

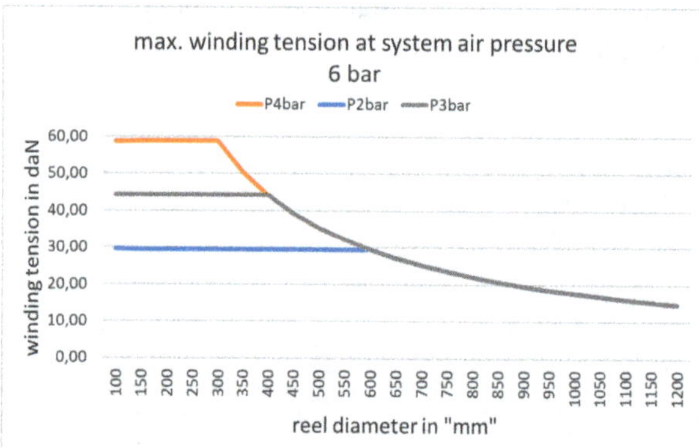

Figure 7.115 Maximum winding tension at system air pressure 6 bar

Winding with an adapted winding tension

To produce reels that can be sold, a setting of the winding tension must be found that enables a reel structure that is not wound too strong near the core and not too loose in the outer area in order to avoid telescoping. To do this, we must observe a few principles:

- The thinner the film, the more sensitive it becomes to warping and poor flatness
- The narrower the reel width, the more sensitive it is to telescoping and core collapse, as well as distortion and poor flatness
- The winding tension of the center winder must decrease as the diameter increases, but the friction pressure must still increase.

The possible diameter and therefore also the possible length (linear meters) of a reel are subject to these conditions in terms of film thickness and/or reel width.

Limits of possible winding tension

In Figure 7.116, the area in which the reels can be produced is shown in green. The maximum permissible tensile force is shown as the top red line, and this must be reduced as the reel length increases. The lower red line shows the minimum required tensile force, which increases with increasing length. Up to a reel diameter of approx. 350 mm, the tension setting is of minor importance. However, as reel diameters increase, this becomes increasingly significant, and the scope becomes increasingly narrower. This is particularly true for thin films and narrow reel widths.

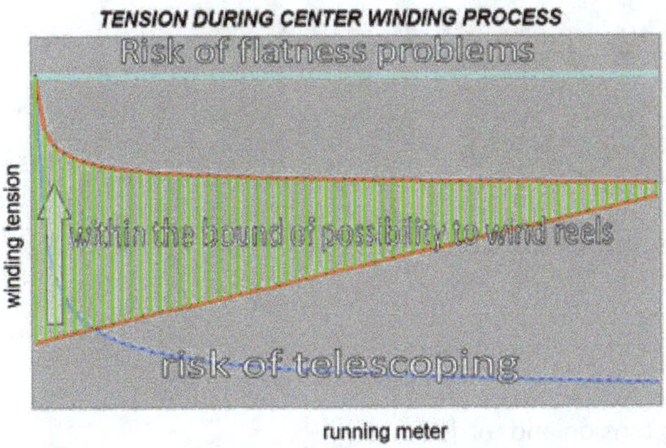

Figure 7.116
Tensile force during winding

The upper blue line marks the course at constant tensile force, whereas the lower blue line shows the reduction in tensile force at constant torque. The intersection with the lower red line shows the "point" when the danger of telescoping of the reels begins. To wind the reels in accordance with quality, we must find a line that is in the green range up to the desired diameter. In order to achieve this, the tension control is equipped with a potentiometer, which allows us to continuously preset the pressure on the friction elements.

Difference between shoulder frictions and friction winding shafts

The difference between winding with a shoulder friction and with friction winding shafts results from the way the winding tension is set up. This is shown in Figure 7.117. Due to the design, the torque in the case of shoulder friction depends on the pressure on the friction disks and their surface (green bars in the diagram above). The torque therefore remains the same, regardless of film thickness and reel width, because the friction surface per reel does not change. With a friction winding shaft, the torque is also generated via the friction surface, but this depends on the reel width.

Because the effective length of the friction strips increases with wider reels, the friction surface and thus the torque increase accordingly (brown bars). Due to the constant torque, the tensile force per mm^2 of film decreases in the event of shoulder friction, both with thicker films and with wider reels (blue line). This means that when setting the winding tension, the width of the individual reel, the number of reels, and the film thickness must be considered.

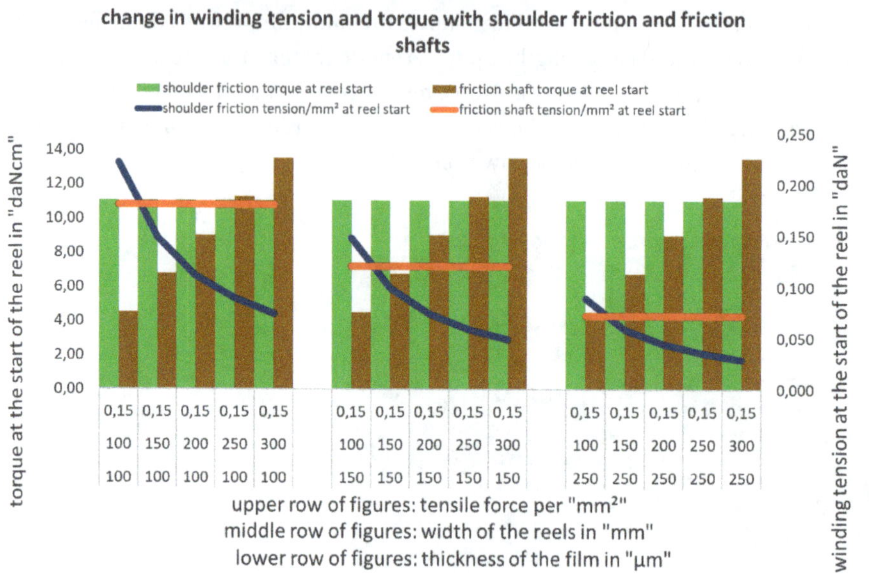

Figure 7.117 Change in winding tension and torque

With the friction winding shaft, the tensile force per mm² of film only depends on the film thickness, but not on the film width (red line). Therefore, when using a friction winding shaft, only the number of reels and the film thickness need to be considered.

The comparison in Table 7.6 shows that the advantages of a friction winding shaft clearly outweigh the disadvantages, especially when changing reels. The reduced downtime when changing reels significantly increases the productivity of a slitter rewinder and this has a cost-saving effect. The amortization of a friction winding shaft is therefore achieved in a very short time.

Table 7.6 Advantages and Disadvantages of Shoulder Frictions and Friction Shafts

Shoulder frictions	Friction shafts
Investment costs are lower	Investment costs are higher
Workshop maintenance costs are less common	Workshop maintenance costs are more common
Cleaning easier; less sensitive to cutting dust	Cleaning more difficult; more sensitive to cutting dust
Higher load capacity (reel weight)	Lower load capacity (reel weight)
Finished rolls cannot be pushed off the winding shaft	Finished rolls can simply be pushed off the winding shaft
To re-equip the sleeves, the frictions and intermediate rings must be pushed on with the sleeves	Sleeves can be positioned quickly using laser beams (significant time savings)

7.4.20 Ionizing Rods and Clamping Bars

Two important details that should not be missing from any winder are ionizing rods in the area of the unwinder and the rewinder, as well as clamping stripes in the area of the rewinder, which are intended to make it easier to stick the individual webs to the empty tubes.

Ionizing rods

In principle, plastic films are very good insulators against electrical currents. However, this insulating property leads to plastic films becoming electrostatically charged when unwinding from a roll due to the webs being separated and when being deflected around a roller. This electricity is caused by friction and cannot be avoided. This charge must be dissipated, and this is achieved with so-called ionizing rods; if this is not considered:

- Dirt particles will be attracted
- A well-unloaded plastic film can be wound up more easily
- The risk of telescoping is reduced
- The danger of the electrical charge jumping to people is averted; this can cause hair to stand on end (Figure 7.118), sparks to be produced, or styrofoam pieces to stick to your hands
- Movements that can lead to accidents are prevented.

Figure 7.118
Effect of static electricity from a non-dissipated ionizing rod

Clamping bars

Figure 7.119 shows an example of the arrangement of clamping bars on various rewinders. The task of these clamping bars or clamping stripes is to maintain the tensile conditions within the winder even while changing the reel. Therefore, the clamping bar must be triggered immediately after stopping and before cutting the webs.

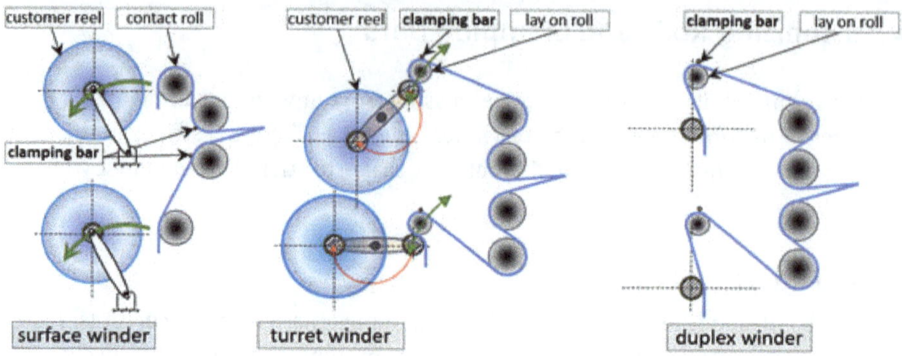

Figure 7.119 Arrangement of clamping bars

When working without this clamping bar, the winding tension is eliminated. The film web would be pulled back slightly by the unwinding pull of the unwinder and the cutting tension, and would hang down in an uncontrolled manner. When glueing the film web to the empty cores, additional handling would be required, and the position of the stripes may be slightly shifted.

Therefore, when using the clamping bar, the film webs are slightly under tension even during the reel change, and glueing is made easier. The positioning of the clamping bars in the machine must be done individually depending on the course of the web path.

7.4.21 Surface Winding and Center Winding

Surface slitter rewinder

In Section 7.4.17, the two possible types of winding were considered, namely center winding and surface or circumferential winding. Both types of reel slitting and rewinding machines are common for calendered films, depending on the type of application.

A surface winder is very suitable for thin films, such as stretched films, which are often printed, as this type of winding can be used to produce rolls with a larger diameter without negatively affecting the flatness. Rolls with lengths (running meters) of 8000 m and more can be produced without any problems. This would be the case, for example, for crosswise stretched films, which are produced on the calender with a thickness of 160 µm and then stretched to a thickness of 50 µm. Producing this with a center winder would be very problematic, as Figure 7.120 shows.

In this case, a 1200 mm wide calender roll would be stretched to a width of 3840 mm. This is very wide for a center slitter rewinder, but a surface winder of this width is technically no problem. On the surface slitter rewinder, the lower, slanting red line would run approximately horizontally (shown as a red, dotted line for clarity), with the risk of telescoping being very low.

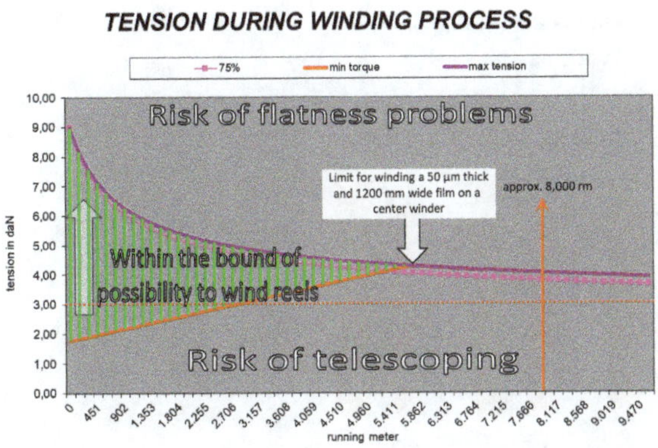

Figure 7.120
Tensile force during winding with the center winder

For cost reasons, surface slitter rewinders are generally used for a smaller number of reels, i.e. for wider finished reels and larger roll diameters. Ideally, each reel is pressed against the contact roll via a pair of swivel arms.

When changing reels, the individual reels must be moved to the weighing and labelling station via a transport device. The duration of a reel change takes more time than with a center slitter rewinder. This is another reason why surface winders are used less often when the requirements can be met with a center winder.

Center slitter rewinder

Center winders use two methods to produce customer reels:

- The double or duplex winder with two winding shafts one above the other
- The turning or turret winder with two turning winders and two winding stations each.

With both types there is no difference in the tensile force and winding tension for the actual winding process.

As explained in Section 7.4.18, the friction winding shaft has already been worked out to have procedural advantages over shoulder friction, which is why the use of friction winding shafts is assumed in this section.

The load limit for friction winding shafts is around 100 kg per meter, which is why support in the middle area is necessary for wider machines. If we assume a maximum diameter of 600 mm for the customer reels, then up to a width of around 1400 mm can be worked without center support. (As a rule, the winding shafts are only occupied by about half the width of the jumbo roll.)

As can be seen from Figure 7.121, finished reels can be easily produced on center winders with a diameter of 600 mm starting from a film thickness of around 100 µm and a finished roll width of 100 mm. This diameter limitation is only a handicap for 3″ cores on a friction winding shaft when customers require larger diameters.

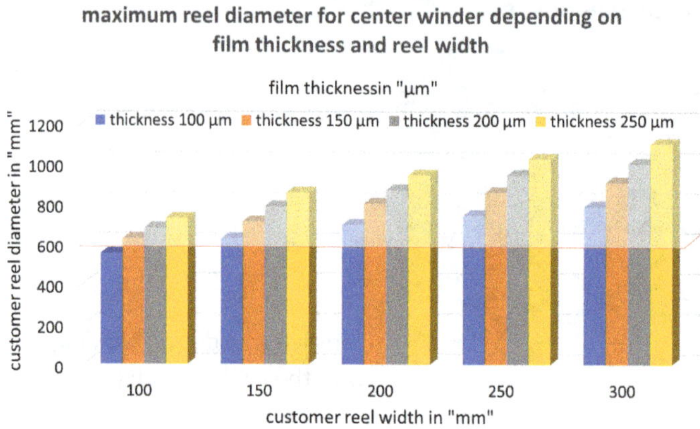

Figure 7.121 Maximum diameter for the center winder

Then the entire design of the cutting machine has to be adjusted and the question arises as to whether this is economically worthwhile. The technical possibilities are not the problem here. The center winding is more suitable for narrower reels than the surface winding. The investment costs are lower regardless of the number of reels to be slit per set, apart from the costs for the slitting section, which would also be incurred with a surface winder. If the budget for a reel slitter rewinder is tight, single knife holders can be dispensed with up to a web width of max. 1400 mm (see Section 7.4.9).

Concluding Remarks

Finally, below are a few tips on which reel slitting and rewinding machine is recommended for which purpose.

Devices that are recommended in any case:

Every slitter rewinder should have the following devices in order to be able to deliver reels of good quality:

- Adhesive table on the unwinder
- Ionizing bars on the unwinder and in front of the rewinders
- Clamping bars in front of the rewinders
- Precise tension and pressure control
- Well-coordinated acceleration and deceleration values, as well as precise speed control
- Push-off device for finished reels
- Pick-up mandrel for the pushed-off reels.

Desirable devices for increasing throughput and quality assurance:

- An automatic transport of the reels to a preferably autonomous weighing station with labelling, packaging, and palletizing of the reels
- A surface inspection system (as is usual on the calender) for the precise localization of reels with quality defects for quality-compliant further use (reworking or sorting out).

Shoulder frictions or friction winding shaft?

The advantages of a friction winding shaft clearly outweigh the disadvantages. Shafts with shoulder friction can only be an alternative for rare applications or for cost reasons.

Top knife on a shaft or single knife holder?

What matters most here is the width of the machine. With a machine width of up to 1400 mm, you can work with a continuous top knife shaft without major problems if your budget is tight.

Here too, automatic positioning of the top and bottom knives is possible. The disadvantages that cannot be avoided are:

- All top knives must have the same diameter
- Replacing the top knives takes more time
- A single knife cannot be exchanged without affecting the other knives
- A "garage" must be created for the unused knives
- The wrap around the bottom knife shaft must be a little bit larger to ensure slitting at the contact point. With sensitive films, this can lead to the edges of the bottom knife being marked if the lower knife shaft is not fully equipped.

In principle, single knife holders are recommended; for wider machines they are required, or production must be carried out with a "wrap cut". This leads to significantly longer changeover times, as automatic knife positioning is no longer possible, because in this arrangement the lower knife shaft must be fully equipped, unless it is possible to stagger the reel width in (for example) 10 mm increments. The knife overlap would also be different across the width, and this would lead to increased wear on the top knives.

Surface slitter rewinder or center slitter rewinder?

This question can only be answered with the involvement of sales or marketing, because this depends on the tasks to be performed:

- As already explained in the previous sections, the surface slitter rewinder is particularly suitable for thin, wider films and large diameters
- The number of reels per set should not be too high in order to avoid lengthy machine downtimes when changing reels

- The film thickness should not exceed approx. 300 μm
- With a center winder, the number of reels per set can be higher
- Pushing onto a mandrel (Figure 7.122) is always possible
- The following work such as weighing, labelling, packaging, and palletizing are also possible.

Figure 7.122 Center winder with mandrel (Source: [6])

However, it should be noted that the diameter of the reels must not be too large to avoid telescoping or flatness defects. An overview is compiled in Figure 7.121. The diameter depends on both the film thickness and the reel width.

Center slitter rewinder as a duplex winder or turret winder?

When it comes to this question, the sales and marketing requirements must also be considered, as the required cutting capacity depends on them. The minimum order size plays a particularly important role in the decision-making process, because machine capacity plays a significant role due to machine downtime during an order change and can significantly reduce capacity. Experience has shown that when changing orders, a period of up to 10 min or more can occur without production due to the necessary work, such as:

- Changing pallets
- Generating the reel list
- Operating data collection
- Changing the jumbo reel and disposing of the remains of the preliminary product
- Adjusting the top and bottom knives according to the reel width
- Setting the laser pointers for core loading and other things.

These factors apply to all types of slitter rewinders. The remaining criteria for deciding whether to use a duplex or turret slitter rewinder are (for example):

- Customer reel width
- Number of reels per set

- Customer reel diameter
- Machine slitting time
- Film thickness slitting speed.

Figure 7.123 shows the influence of these factors based on a 250 µm thick film. The blue bars show the capacity of a duplex slitter rewinder, while the green bars show the increase in capacity of a turret slitter rewinder. The sum of the two bars (blue line) represents the capacity of a turret slitter rewinder. It is clearly visible that the turret slitter rewinder makes an increase in output possible, especially with larger diameters and a moderate number of reels per set.

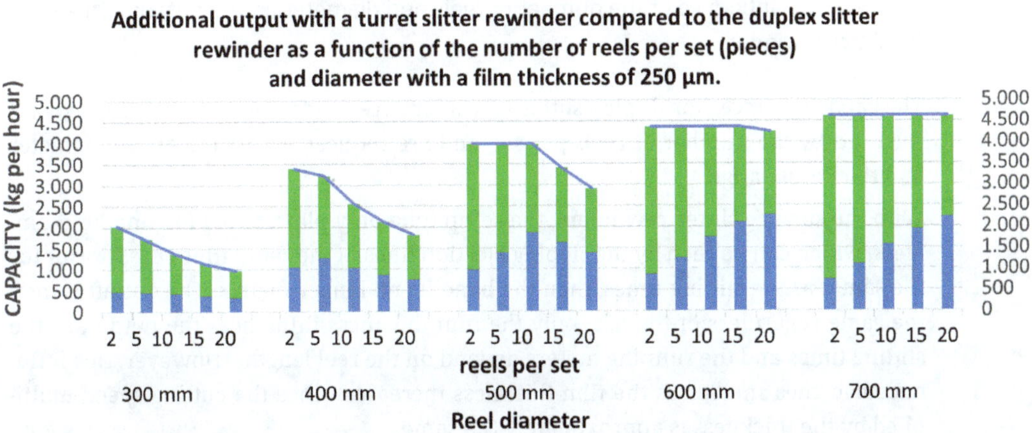

Figure 7.123 Duplex or turret slitter rewinder

Figure 7.124 and Figure 7.125 show the capacities of the two systems over a thickness range of 100 µm to 600 µm.

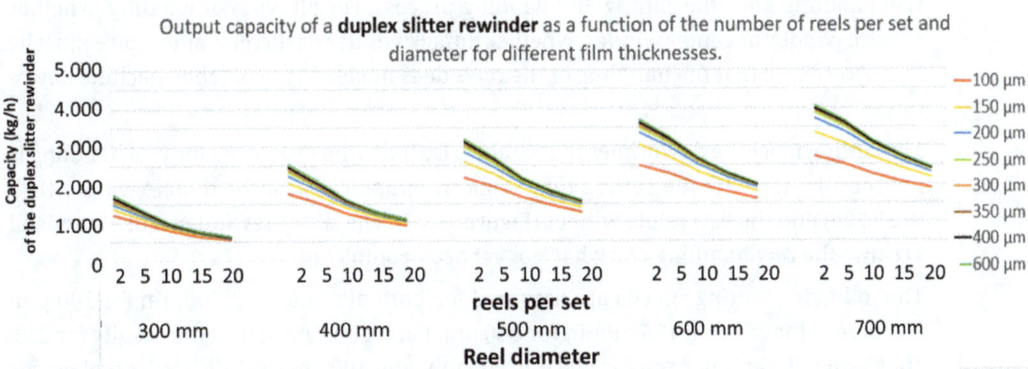

Figure 7.124 Influence of the number of reels and diameter on the output of the duplex slitter rewinder

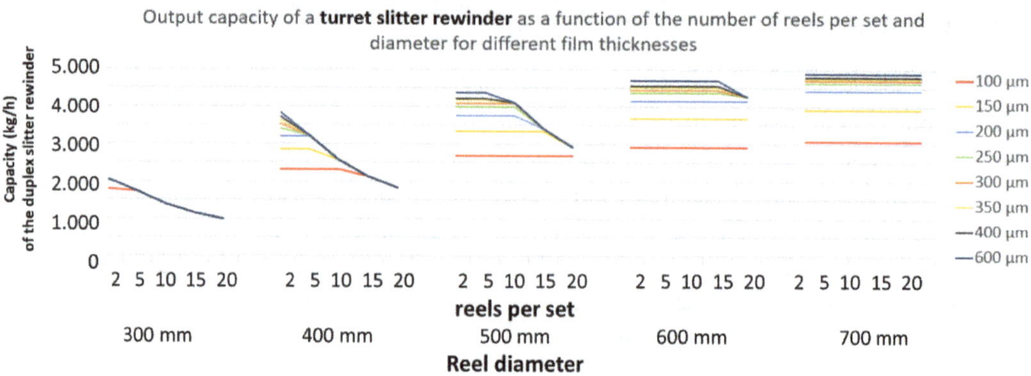

Figure 7.125 Influence of the number of reels and diameter on the output of the turret slitter rewinder

The capacity curves for duplex slitter rewinders are fundamentally very strongly influenced by the number of reels per set, and are reduced across the entire diameter and thickness range.

With the surface slitter rewinders, the setup time only plays a role for smaller diameters, which can be seen by the sloping ejection lines. If the setup times are shorter (or the same) as the slitting times, then the benefits no longer matter. The output is then the same regardless of the use, only the film thickness influences the output, as the slitting times and the running meters depend on the reel lengths. However, this influence becomes smaller as the film thickness increases, since the cutting speed multiplied by the thickness is approximately the same.

If due to these considerations, the higher investment costs are amortized more quickly (return of investment), a turret slitter rewinder is preferable to a duplex slitter rewinder.

Reel handling:

To fully make use of the capacity of a modern slitter rewinder, attention must be paid to reel handling after the slitting and winding process. For all types of winding, whether surface winder or center winder, whether duplex winder or turret winder, care must be taken to ensure that the handling of the reels does not lead to avoidable machine downtime.

The cutting machine downtimes that could arise if the time required for reel handling is not coordinated with the cutting time is shown graphically for the double winder (Figure 7.126) and the reversible winder (Figure 7.127). The slitter rewinder is at a standstill because the reel handling work has not yet been completed.

The following cutting speeds are assumed for both diagrams: 325 m/min for 100 µm; 200 m/min for 250 µm; 125 m/min for 400 µm. Each diagram shows the results for 2, 5, 10, 15, and 20 reels per set, for 400 µm, 250 µm, and 100 µm thicknesses, as well as for diameters of 300 mm to 700 mm in 100 mm increments.

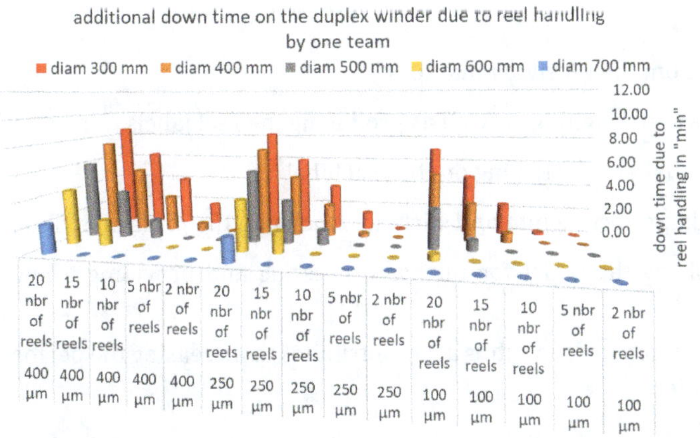

Figure 7.126 Downtime on the double winder due to reel handling

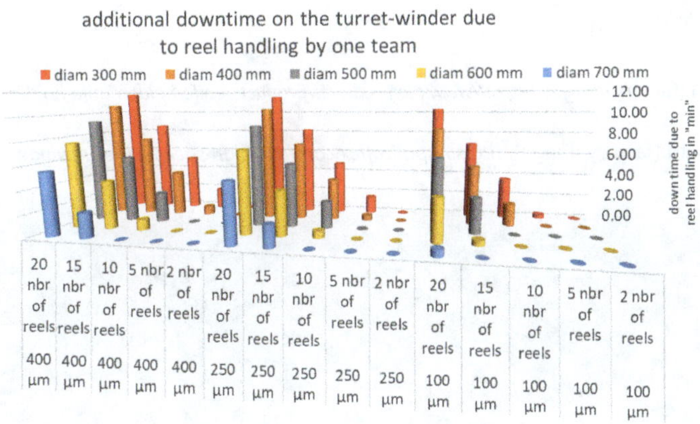

Figure 7.127 Downtime on the reversible (turning) winder due to reel handling

In order to avoid such machine downtime, the slitting machine team tends to reduce the speed. Then the slitting time is extended and coincides with the reel handling time. Therefore, reel handling must be given the necessary attention (Figure 7.128).

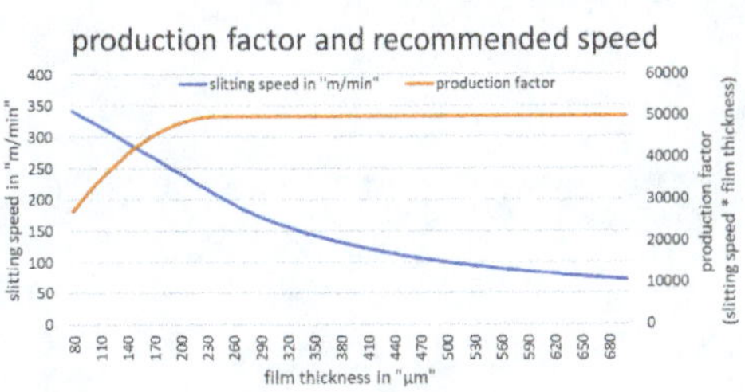

Figure 7.128 Production factor and reference speed

The work of reel handling is:

- Removing the reels from the receiving mandrel
- Feeding to the scales and weighing, as well as producing the reel labels
- Sticking on the sleeve label and sticking on the reel label
- Wrapping the reels in protective film and stick the label onto the protective film
- Palletizing—transporting the pallet for further processing and providing a new pallet.

This work can easily be automated, and thus also contributes to increasing the performance of a slitter rewinder.

References for Chapter 7

[1] Kohlert, C. [lecture]: Klöckner Pentaplast

[2] Kompressorenteile, *DIENES Schneidtechnologie*

[3] Bilstein, *https://bilstein.website*

[4] Rotary slitter blades, Maxcess

[5] DIENES Messerhalter PSG 1 [factory image], *https://www.dienes.de/en/products/knife-holder/knife-holder-psg-1/*

[6] SP Ultraflex Systems Pvt Ltd [factory image], *https://spultraflex.com/our-high-speed-slitter-rewinder-products.html*

8 Digitalization

8.1 Digitalization and Industry 4.0

Frank Kleinert

8.1.1 Historical Classification

With the introduction of computers, the automated monitoring of processes, and the use of digital data, a digitalization process has emerged over the last 50 years that started in industry but also developed in the private environment. In the 1970s and 1980s, ERP (enterprise resource planning) systems were developed, for example by the company SAP, which primarily support controlling, purchasing and sales. At the same time, MES (manufacturing execution system) solutions were programmed, with a focus on production processes (Figure 8.1).

The spread of the faster wireless standard 5G or LTE, fast company networks, and the emergence of global cloud providers with high computing power have supported the spread of digitalization. Figure 8.1 is a schematic representation of the digitalization of industry, where the term "Industry 4.0" stands for the networking and communication of machines/ERP/MES systems. The focus of Industry 4.0 is on the intelligent networking of machines and systems using existing internet technology. Use cases can be very broad:

- Flexible production: Networking of production processes in the company but also digital linking of suppliers and customers.
  Goal: Optimization of processes and procedures.
- Optimized logistics: Production planning, inventory management, and delivery routes are synchronized based on historical data.
  Goal: Optimization of production cycles.

- Use of data: Digital data on the quality and condition of the system is recorded and evaluated centrally.

  Goal: An increase of efficiency and service.

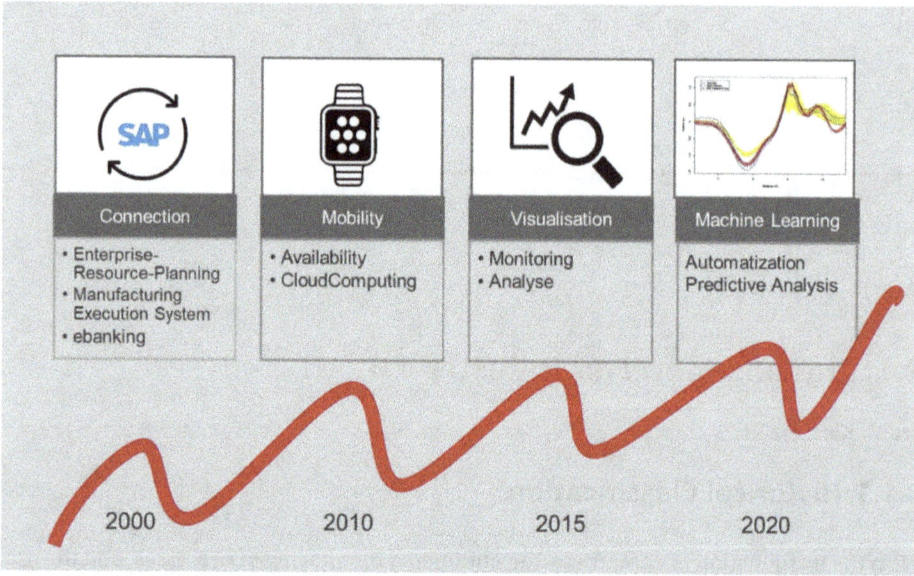

Figure 8.1 Trending development of Industry 4.0 and machine learning (Source: [1])

8.1.2 Typical Challenges When Introducing Industry 4.0

Digitalization according to the principle of Industry 4.0 has various challenges, with typical hurdles in medium-sized companies being:

- Existing IT infrastructure
- Lack of standardization in production
- Knowledge and capacity of personnel
- Recognizing applications with clear benefits
- A lack of data strategy in management.

A typical IT infrastructure was established in the 2000s. This was based on the understanding that data is gradually consolidated in the levels and the generated data is promoted up the pyramid (Figure 8.2). There are defined interfaces between the levels, which typically transmit limited data. Details about machine settings or the quality data recorded in-line are often lost or remain unstructured on servers.

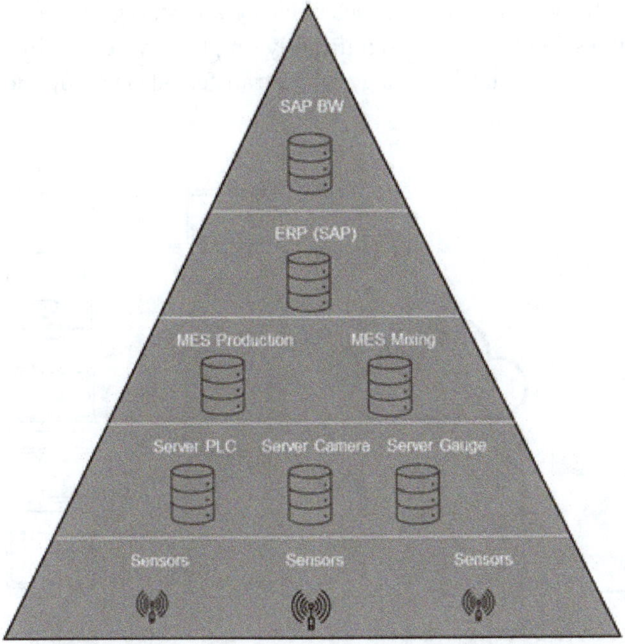

Figure 8.2 Typical IT infrastructure with sensors, servers, SCADA, MES, and ERP systems (Source: [2, 3])

The IT systems of the last 30 years therefore have to overcome five challenges (Figure 8.3). Solutions that allow end-to-end networking are ideal for new investments. Usually, a company will use a combination of existing infrastructure with the coupling of cloud solutions so that the existing systems can continue to be used.

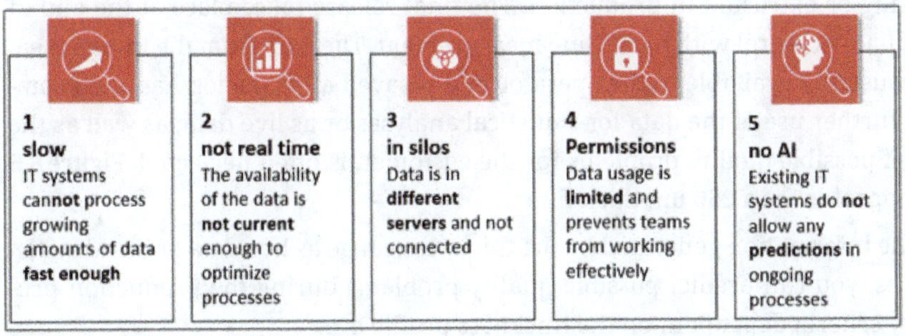

Figure 8.3 Challenges for the IT infrastructure in the transition to Industry 4.0

Data is transferred to the cloud as required and combined with other data so that existing servers and their challenges (Figure 8.4) can be resolved. This means that the existing IT infrastructure remains in place and the required added value is created, which can be written back into the existing MES/ERP system if necessary. Hybrid IT systems combine the advantages of both systems, such as speed, security, and storage capacity. Rolling out to multiple machines and locations is significantly simplified when sensors and interfaces are standardized, and cloud technology can be transferred via "copy and paste".

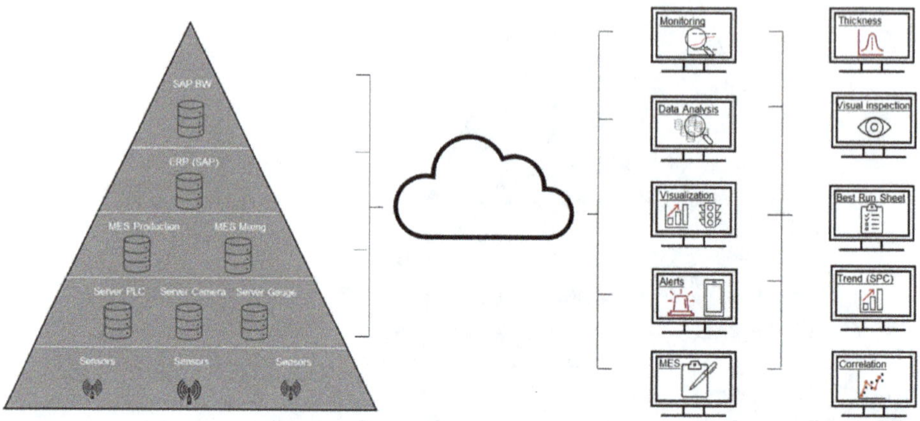

Figure 8.4 Hybrid IT infrastructure by coupling local and cloud-based IT (Source: [2])

8.1.3 In-line Film Thickness Monitoring

Use cases for Industry 4.0 are mentioned in many literature sources. Since digital twins, prediction of maintenance intervals, and automated inventory monitoring are often described as examples, the prediction of quality problems will be discussed here.

The thickness of the film is usually monitored in continuous processing processes such as calendering or extrusion in production. The final release takes place at the end of the roll by quality control with an off-line measurement. The data from the in-line measurement is usually available to the operator and is saved after the job has been completed. The further use of the data for statistical analyses or as live data, as well as the prediction of possible quality problems for the customer, is often neglected. Figure 8.5 shows a "geoprofile" of a 250 µm thick film.

If you use the information generated by the thickness gauge as live data and add statistical analyses, you can predict possible quality problems during the production process. Quality criteria dependent on the thickness profile are:

- The thickness specification limit
- Detecting piston rings
- Hard and soft edges
- Possible telescoping of customer reels.

Figure 8.5 Geoprofile display of a film 2000 m long and 1700 mm wide; blue and orange areas are deviations from the mean value

Figure 8.6 shows the normal distribution, the box plot, and the *Q–Q* plot of a production that has been identified as critical.

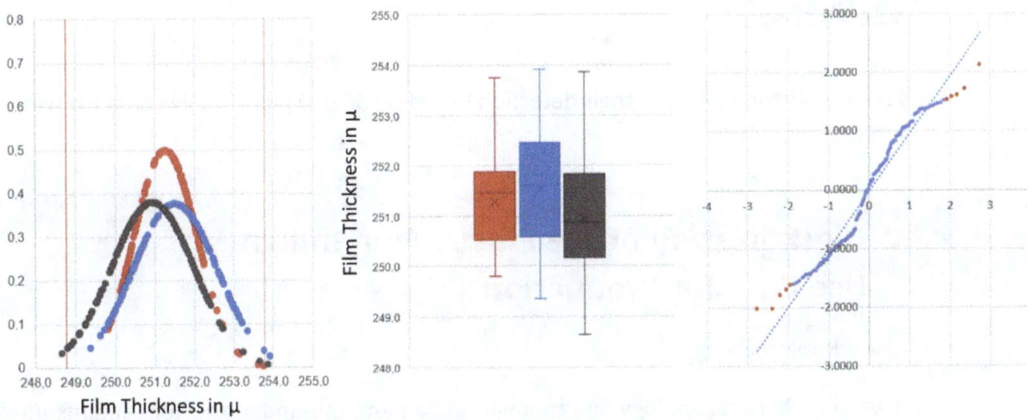

Figure 8.6 Statistical analyses to describe the normal distribution, to visualize the standard deviation, and to detect measured values outside the normal distribution

Based on the data analysis, a visualization of the critical areas of a film is created (Figure 8.7). At the same time, the statistical analyses provide indicators for predicting quality problems in the subsequent step of the process.

If you link the data from the in-line measurement and the data from the MES to compare the target specifications, all that is required is a visualization for the operator and the logging of the statistical data for each role or order.

Of course, this type of quality prediction using in-line data is also possible for other quality deviations such as:

- Burns
- Air bubbles

- Gels
- Contamination such as metal
- Color variations.

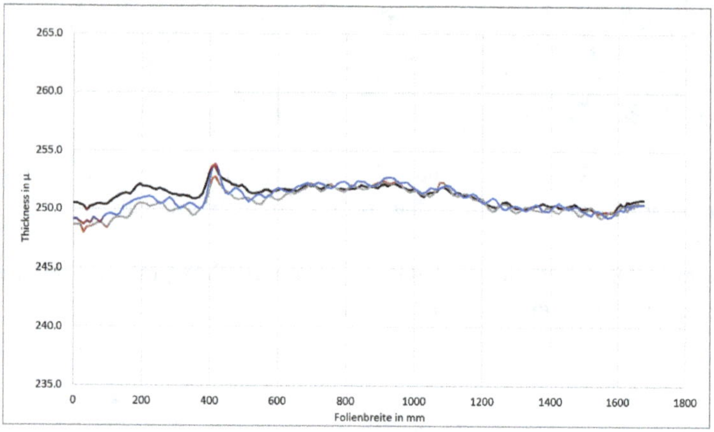

Figure 8.7 Piston rings and their detection based on 30,000 pieces of live data from inline measurement

8.2 Digitalization of the Plastic Film Industry: Use Case for Production

Michael Kohlert

In a world where technology and innovation go hand in hand, no sector remains untouched by the wave of digitalization. The plastic film industry, once a traditional business that relied on proven manufacturing processes, is now on the cusp of a revolution.

Digitalization has not only changed the way plastic films are produced and processed, but also how they are optimized. This article highlights the profound changes that digitalization is bringing to the plastic film industry, and shows how companies can use this change to become more competitive and meet the future demands of the market.

8.2.1 Introduction

The plastic film industry produces millions of tonnes of waste every year. Machine failures that lead to downtimes and waste of raw materials, incorrect machine settings, misdirected use of materials, low technical know-how of the system operators, and fluctuations in material quality due to inadequate incoming goods inspection cost the plastic film industry billions of euros.

"The intelligent factory", "predictive maintenance" and "big data" are buzzwords for Industry 4.0 these days. Around 1980, programmable microcontrollers were a step in this direction, followed by the transfer of data to databases and the extraction of knowledge from large amounts of data—new ways to make information usable in the process (Figure 8.8).

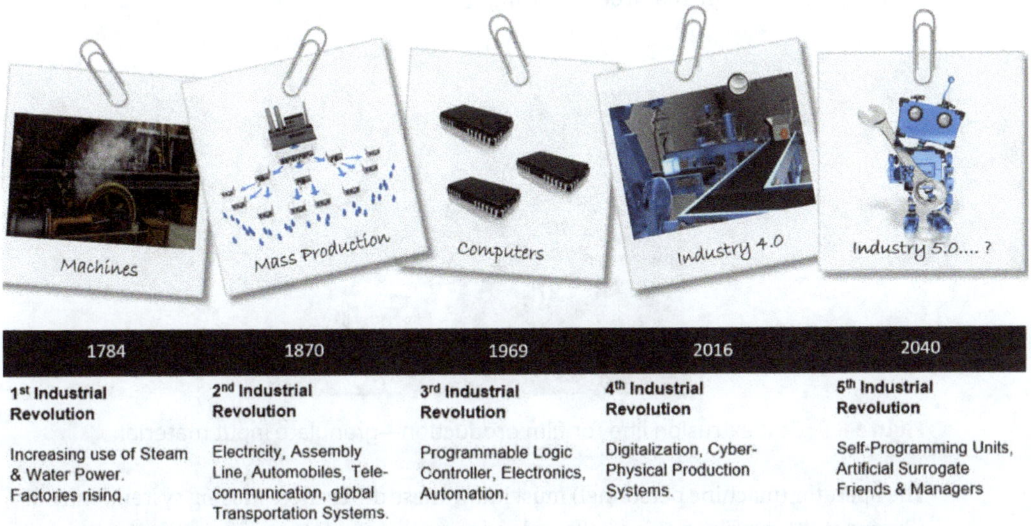

Figure 8.8 Time series of industrial revolutions

With today's methods, real-time monitoring and evaluation of data is becoming increasingly important to minimize costs and increase system efficiency. Visualization aids and statistical controls are used to make the work of the plant operator faster and easier. These applications use advanced machine learning algorithms to identify potential problems in machines, allowing employees to take corrective actions and avoid production downtime. The goal of a self-controlling machine ("Self-X") to achieve optimal process stability is getting closer: an extrusion system that monitors all parameters in seconds, calculates trends, reacts proactively to deviations, and regulates itself.

The plastic film, sheet extrusion, and cardboard industries are increasingly focusing on Industry 4.0 pilot projects adapted to the location's own needs. These projects include big data, human–machine interfacea, predictive maintenance, data mining, mobile device management, and virtual storage management, to name a few. In this section the focus is on a condition monitoring system that takes the form of a human–machine interface that assists the production employee in process optimization.

8.2.2 Human-Machine Interface for AI Recommendations

An example extrusion system (cast, blown film) with a production volume of 1100 kg/h produces 5–15% of rejects. The granules (PE, PP) and additives conveyed from the metering into the extruder are heated at 100–220°C, depending on the material type. The liquid gel is pressed through a nozzle/ring onto a glazing roll (calender stack) or into a bladder, cooled via roller and traversing systems to the reel winder, and handed over to an AGV (automated guide vehicle) or forklift as a finished reel. The machines differ partly in their design and process settings (Figure 8.9).

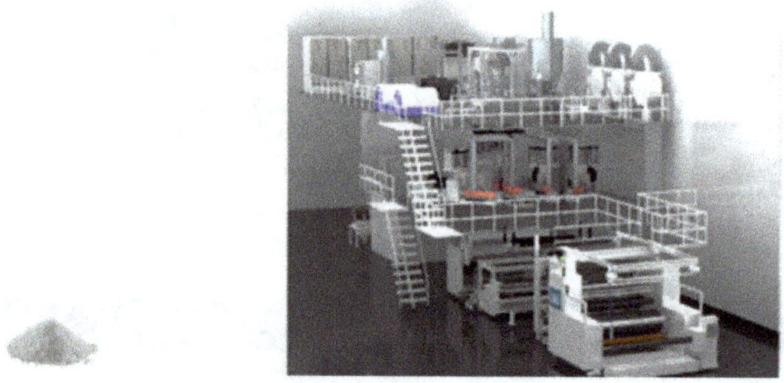

Figure 8.9 Cast extrusion line for film production—granulate input material

The operator (machine personnel) must know these differences (dosing system, extruder (fluxing unit), nozzle, glazing rolls (calender stack), winder, and scale), understand the specifications (shrinkage, color, thickness, opacity, basis weight, and moisture) of individual products, quality rules (six-sigma, variance, standard deviation) according to customer specifications, and monitor the entire machine over a length of 30 m and a height of 20 m using sensors.

Changes are usually noted in writing and maintained in other local systems, such as the manufacturing execution system (MES) and computer-aided quality (CAQ) system. With larger machinery, jumpers are deployed between the machines to minimize personnel costs per kg. However, this also increases the complexity of the cross-machine rules that must be adhered too.

The machine operator ensures the smooth production of the order types to be manufactured according to the order planning specifications. The product mix leads to static and dynamic material, parameter, and aggregate changes that are carried out by the machine personnel and the setup team. Quality must be monitored constantly and sometimes takes up 30% of the actual work. If the operator recognizes a defect early on that is a result of quality or material deviations, countermeasures can be taken. Due to the high complexity, complete monitoring by the operator cannot be guaranteed without the necessary sensory support. Today's measurement technology can sometimes monitor individual characteristics (defects, thickness, and color), but the reason for many scrap problems lies in the interdependence of several problem variables occurring at the same time. This multivariate dependency cannot be recognized by an untrained system operator and can hardly be calculated without analysis tools in a reasonable amount of time during day-to-day business and recurring new problem areas. Moreover, an evaluation by the production manager and others can sometimes take days. The result is then already outdated because new operational problems are faced.

In this situation, a software application was developed that, integrated into the network, has access to all subsystems and continuously monitors them. The program was developed in collaboration with production for production. An interface (HMI) offers the system operator recommendations to improve the operation of the extrusion system.

8.2.3 Data Collection and Storage

A film extrusion machine (cast film or blown film) is controlled by up to five programmable logic controllers (PLCs), mostly from different manufacturers of the various units. They log temperatures, pressures, speeds, and other performance parameters using a variety of sensors (Figure 8.10).

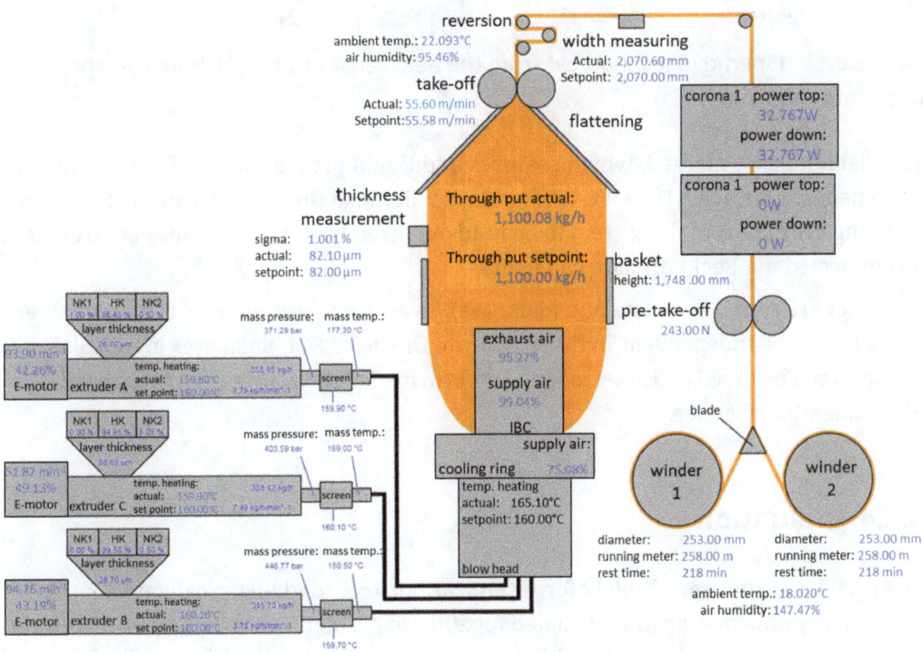

Figure 8.10 Blown film extrusion system—overview of the control data (Source: Maria Soell GmbH)

Each machine records up to 2000 parameters per minute, generating 7 GB of data per day. In comparison, a car has just over 200 sensors. This data from a production machine can now be used for predictive maintenance. However, the company's infrastructure must be prepared for reading and storing the data (Figure 8.11). In most cases, there are established structures with little standardization. Basically, three data blocks are always required—order data (ERP, MES), control data (PLC), and quality data (CAQ, measurement technology).

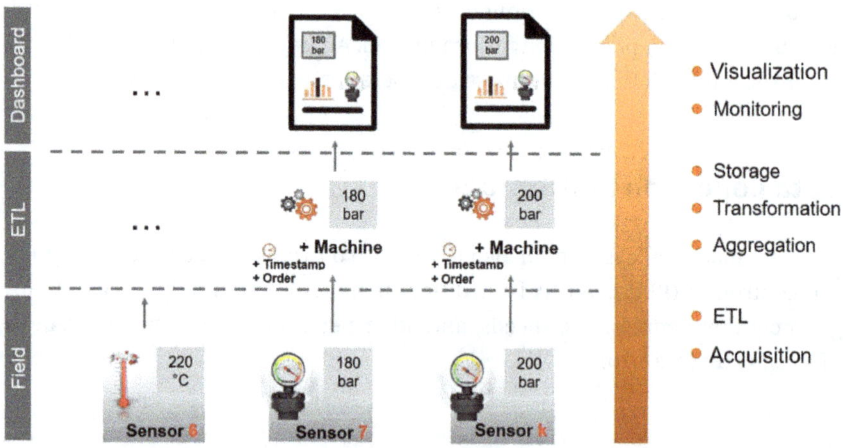

Figure 8.11 "Operative technology" from the field sensor to the dashboard—shop floor data acquisition

This data must be merged, which requires additional programming effort due to heterogeneous interfaces. The OPC-UA standard will make things easier in the future, by allowing every part of the system to be read out directly. Until then, older controls can be connected via Beckhoff data collectors.

The dates are stored on a data hub (database). To achieve decoupling of the system hierarchy and to be independent in the long term, Oracle or SQL databases are ideal. Here the dates can be saved and accessed in a uniform format in the three predetermined data blocks mentioned above.

8.2.4 Evaluation

The raw data is not saved solely for visualization. Recorded historical data combined with quality information are evaluated together to predict future machine downtimes based on statistical analyses. For this purpose, 2 years' worth of data was evaluated.

The sensory timestamps were classified based on the quality data. The sensory information is divided into data during a scrap process (class 2) and data from good production (class 1). Furthermore, a further, more dynamic division of the data was carried out. For this purpose, trajectories were formed that were available in two classes.

First, the sensors were evaluated separately to identify interdependencies. As an example, Figure 8.12 shows the evaluation of sensory information via the mean of a confidence interval, as well as the minimum and maximum for each sensor. It turns out that the confidence interval is subject to only very slight fluctuations between the data from two sensors. Based on this analysis, it was recognized that scrap production cycles are very difficult to identify using individual sensors, even if singule values can be found outside the confidence interval. This multivariate approach to anomaly detection uses the projection of high-dimensional data sets and reduces them to an average of two dimensions using a non-linear projection method.

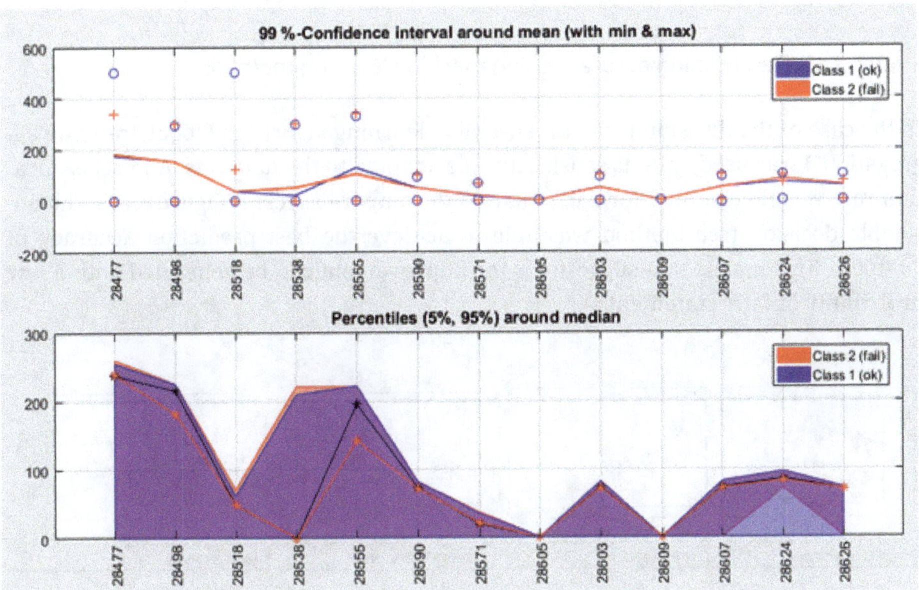

Figure 8.12 Confidence interval of sensory data from an extrusion machine

8.2.5 Data Processing and Analysis

The univariate analysis and projections described previously are not sufficient for a suitable determination of the machine condition. Classification methods from the field of machine learning methods are more promising here, by allowing predictions about the condition of the machine to be determined based on minute-by-minute sensor data. Historical data from the past 24 months was used for each machine (Figure 8.13). The data set was randomly divided into 60% training data and 40% validation/test data.

Four different methods were tested on the underlying data: nearest-neighbor classification, decision trees, neural networks, and naïve Bayes classification.

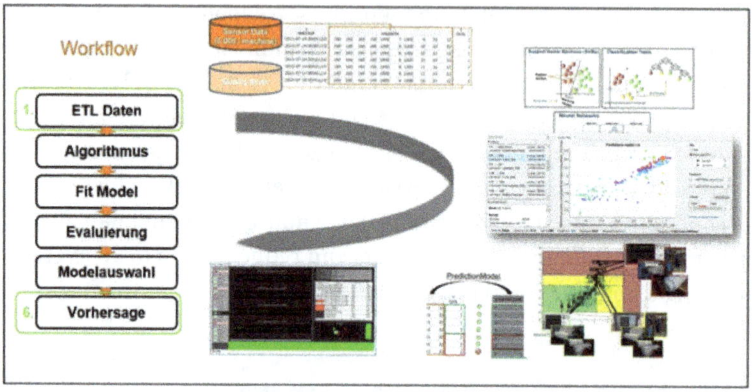

Figure 8.13 Determining accuracy using machine learning methods

In the case of the decision trees, an ensemble learning algorithm ("bootstrapping aggregation") was used, and its sensitivity was adapted to the data sets. A misclassification rate was determined for each method to compare success (Figure 8.14). The ensemble decision tree method was able to achieve the best prediction accuracy of 95–100%. This means that suggestions for improvement can be generated with a low probability of false statements.

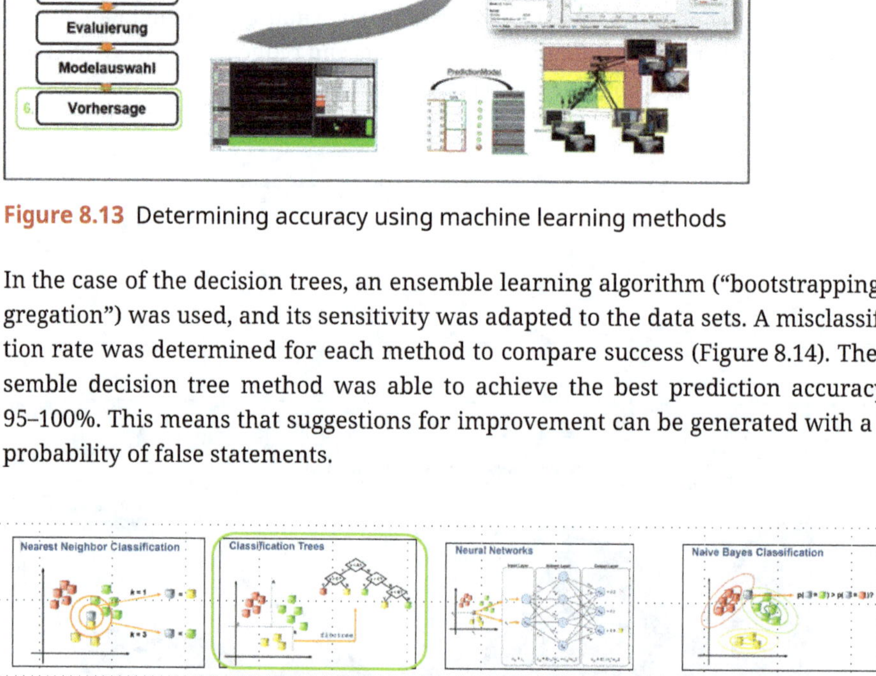

	NearestNeighbour	TreeBagger	NeuralNetwork	NaiveBayes
	Misclassification % (Mean)	*Misclassification % (Mean)*	*Misclassification % (Mean)*	*Misclassification % (Mean)*
M151	24%	2%	8%	10%
M152	44%	5%	23%	13%
M153	23%	2%	13%	13%
M156	12%	2%	3%	9%
M157	11%	1%	10%	8%
M158	29%	2%	14%	17%
M159	21%	0%	3%	2%
M181	1%	0%	1%	2%

Figure 8.14 Misclassification rate by learning method and asset

8.2.6 Visualization in Production

The developed models were adapted and programmed for prediction in a real-time application. The prediction function has been integrated into the application in the form of a traffic light and a message (Figure 8.15). If a machine failure is predicted, the machine operator receives a recommendation to shut down the machine in a controlled manner and so reduce unnecessary scrap time.

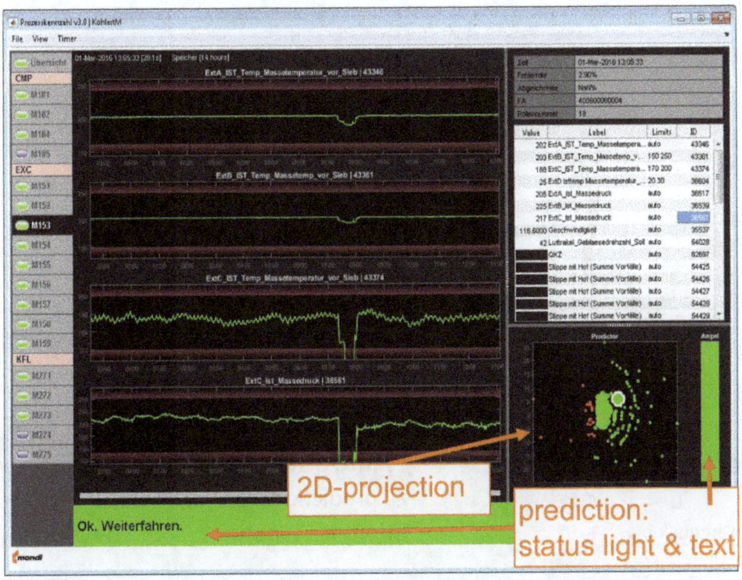

Figure 8.15 Application for predicting scrap, created with MATLAB

With this approach, the quality status of each time stamp was considered; as an example, production data for good production can be seen in the green marked area, while reject production is in the red marked area. The static condition of the machine can therefore be displayed very easily.

The predictive models are stored locally so that the application can retrieve this data at any time. The models are fed with new data every day and carry out an automatic update. A training program collects the last 3 months of data and performs new calculations. The prediction quality is then compared with the previous models. If the calculation leads to an improvement, the new models are used. The models discard old and new data that does not stand up to renewed data checking. Three models are used here to evaluate the data, which classify the data in the form of an agent system with majority voting.

The AI's statement about whether the machine needs to "continue running" or "shut down in a controlled manner" was developed together with the production employees. The system operators have little experience with statistical analysis and machine learning methods. It is therefore particularly important that this group of people accompanies the project from the beginning and is supported at every step.

Figure 8.16 shows a system from St Petersburg University of Technology that uses AI and trajectories to improve predictions. When looking at the movement sequences of the states of a machine, the direction can be used to make a statement about future states.

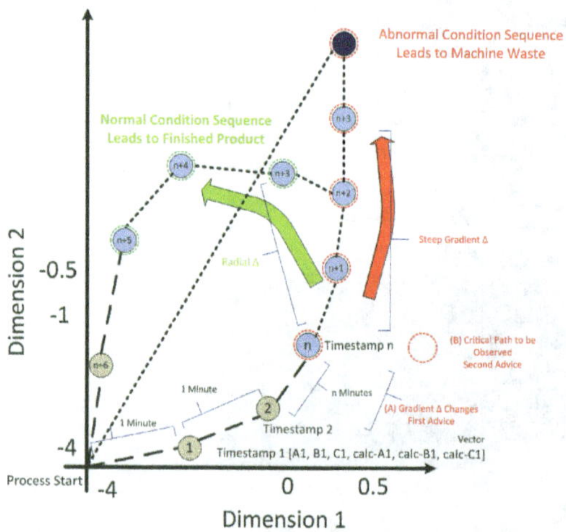

Figure 8.16 AI prediction using trajectory approach (Source: TU St Petersburg)

The trajectory approach allows a time prediction of several hours. This approach should be improved in the future with simulation. This would allow the machine to adapt parameters that change in line with the state.

It would be recommended to enrich the training level of a plant operator with six-sigma training. In the near future, knowledge of statistical process control will become more important for the profitable use of a variety of machine data in various subsystems. Digitalization on the production machine places high expectations on user knowledge.

8.2.7 Outlook

In the course of digitalization, it makes sense to develop a holistic strategy and roadmap for your own company. The priorities can thus be set and communicated annually via investment programs. Augmented reality will take an interesting approach in the future, and possible considerations include:

- Using AR to enable maintenance support, optimize workflows, and facilitate employee training through real-time data and visualizations
- Security and privacy, including encryption and user authentication, are essential when implementing AR
- AR implementation processes include design, construction, testing, and production, supported by comprehensive employee training.

8.2.8 Summary

Industry 4.0 creates new approaches to optimizing production processes:

- Real-time monitoring and adjustment of automated production lines with IoT sensors [4]
- Use of cloud solutions (e.g. Amazon Web Services) and platforms (e.g. Siemens Mind Sphere, [4]) for data analysis and visualization
- Use machine learning to recognize patterns and identify problems early
- Integration with enterprise systems for end-to-end monitoring and optimization
- Using cloud-based collaboration platforms for distributed teams
- Implement security solutions and secure storage methods to protect data
- Use of twin technologies and simulation software for virtual testing
- Application of 3D printing for rapid prototype development
- Using advanced data analysis systems and AI for efficient data processing
- Implementation of IoT sensors for real-time monitoring and alerting for the evaluation of mass pressure fluctuations in the extruder
- Use of automated quality inspection systems and image processing systems, such as computer vision in in-line operation of calendering systems
- Application of digital document management systems and automated compliance tools
- Establishment of e-learning platforms and knowledge bases for ongoing employee training.

In the plastic film and cardboard industry, these adapted processes lead to higher product quality while at the same time reducing rejects, downtime, and energy consumption.

However, the focus here has been on using machine learning to recognize patterns and identify problems early. The Industry 4.0 approach explained in this chapter lies is based on automatically selected sensory data and employee knowledge, which has been used to develop an assistance system with a practical application in the plastic film industry. Recommendations for production employees were developed from predictive models using historical data using machine learning algorithms, with thousands of sensory data points (mass, pressure, temperature, etc.) being recorded per minute for each production machine, with the quality status being automatically recorded for each product. To obtain the best prediction, all data was subjected to extensive cleaning, then coupled with the quality data, and various algorithms were evaluated based on historical data. The most robust model with the highest accuracy was selected and used to generate recommendations. In addition, the methodology was tested on other machines (plate extrusion and cardboard machine) and is currently in the evaluation phase.

References for Chapter 8

[1] Chistyakova, T. B., Kleinert, F., Teterin, M. A.: "Big data analysis in film production", in: *Cyber-Physical Systems: Advances in Design & Modelling*, 2020, Springer, pp. 229–236

[2] Kleinert, F.: "Prediction of quality issues in plastic film—The challenge between OT and IT", AI and Big Data, Leaders Dialog, Zürich, 2023

[3] Kleinert, F., Ewald, M.: Real-time data analysis to predict quality—Experience Industry 4.0. Processing PVC—Technology and Applications, SKZ (The Plastics Center), 2023

[4] Mind Sphere—How the IoT operating system works, Siemens, *www.digital-manufacturing-magazin.de*

Additional Sources for Chapter 8

Kohlert, M., Drewes, S.: Produktionsoptimierung in der Cloud, 2017, *www.all-electronics.de*

Kohlert, M.: "Industrie 4.0 in der Kunststofffolien- und Papierindustrie", in: Proceedings of the 2nd Technical Event of the IEEE CIS German Chapter, Institute of Integrated Sensor Systems, Kaiserslautern, 2017

Kohlert, M.: "Industrie 4.0 in der Qualitätssicherung", in: Folienextrusion—Trends bei Rohstoffen, Verarbeitung und Anwendungen, Institut für Kunststoffverarbeitung, Aachen, 2017

Kohlert, M.: "Advanced process analysis and on-line evaluation for computer-aided monitoring in polymer film industry", Vortrag Gehalten im Arbeitskreis der Hochschullehrer für Messtechnik e.V., Ilmenau, 2015

9 Recycling of Polymer Films

Christian Kohlert

Recycling is playing an increasingly important role these days due to unnecessary resource conservation, unnecessary material and energy consumption, environmental pollution, and disposal of used goods. Figure 9.1 shows the distribution of CO_2 emissions across individual areas of life, showing that packaging generates around 1% of total emissions. Even if other emissions have greater savings potential than the packaging itself, packaging should still make its contribution to the savings potential.

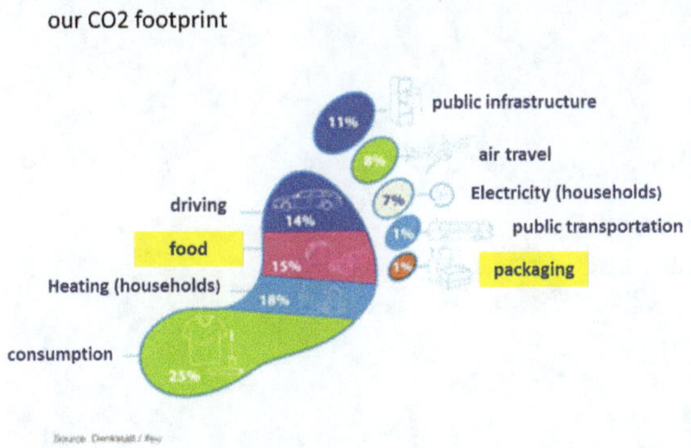

Figure 9.1 Contribution of packaging towards overall CO_2 footprint

The most sensible options for reducing this impact in the film industry are:

- Preventing waste generation
  - Optimization of process steps
  - Maximizing yield from raw materials
- Minimizing production waste
  - Waste reduction through qualitative storage and transport (including cleanliness)
  - Internal recycling in the production process
  - Milling and return to the production process

- Recycling
    - Collecting and sorting for external recycling
    - Collecting and down-cycling within individual companies
    - Down-cycling into park benches, noise barriers and highway walls
- Combustion
- Landfill

Recycling is now possible in the industry for monofilms or separable multilayer films. It becomes problematic with thin multilayer films made of different polymers, which can hardly be processed into high-quality recycled material due to different melting points and viscosities. In such cases, the usual option left is down-cycling into lower-value production goods. However, the latest findings show that multilayer films in the µm range can be reactively extruded into high-quality granules and filaments for 3D printing.

Mixed polymers may be of high quality, but re- or upcycling is not trivial due to the complex relationships [1]. Of course, the usual thermal recycling (combustion) is not the best solution. However, with the help of adapted compatibilizers, granules can be produced from mixed polymer waste (e.g. co-extruded multilayer film), which can be used again in a film product or similar products depending on the requirements.

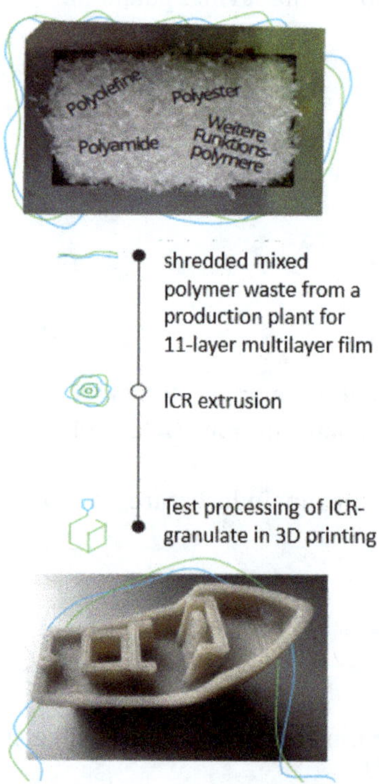

Figure 9.2
Processing mixed polymer waste into material for 3D printing (Source: [1])

Depending on the composition of the mixed polymer waste, advantages such as higher strength, better heat resistance, and better barrier properties compared to (for example) pure polyolefin films can be achieved by using one-piece recyclate through fine distribution and binding of the "foreign polymer particles" to a suitable matrix.

Through individually tailored compounding and reactive extrusion (ICR extrusion), polymer waste can be processed into high-quality granules. These can then be prepared for further processing, for example extrusion, injection molding, 3D printing, or additive processing. Figure 9.2 shows how mixed polymer waste from an 11-layer blown film system consisting primarily of PA/polyolefin compound with a high polyolefin content was processed after processing in a 3D printer.

References for Chapter 9

[1] Company material, GMBU e.V. Halle, 2022

10 Future of the Production of Polymeric Films Using Calender Technology

Christian Kohlert

As shown in Section 2.1 and Figure 2.17, the development of calender technology has shown various development stages, from the evolution of textile to rubber and plastic calenders, the modernization of construction and electrical engineering, and the development of new coating technologies and calenderable polymers.

The future of calender technology will bring with it increased refinement of film surfaces "in-line", i.e. in the manufacturing process, via surface treatment (plasma) and surface coatings with the help of nanotechnology. This creates film properties that cannot be attributed to the polymer, such as conductivity, printability, or UV stability. Due to the stagnating PVC film production in western industrialized countries, new products are being developed for the existing calenders, such as WPC (wood–polymer composites) and other biologically based or degradable polymers. The use of nanotechnology by spraying nanosols opens previously unimagined application possibilities for calendering technology, from active surfaces (e.g. antimicrobial, anticorrosive, and aromatic) to "intelligent" surfaces (e.g. time indicators, temperature/time indicators, and electrically luminous) to special effects such as electrically dimmable or anti-counterfeiting surfaces (see Figure 7.9).

Regarding the actual calendering technology, a combination of the mathematically described sub-processes (Section 4.3) will lead to semi-automation to full automation, and will further result in learning and optimization options controlled via VR glasses. Subsequent developments then move towards Industry 4.0 for IT-supported process improvements. This uncovers potential savings in terms of personnel, material, time, and energy reserves, and makes possible a corresponding reduction in terms of environmental impact and an improvement in terms of economic potential. This shows that the highly productive calendering technology has developed steadily over the last few decades and has retained its place as a film manufacturing process in the 21st century.

For small quantities and many changeover processes, extrusion technology is playing an increasingly important role, especially for non-high-viscosity films such as polyolefin and PET, as this process requires smaller investments and can be implemented more energetically and cheaper in terms of changeover thanks to the closed machine. Multilayer technology will also continue to develop with the recycling options described in this book (Section 9), as the product requirements in terms of strength, barrier properties, and production efficiency can be addressed in a way that protects the material.

Index

www.ingramcontent.com/pod-product-compliance
Lightning Source LLC
LaVergne TN
LVHW081928200926
848484LV00009B/2395
* 9 7 8 1 5 6 9 9 0 4 4 0 4 *